Troubleshooting in the Chemical Process Industries

Troubleshooting is an activity on which process engineers spend a relatively large portion of their time throughout their career. And yet, it is a topic not adequately covered in chemical engineering curricula. While the fundamental principles surrounding the field of chemical engineering are taught, they are presented as fragments and, like pieces of a puzzle, must be put together when analyzing a potential problem. *Troubleshooting in the Chemical Process Industries* offers readers practical guidance on problem-solving in the chemical process industry from a process engineering perspective. It links fundamental principles of chemical engineering together, considers the data gathering and analysis steps in the problem-solving process, and details how they can be effectively used to identify root causes of and resolve operating problems. The book highlights the importance of troubleshooting as a response to abnormal process behavior that threatens or prevents achievement of targets, such as production volume, product sales specifications, and environmental and process safety performance.

- Considers troubleshooting in three categories: for existing facilities, during retrofit of an existing facility, and for new plants
- Covers plant performance monitoring and discusses the source of process data relative to instrumentation and associated errors
- Discusses process simulation in troubleshooting, strategies for dealing with the empirical aspects of processes, and manipulation of raw process data into meaningful performance information
- Explores the perspectives of process engineers and process operators, comparing and contrasting both functions, considering human factor issues
- Features industrial case studies

With this practical book, process engineers are given the know-how to solve real-world problems they are likely to encounter in industry through the application of sound process engineering methodologies.

André Bernard is a Senior Process Engineer in the Central Engineering Department of NOVA Chemicals, now Borouge International. He has over 35 years of experience in plant operations and process design. He has published several technical papers over the years and holds BS and MS degrees in chemical engineering from l'École Polytechnique de Montréal in Canada. He is a Fellow of the American Institute of Chemical Engineers (AIChE) and a registered professional engineer in the provinces of Ontario and Québec in Canada.

Troubleshooting in the Chemical Process Industries

André Bernard

CRC Press
Taylor & Francis Group
Boca Raton London New York

CRC Press is an imprint of the
Taylor & Francis Group, an **informa** business

Designed cover image: © 2026 Shutterstock

First edition published 2027
by CRC Press
2385 NW Executive Center Drive, Suite 320, Boca Raton FL 33431

and by CRC Press
4 Park Square, Milton Park, Abingdon, Oxon, OX14 4RN

CRC Press is an imprint of Taylor & Francis Group, LLC

ISBN: 978-1-032-58342-6 (hbk)
ISBN: 978-1-032-58343-3 (pbk)
ISBN: 978-1-003-44967-6 (ebk)

DOI: 10.1201/9781003449676

Typeset in Times
by Apex CoVantage, LLC

To my wife Sandra Hunt, for her support and encouragement over all those years. Her love has fueled my sense of purpose.

Contents

Preface

Troubleshooting is an activity for which process engineers spend a relatively large portion of their time throughout their career. And yet, it is a topic that is neither adequately covered nor included in the curriculum of chemical engineering programs. While all the fundamental principles surrounding the field of chemical engineering are well established, they are presented as fragmented topics that, like pieces of a puzzle, must be put together when analyzing a potential problem. Unlike in a school, operating problems in a chemical plant are not formulated as a black and white question with associated data. It is the responsibility of the process engineer to frame the question (problem definition) and collect the data required for evaluating the process (data gathering). I sadly noticed throughout my career that many young engineers are in a hurry to throw out their textbooks and deny the knowledge they work so hard to learn while pursuing their degree. Whether eager to establish themselves as experts or mimicking the behavior of senior coworkers, they often relegate good process engineering to intuition. However, while intuition may help, troubleshooting chemical processes goes beyond that and good intuition comes with experience. In too many instances, I witness troubleshooting exercises where people would gather in a conference room, brainstorm potential causes, and often become more focused on the cure rather than the source of the problem. Many of those sessions reminded me of a metaphor used by Albert J. Bernstein and Sydney Craft Rozen in their book titled *Neanderthals at Work*, where they compare such meetings to a "band of people who do the equivalent of casting runes and consulting the gods".

As a young engineer, I quickly realized that you cannot compete with experienced process operators who have spent several years operating plants. I was fortunate to be pulled aside by a senior engineer in the early days of my career and be told that while I was inexperienced, all the theory I learned was fresh in my brain and the best way I could help was using it. So how could I help? I made a parallel between doctor/patient and process engineer/process operator interactions. Doctors first listen to their patient describing changes typically related to discomfort with their bodies (problem definition). Throughout the conversation, the doctor collects a list of symptoms (data gathering) and may request further testing so he could correlate the symptoms to potential illness (root cause), from which he can propose a treatment (cure/solution). The process engineer/process operator relationship is somewhat similar. The process operator is typically first to observe and experience unusual plant behaviors, which like an illness cause him difficulty (discomfort) in sustaining plant performance and would typically require process engineering support. The time factor distinguishes the environment in which both positions execute their work. Process operators have limited time for troubleshooting as they must operate the plant and cannot afford distractions when a unit is misbehaving. Process engineers, on the other hand, could dedicate more time to analyze a problem.

I started my career in the late 1980s in a plant that was using an analog control panel. Process control was very basic and changed unusually as it involved hardware alterations. Process data were limited to roller charts, hard copy log sheets, and process operator logbooks. The first desktop computers were introduced six months after I started my career. Heat and material balances (H&MB) and process calculations were performed with hand calculators. Gathering long-term historical data on process variables would have been time consuming and was rarely done during troubleshooting. The data collection was mostly limited to the hours surrounding a change in plant behavior, with few, if any, associated process calculations. Around that period, many plants were transitioning to distributed control systems (DCSs), with associated data historians. This transition revolutionized the availability of process data that could now be accessed from both desktop and laptop computers. From my experience, the potential benefits of data historians have been underutilized. I observed many process engineers displaying trends of raw process variables in the time domain where they try to find correlations between variables. In fact, this practice, when compared to the paper roller charts, only provides the ability to extend the time span and overlapping of more variables on the same plot. From my perspective, the power of data historian is in providing the capability of doing heat/material balance and process calculations associated with the performance of operating units. As process engineer, we should be aware of the dependency of various process variables. Therefore, plotting those variables against each other may paint a more accurate picture on the health of a process unit. Effective troubleshooting relies on good data analysis. Failing to do so leaves the troubleshooting team without answers or far-fetched theories that could often result in unnecessary plant outage. Those outages are costly and often associated with process safety risks, not counting the fact that a decision of shutting a plant down must be associated with the required corrective actions. There is nothing more frustrating than taking an outage without finding any deficiencies.

There are many troubleshooting work methods/processes used in the industry. Most, if not all, methods involve data gathering and data analysis. The objective of this book is not to develop a new method but rather reinforce the data gathering/data analysis piece through application of good process engineering methodologies. The process operator/engineer interface is also discussed.

Acknowledgments

I tried to share in this book what I have learned and applied over the years, which made me successful in troubleshooting. A lot of this knowledge came not only from literature but also from the teaching and mentoring I received by many. The journey goes as far back as high school, where my track coach Pierre Larose, who sadly passed away at the young age of 37, taught me that success comes with discipline and preparation. His guidance went beyond sports, where he had his way in making me realize the importance of knowledge and education. This influence paved the way to my journey in chemical engineering, an exciting career that brings elements of satisfaction comparable to scoring goals or winning important games when resolving difficult problems. It is said that if you love your job, you will not work a single day. Chemical engineering is a wonderful career when you get to apply the fundamental principles learned in school to real plant problems. I am also grateful to my long-time friend, mentor, and co-worker, David Gent, who nurtured my curiosity and enabled my desire to bridge the gap between theory and practice. In my career I had the privilege of working with highly skilled people like Henry Kister and Dan Summers who, besides being inspirational, were not shy in transferring a lot of knowledge. Their enthusiasm toward the profession of chemical engineering is inspiring. I would not want to forget Richard Hayden, with whom I had the opportunity to work for several years. Richard sadly passed away in 2017, but I think, he would have felt proud to see me writing this book.

Writing this book has been a challenge from a perspective of time commitment and it took a toll on my personal and family life. I am extremely grateful to my wife Sandra Hunt for her understanding and support.

I would like to thank Bob Esposito for reaching out and initiating the idea for this book. I am grateful for the precious advice and guidance from my editors Allison Shatkin and Ariel Finkle. Special thanks to Henry Kister, Phil James, Ross Taylor, Sanjeev Kapur, Ed Mohundro, and Dan Summers for reviewing the original plan for this book and reviewing some of the sections. Their advice and suggestions are greatly appreciated.

1 What Is Troubleshooting and the Work Processes

1.1 WHAT IS TROUBLESHOOTING?

Troubleshooting is a term that is widely used in the Chemical Process Industry (CPI) and refers to problem solving. A problem in a chemical plant can be defined by the combination of one or several anomalies which are threatening the ability of the facility in meeting its objectives. While frequently informal in the initial stage of recognition, it often becomes part of a systematic work process as the problem grows in severity and visibility. Whereas there are many different work processes in use, they are typically aimed at defining the problem, identifying the root causes, and establishing and implementing a remediation plan.

Before further discussing troubleshooting in the CPI, it is important to understand the functionality and objectives of a chemical plant and know what would be defined as problems in need of resolution. From a high-level perspective, industrial chemical processes involve the transformation of raw materials into product streams of higher value. A chemical process can be subdivided into a few major activities, primarily reaction, separation, and/or blending. A reaction is a step by which the chemical structure of raw materials is modified into higher value products. The stream exiting the reaction section is a mixture containing the desired product along with coproducts and impurities that without separation have no practicality and commercial value. Some processes include blending steps, where additives are introduced to confer desired properties on the final products. Energy generated or imported in the form of heat and/or electrical power is associated with the functionality of the process. The production output of a chemical plant is limited by the scale and capability of the process units involved and is typically defined by the unit having the smallest capacity, often referred as bottleneck. Chemical plants are designed and operated for meeting product sales volumes and specifications while complying with environmental and safety regulations. Production planning and overall plant optimization go beyond the process itself. Feedstock flexibility, availability and costs (contract vs. spot market), single versus multiple products, product sales orders, pricing, and management of inventories add to the complexity of maximizing profitability of a chemical facility.

In general, the need for process troubleshooting arises when anomalies within the boundary of the plant cause deviations from expected process performance. The result of these deviations translates into consequences ranging from production cost impacts, rate limitations to a total plant outage. A key step in troubleshooting is to recognize and effectively diagnose the anomalies and their impact on process performance. Poor diagnostic can result in costly and sometimes unnecessary plant outage. Loss of performance due to fouling, for instance, may restrict the capacity of a unit

DOI: 10.1201/9781003449676-1

and increase energy demand. However, if it does not pose an immediate threat for the plant in meeting committed sales volumes while remaining compliant relative to regulations, cleaning of the unit may be deferred to the first window of opportunity that would minimize the impacts on the plant profitability. On the other hand, stream's cross-contamination caused by a leak that has reached a magnitude where the plant can no longer meet product specification would require an immediate unit outage. In the above examples, the anomalies would translate into observable consequences on the process, higher utility demand caused by a fouled exchanger, and off-specification product due to a leak. The observed consequences typically frame the problem definition of a troubleshooting work process. However, recognizing the problem does not necessarily pave the way to the root cause. Identifying the root cause of a problem is the most critical step in troubleshooting as it also offers the path toward a resolution and/or cure.

Troubleshooting activities are not limited to ongoing plant problems as they extend to after the fact events such as process upsets and/or process safety incidents. Those events are typically reviewed as part of formal incident investigations, which follows systematic troubleshooting work processes. The objectives being identification of the root causes, understanding the sequence of events, and recommendations intended to prevent reoccurrence or minimize the consequences of similar events.

1.1.1 Optimization versus Troubleshooting

Optimization differs from troubleshooting as it is an exercise intended for maximizing the capability of the process while minimizing cost. A process might not be optimized but still meet its intent. As discussed earlier, the need for troubleshooting arises when abnormal unit performance is threatening or preventing the plant from meeting its objectives. There are different levels of optimization ranging from informal and unstructured to formal and integrated with business processes.

Informal optimization is typically done at the unit/area level. In large facilities, chemical plants are normally subdivided into different process areas, which are operated by different groups of people. Each area or department operates under their own budgets and sets of key performance indicators (KPIs). Informal optimization occurs on a daily basis by reviewing operating data and making adjustments intended to maximize utilization of the asset and minimize cost. However, improper considerations of cross-boundaries or inter-dependencies between process areas can result in overall plant inefficiencies. In other words, what is beneficial to one area may negatively impact other areas such that the overall plant operates below its full capabilities. Chemical plants are often highly integrated with some of the relationship between areas being counter-intuitive.

Informal optimization often relies on historical data, experience, and intuition of those involves in setting operational direction. Constraint imposed by equipment characteristics is rarely taken into considerations if operating data suggest beneficial performance. However, while in the short term operating equipment or units outside the boundaries of their design characteristics may appear beneficial, mid- to long-term performance may exhibit a significant drop. Use of chemical inhibitors and/or dispersants intended for minimizing fouling of equipment is another example where reducing dosage would show short-term cost benefits but affect the expected run length of the operating unit [1]. Not knowing or understanding the operational boundaries of units and equipment may result in extensive troubleshooting exercises or normalization of deviations, which may cause a higher frequency of unit outages and maintenance workload.

Formal optimization is done at the global plant level and extend beyond unit operation to include outside battery limits elements, such as feedstocks, utilities, chemicals, waste treatment costs, and others. The intent is to maximize profit by optimizing the production, supply and demand chains. Different methodologies are being used in the chemical industry from linear programming (LP), non-linear programming (NLP), and offline/online optimizers. These methods, though different in complexity, are driven by an objective function and a list of constraints. The objective function is a mathematical expression that defines profit relative to cost and revenues. The constraints, while not limited to, often relates to process equipment's capacity and design characteristics.

Big data analytics [2] and artificial intelligence (AI) [3, 4] are gaining in popularity and extend beyond conventional optimization. They leverage on the integration of multiple database, such as process data historian, laboratory data, maintenance/reliability, commercial (feedstocks/products pricing), and others. While some hybrid models might still use first principal models, but on a much reduced frequency than conventional optimizers, big data analytics and AI rely strongly on historical data and statistics for predicting constraints, running case studies, and sensitivity analysis. It also provides early detection on process equipment performance degradation. Machine learning algorithms is another aspect that distinguishes AI from conventional optimization.

The skill level required for developing LP, NLP, or optimization models far exceeds than what is required for just running these applications. Big data and AI add to the development complexity. Therefore, it is not unusual that the team developing the applications is outsourced or, if internal, disbanded after implementation/commissioning. Given that chemical plants often experience changes over the years, these applications could become obsolete if not properly maintained resulting in poor plant performance, which may require troubleshooting.

Disregarding big data and AI, troubleshooting crosses path with optimization in terms of process unit constraint. These constraints are expressed either as constant or as mathematical expression in an optimization model and are typically based on expected design performance. Process unit anomalies may cause the constraint boundaries to shift and make the optimization solution unachievable. Effective troubleshooting would identify the new constraints and allow updating the optimization models accordingly. Big data and AI, on the other hand, could potentially predict the dynamic nature of those constraints, if sufficient and relevant historical data are available.

The empirical nature of chemical reactions responsible for fouling process units is typically disregarded in optimization models. While the reaction section of a chemical plant uses kinetics models, the recovery area is typically modeled as a non-reactive process. As such, there are conditions which would appear more optimal from a thermodynamic standpoint, but in reality, due to the presence of reactive impurities, would result in premature fouling of the unit causing higher downtime frequency. Prevention or minimization of those fouling reactions could be handled by limiting key process parameters (i.e., optimization constraints), such as temperature, residence time, and/or using chemicals inhibitors and/or dispersants. The dosage and selection of those chemicals involve a level of empiricism and are typically excluded from optimization models.

1.1.2 Troubleshooting Work Processes

While the focus of this book is not on troubleshooting work processes, it is important to review some of the problem-solving methodologies typically used in the CPI so that process engineering data can be effectively integrated to the selected technique. Through their education, engineers' understanding of fundamental concepts is tested in the form of written questions requesting clear answers and includes all the data required for resolving the problem. Solving such problems typically follow the same thinking process; identify known and unknown variables, data correlations (tables, charts, and mathematical formulas), and executing the calculations that will provide the answer requested by the question. Real plant problems do not appear that way and require a structured methodology for effective resolution. While good intuition and creativity are good assets to process engineers, they are only one element in the troubleshooting toolbox. Relying solely on those elements could result in biased opinions unsupported by facts and will likely result in unsuccessful problem resolution.

Some of the troubleshooting methodologies documented in literature arise from process risk assessment and incident investigation techniques. From the perspective of troubleshooting, process incidents are somewhat similar to underperforming operation as they are both the result of anomalies resulting in undesired and observable consequences. Risk analyses, while following similar methodologies, use backward thinking from hypothetical root causes to expected consequences.

Regardless of problem-solving methodologies, the boundaries of a chemical process whether or not meeting its intended objectives are limited by thermodynamic and transport phenomena consistency, and process equipment characteristics. Understanding this concept limits the number of potential root causes realistically possible.

To help in discussion on troubleshooting work processes, a case study was developed and forwarded to a small population of engineers and process operators with different level of experiences. They were instructed to review the data and describe how they would resolve the problem with emphasis not so much on getting the right answers but rather on the steps they would follow and on their recommended path forward.

CASE STUDY: DISTILLATION TOWER FLOW

The System

A distillation tower is used for separating a light fraction (identified as A) from a heavy fraction (identified as B). As shown on Figure 1.1, the tower has two feed points (Feed #1 and Feed #2), which originate from different upstream units.

Low-pressure steam provides heat to the reboiler and a refrigeration stream cooling to the overhead condenser. Reflux is operated on flow control and the distillate is withdrawn from a level (reflux drum) to flow (distillate product) cascade control loop. Low-pressure steam to the reboiler is regulated from tray 43 temperature, which remotely set-point the low-pressure steam flow controller. The low-pressure steam can also be controlled strictly on flow control. The

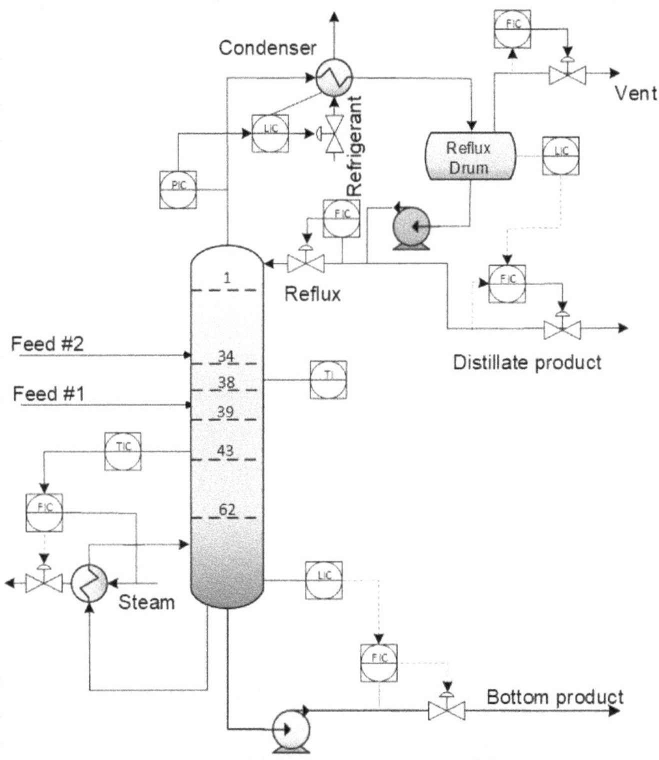

FIGURE 1.1 Case Study 1 – Distillation Tower Flow Diagram.

Reproduced with Permission from figure 1.1, Distillation: Avoid Problems during Tower Startup, André Bernard, Chemical Engineering, June 2012

bottom product is withdrawn on a level (sump) to flow cascade control loop. The tower has a total of 62 trays.

The tower is a downstream unit with Feed #1 originating from a pre-fractionator that is not always in service.

The Problem

Following a plant outage, the tower can no longer meet the bottom product stream specification relative to the light component A. The sales specification of the bottom stream is 400 ppm weight, and the analyzer has been varying between 115 and 25,240 ppm weight, averaging 16,200 ppm weight for the first 15 hours of operation. The distillate sales specification relative to the heavy component B is met and the tower is responding to reflux. Figure 1.2 provides trends for some key operating variables for the first 15 hours of operation. Table 1.1 provides a simulation comparison of the expected performance between the pre-outage and the post startup conditions.

The Situation

The front end of the plant is currently in operation with raw materials/feedstock supplying the unit. The bottom product stream from the tower is stored in an off-specification storage tank, which is recycled to the front end of the plant. This situation cannot be sustained as the off-specification storage tanks will eventually be full, at which point the plant will have to be shut down.

You are brought in as a process engineer and requested to analyze the problem and recommend a path forward.

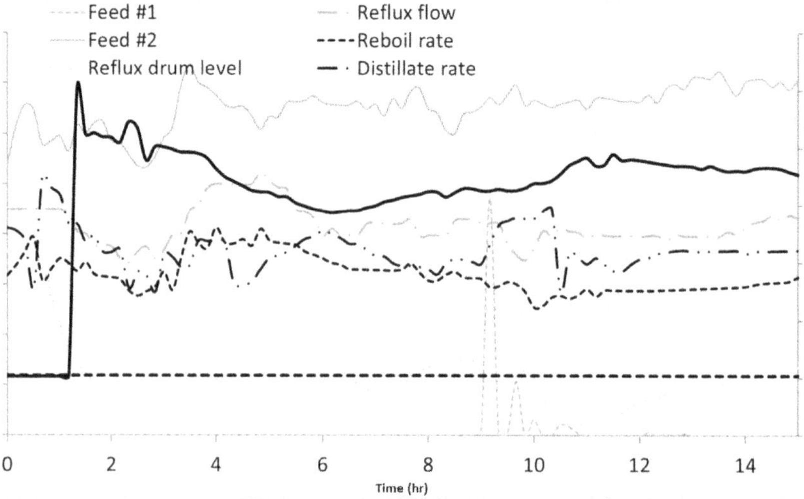

FIGURE 1.2 Key Operating Variable Trends.

TABLE 1.1
Pre- and Post-Outage Simulation Data

Description	Feed #1	Feed #2	Distillate	Bottom	Vent	Reflux
Pre-Outage						
Flow (klb/hr)	59.5	107	85.4	81.1	0	95.6
Pressure (psig)	122	122	290	125	N/A	290
Temperature (°F)	143.9	83.5	62.8	182.2	N/A	62.8
A (wt. frac.)	0.2317	0.6691	1.0	0.0001	N/A	1
B (wt. frac.)	0.7683	0.3309	0.0	0.0	N/A	0
A (mol. frac.)	0.3169	0.7351	1.0	0.0001	N/A	1
B (mol. frac)	0.6831	0.2649	0.0	0.9999	N/A	0
Post-Outage						
Flow (klb/hr)	0	215.7	88.2	127.5	0	87.2
Pressure (psig)	N/A	122	290	125	N/A	290
Temperature (°F)	N/A	73.5	62.9	125.1	N/A	62.9
A (wt. frac.)	N/A	0.3797	0.9970	0.0001	N/A	0.997
B (wt. frac.)	N/A	0.6203	0.0030	0.9999	N/A	0.003
A (mol. frac.)	N/A	0.4960	0.9977	0.0001	N/A	0.9977
B (mol. frac.)	N/A	0.5040	0.0023	0.9999	N/A	0.0023

Five common themes came out from the respondent who looked at the case study:

- Establishing a list of anomalies
- Data validation and system checks
- Listing potential causes/consequences and testing for proving or disproving those potential causes
- Understanding process unit characteristics
- Looking at historical data

New chemical engineering graduates without any plant operating experience while showing the abilities at identifying the process anomalies, framing a problem definition, and looking at pre- and post-outage differences were struggling in analyzing data presented in the form of historical trends. Their desire to find an answer led to speculation on causes, not supported by facts. New chemical engineering graduates who had been exposed to plant operation through internship showed more abilities in analyzing process variable trends and already demonstrated the ability for establishing a list of causes and effects along with testing required for proving or disproving those potential causes. In fact, little differences were found between their answers and those who already had 2–3 years of experience in the field. This suggests that new chemical engineering graduates without previous plant operation experience would reach the level of young engineers having 3–5 years relatively quickly once immersed in plant operation. Understanding of process unit characteristics was, however, missed by all respondents having less than 5 years of experience. More seasoned engineers with 10-plus years'experience captured those elements, such as comparison of operating and design conditions to design basis of the equipment and relevant rating calculations. It can be speculated that younger engineers would assume that the equipment is properly designed. Whether by ignorance or by fear of sounding arrogant, they would not challenge that operating conditions imposed on the unit may fall outside the design envelop of the equipment. Previous troubleshooting exposure to similar problems and design experience would raise awareness relative to process unit capabilities, potentially explaining why seasoned engineers would pay more attention to these criterions. As stated earlier, the root causes of problems in a chemical plant are typically related to inconsistent thermodynamic conditions, operating equipment outside its design envelop, and/or deterioration of equipment resulting in unexpected performance. Process thermodynamic and unit operation characteristics are concepts that should be fresh in the mind of young engineers. Use of those concepts is something that should be enabled and nurtured in the development of young engineers.

Response from process operators while falling within the same themes described earlier involved direct interaction with the unit. Instrumentation checks through instantaneous mass balances and calibrations request. Setting operating conditions to pre-outage values then moving one process variable at a time to prove/disprove potential root causes. Requesting technical support if none of the above strategy was successful in resolving the problem. The time factor is one of the main elements that distinguish process operators from engineers in executing problem solving and decision-making strategies. For instance, responding to process upsets requires a process operator to respond to an overwhelming number of alarms immediately, making process moves required for bringing the unit into a safe state, followed by recovery or re-start procedures. The above case study does not pose any safety treat, allowing time for varying conditions and performing live testing on the unit. Regardless, their interaction with the unit is in real time, while engineers analyze data offline. Process operators run the unit and, as such, have extensive exposure to its behavior and develop deep understanding of process variable interactions giving them strong intuition. While this latter element may prove useful in problem-solving session, caution must be applied as it could also result in premature conclusions not supported by facts.

The respondents' answers to this case study, while providing useful information, cannot re-create the intensity of the moment and eagerness to resolve this problem for avoiding a costly unit outage. This case study was based on an actual problem documented in literature [5], which involved a depropanizer during startup. Information from the paper is being used in the following discussion for complementing answers received from the respondent.

Regardless of formal or informal methodology, the troubleshooter must first understand the intended purpose of the process unit or system in question, frame a problem definition, what is the problem and why does it need to be resolved. The purpose of the tower in the above case study is clear and involves separation of a light fraction A from a heavy fraction B. The problem is typically the observed anomaly and the reason for resolving it. The consequences can be summarized by the following statements for the case study:

Problem definition: The tower cannot meet the bottom stream specification.

Consequences:

1. Interruption in supplying downstream products to customer due to off-specification of an intermediate stream.
2. Potential unit outage due to limited off-specification product storage capacity.

In the actual case, the bottom stream from the tower (i.e., depropanizer) is an intermediate stream which undergoes further processing for meeting sale specification of a C4 product. As such, the tower bottom has an operating specification, which when exceeded cause the final C4 product stream to exceed sales specification relative to light component *A* (i.e., C3s).

The next step is to gather available data and facts associated with the operating unit exhibiting the anomaly. Figure 1.1 provides information that would have been extracted from process flow diagrams (PFDs) and pipe and instrumentation diagram (P&IDs). As illustrated, the system boundary is limited to the tower, which, as the troubleshooting exercise evolves, may or may not require to expand on sub-systems interacting with the tower. Raw process variable trends from the plant data historian are provided in Figure 1.2 and simulated material balances from pre-outage and post-startup conditions are provided in Table 1.1. Additional information believed needed for effective troubleshooting was intentionally left out in the problem description submitted with the expectation that it would have been requested or identified as essential by the respondents. For instance, while some respondents referred to tower rating calculations, none asked for or mentioned the need to know the type and characteristics of internals present in the tower. Availability or recommendations to conduct non-intrusive testing such as gamma scanning for confirming that the tower internal hardware is in place and assessing potential flooding and/or distribution problems was missed as well.

The following observations complement the information provided in the submitted case study and are required for effectively analyzing the problem.

- The tower is equipped with conventional two-pass trays above Feed #1 and high-capacity trays below.
- Tray operating windows illustrating the tray operating limits are available.
- A gamma scan was requested by process engineering once the problem was reported. Given the variability in operating conditions illustrated in Figure 1.2, process engineering requested operation to maintain constant and stable conditions for the duration of the gamma scan. This period of stable conditions also provided data for process simulations.
- Design process simulation of the tower was available.

The next step is to organize the above information systematically for relating the symptoms of the problem to the potential root cause. This can be accomplished by different methodologies or work process as described below.

The 5-Why Method, originally developed by Sakichi Toyoda in the 1930s and popularized by Toyota in the 1970s [6], is a simple problem-solving methodology where the troubleshooter starts by questioning why a deviation in performance is occurring. The method is called 5-why as it is expected that answering the question why five times would lead to the root cause. Despite the name, the number of iterations is not fixed nor important, what matters is that the answer to the last iteration leads to the root cause, or, as it will be discussed later, to refinement of the problem definition. A distinction is made here between the root cause and problem definition. A root cause provides clarity on the mechanism that caused the deviations and, as such, a path back to normality. The last iteration of the 5-Why method could lead to a probable cause for which the mechanisms causing the deviation as the remedial actions still remains unclear. In this case, the answer from the last Why-iteration would become a problem definition requiring more research or a different approach to problem solving.

While there are some elements of the 5-Why method, none of the respondents in the case study formally used this methodology. The following are iterations of the 5-Why method applied on the case study using the problem definition stated earlier as starting point.

1. **Why** *is the tower bottom stream not meeting specification?*
   - *Little to no fractionation in the stripping section.* It is known from the problem description that the rectifying section is fractionating as the tower is known to respond to reflux and the distillate stream is meeting specifications.
2. **Why** *is there little to no fractionation in the stripping section?*
   - *The tray downcomers are unsealed.* The gamma scan reveals very little froth height in the stripping section. The gamma scan does not suggest any tray damage. Unsealed downcomers results in vapor blowing through the downcomers, which bypass the active area of the tray causing poor vapor/liquid contact and poor fractionation.
3. **Why** *are the tray downcomers unsealed in the stripping section?*
   - *The stripping section is equipped with high-capacity trays having downcomer configuration sensitive to loss of seal at low rates.* However, simulation and rating calculations suggest an operating point close but above the downcomer seal point.
4. **Why** *are the downcomer unsealed if the operating point is above the seal point?*
   - The answer to this question relies on an unproven theory or assumption related to a path to low rate, which is discussed in more detail in the historical data methodology.

Using the 5-Why method here would suggest that the root cause was a path to low rate resulting in unsealed downcomer. If true, increasing rates should be sufficient in sealing the downcomers and regaining fractionation efficiency. Validation of this theory is further explored through analyses of historical data.

Historical data methodology, also referred as epidemiologic approach [7] in risk analysis, involves the collection of historical data relevant to the system being investigated. Besides processing raw data from plant data historians, this would also include gathering of laboratory analysis, previous performance test reports, and in the case of process upsets, incident reports. This gathering of data should be complemented with process engineering calculations that determine the operating point relative to the design envelop of the equipment and also ensure that the conditions imposed on the system are thermodynamically consistent. Commonalities and differences between events should shed light on the potential root causes and remediation actions.

In this case study, once operation stabilized operating conditions, a gamma scan of the tower was conducted, and process data was collected from the plant data historian for material balances and process simulation purposes. Tray vapor/liquid hydraulic loads were calculated from process simulations and plotted on the tray operating windows for the high-capacity tray and rating calculations done for the conventional two-pass trays. Process simulation results suggested that the tower be capable of meeting overhead and bottom product stream specifications under the process conditions evaluated. The process loads, while slightly above the pre-outage level, were near the turndown limit of the high-capacity trays. That lower limit was characterized by the minimum liquid load required for sealing the downcomers. Nonetheless, the pre-outage tray load was even closer to that low limit with the tower meeting product specifications The gamma scan, however, re-emphasized the potential for unsealed downcomers as little froth was found on a significant number of high-capacity trays.

The question then was why would the tower perform well before the outage at a hydraulic load closer to the downcomer seal point. This question was answered with a theory believed to have merit but could not be proven with the time constraints and the capabilities at end. This theory was based on the path to low rate. In the pre-outage conditions, historical data confirmed that the high-capacity trays were brought close to the downcomer seal point from high to low rates. The downcomers were already sealed as the conditions approached the seal point, while in the post-startup conditions, they were unsealed. It was suggested that for breaking the momentum of the vapor velocity in the downcomer, a higher liquid load was required.

Fault tree analysis (FTA) [7] is a method used in risk assessment, referred as top-down approach starting from the outcome then backward to the potential causes. The outcome is an undesired hazard, and the intent of the exercise is to determine the expected frequency of failure of a single or combination of components within a system that will result in the hazardous outcome. Each potential cause is associated with probabilities and frames the structure of the tree. Resulting probabilities relative to a combination of causes are accounted using "AND" and "OR" conditions (logic gate). FTAs assume that component failures are binary; they operate or they fail. Performance degradation neither is covered by an FTA nor shows transient behaviors. An FTA would include basic system components, human interface, control, and safety systems.

Availability of an FTA on a system being troubleshot would provide a good starting point and would fall under the data gathering steps of the historical data methodology previously discussed. An FTA could have been executed during the design phase of an operating unit or as part of reoccurring plant process hazard analysis (PHA).

In troubleshooting, the top-down approach of an FTA makes it an appropriate methodology for seeking the potential root cause of an undesired process outcome. The binary nature of the methodology and the inability to treat transient conditions and performance degradation would, however, limit its capabilities. Nonetheless, if not leading to the root cause, it would eliminate several potential causes related to system malfunctions. The methodology would apply well to systems that involve many components and safety instrumentation systems (SISs), such as chemical reactors, fired and rotating equipment. In a troubleshooting context, FTAs could be simplified as there are not requirements for estimating the frequency or probability of system failures as their status is confirmed through actual plant data.

Figure 1.3 illustrates a simplified FTA that could apply to the case study with the top of the tree being the undesired outcome, the tower bottom stream not meeting specification. In terms of binary failure, two potential causes are identified, steam blowing through the reboiler or failure of the control valve regulating the steam flow. Steam blow through the reboiler is characterized by a single failure, steam trap, or condensate valve failure, depending on system configuration. Failure of the steam control valve can however be the result of multiple failures. Assuming that the tower

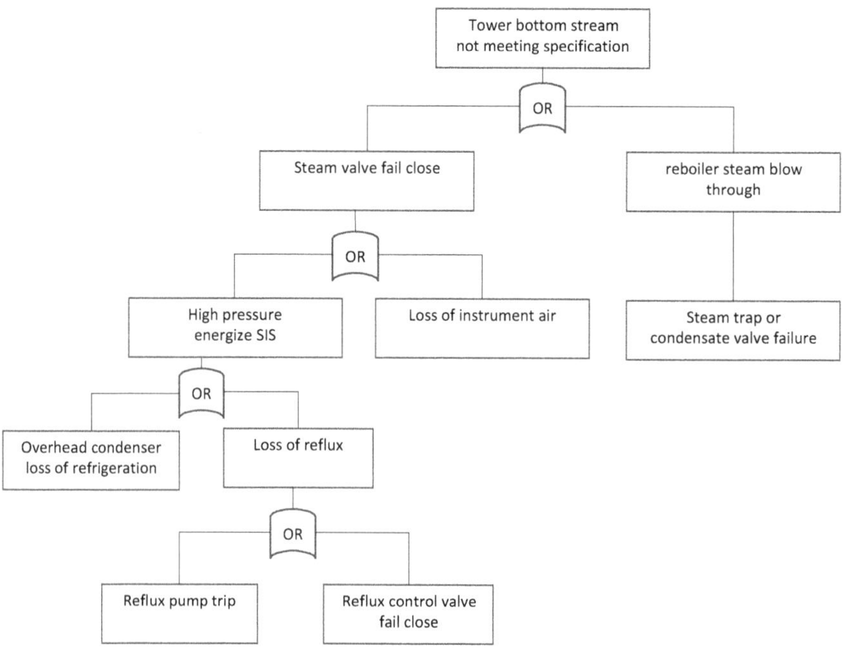

FIGURE 1.3 Case Study FTA.

is equipped with an SIS system protecting the vessel from an over-pressure, activation of this system would involve removing the heat by closing the steam valve or a dedicated trip valve on a high-pressure excursion. Activation of the SIS can be triggered by either a loss of a refrigeration in the overhead condenser or a loss of reflux. The loss of reflux can then be the result of a reflux pump trip or reflux control valve failing in the close position. Another mode of failure independent from the SIS system is a loss of instrument air, which can cause the steam valve to fail close if configure as a fail close valve.

The failing mechanisms described in the FTA are easily ruled out as the tower operating data, temperature profile, and heat balance are indicative of a functional reboiler. The binary nature of the FTA does not make this methodology appropriate in this case as the problem is not related to a component failure but rather to operating conditions' transition outside the capability of the equipment. Removing the binary criteria of the FTA would open consideration to more potential causes. Keeping the top-down approach from outcome to potential causes would result in a cause and effect diagram (CED), also known as Ishikawa or fishbone diagram [8].

CEDs [8] were popularized by Karou Ishikawa in the 1960s in quality management processes. The diagram is also often called fishbone diagram due to its structure with the effect or outcome at the head of the fish preceded by the bone structure, which is divided into six categories intended to segregate factors influencing the outcome. While there could be some variations around the six categories used in the diagram, they would typically refer to people, material, measurements, environment, methods, and equipment. The six categories and outcomes represent the initial starting point of the diagram, which is typically used in brainstorming sessions where it is displayed on a flip chart. Brainstorming participants are then asked to populate the diagram with potential causes, which are assigned to the six categories. A CED is illustrated on Figure 1.4 for the case study. Underlined potential causes are

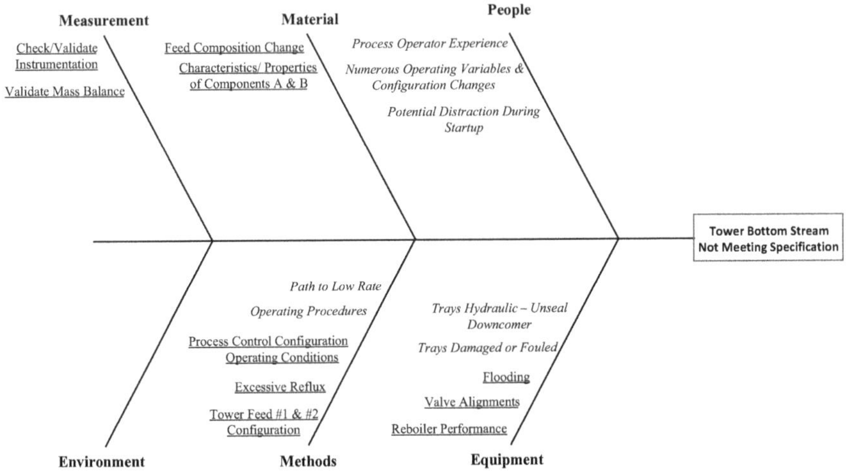

FIGURE 1.4 Case Study CED.

those identified by the respondent and those in italic were added for the purpose of discussion.

It is interesting but not surprising that none of the respondents identified potential causes related to people or human intervention. This category is often overlooked for fear of assigning blame to individuals. However, literature [9] suggests that 50–90% of process safety incidents are caused by human errors. Ian Nimmo [10] indicated that 40% of incidents from studies he conducted, and data extracted from published database were related to human errors. This number increased to 80% when considering the root causes in the Equipment and Methods categories, as illustrated in Figure 1.4. The topic of human factor will be discussed in further detail in a later chapter. For the purpose of the case study, a few probable causes were added to the CED, such as process operator experience, numerous operating variables and configuration changes, and potential distraction during implementation of the startup procedure.

Under Material category, the respondents highlighted the change in feed composition, and some respondents made references to the characteristics and properties of component *A* and *B*. Under Measurements, checks and validation of instrumentation along with material balance were captured. No potential causes were attributed to the Environment category. Cooling to the overhead condenser is provided by a refrigeration stream for which the conditions are unaffected by ambient conditions. On the other hand, a condenser serviced by cooling water would be affected by atmospheric conditions during warm weather. Though not affecting the refrigeration stream and fractionation of the tower, the power of the refrigeration compressor could also be affected by cooling water temperature. However, for this case study there are no potential causes related to the Environment category that would result in the tower not meeting its bottom product specifications.

The Methods category included the process itself and methodology for operating the unit. The respondents highlighted the difference in feed configuration when comparing pre- and post-outage around Feeds #1 and #2. Excessive reflux was highlighted by a respondent due to frequent decay of the reflux drum level suspected associated with the tower being off material balance. Process control configurations such as Manual, Automatic, or Cascade mode selected by the process operator were identified as a potential cause. Operating procedures were not captured by the respondent and added on the list along with the path to the low rate previously discussed.

Under the Equipment category, the respondent identified the reboiler performance, valve alignments, and potential tray flooding. Damaged or fouling and tray hydraulic, such as unsealed downcomers, was added to the list of potential causes.

To Ian Nimmo's point, the human factor goes beyond the People category as most of the potential causes listed on the Methods and some under the Equipment categories involve people decision and interaction with the process. The tower feed configuration, controller operating mode, and valve alignments in the case study were set by process operators.

The CED is an effective and simple methodology when conducting group brainstorming sessions for gathering potential causes and subdividing them by categories. Disadvantages of the method reported include CEDs growing to a point where it becomes confusing with difficulties in illustrating interdependency between potential causes. CEDs do not provide any guidance relative to proving or disproving any of the potential causes listed. This weakness is addressed in the Kepner–Tregoe (KT) problem-solving methodology through a systematic approach extending beyond simple cause and effects.

KT [11] originated from the work of Benjamin B. Tregoe and Charles H. Kepner who were social scientists conducting research on decision-making and problem solving at the Strategic Air Command in the late 1950s. They discovered that effective Air Force officers were characterized by their logical thinking in processing information before acting. Those findings were expanded to the thinking habits of business managers and led to the development of a logical thinking approach toward problem analysis, decision analysis, potential problem (opportunity) analysis, and situation appraisal.

In KT, the problem analysis methodology is divided into six categories:

1. *State the problem,* or as common to other methodologies, a problem definition. While stating the problem may seem straightforward, the KT methodology emphasizes a precise problem definition where you can no longer explain why the deviation exists. For example, in the case study, the tower bottom is an intermediate stream undergoing further processing before qualifying as saleable product. While the saleable product does not meet specification and must be stored in an off-specification storage tank, stating the problem as an off-specification product would expand the scope of the analysis to all the operating units involved from the tower to storage. Since it is understood that for meeting the sale specification, the tower bottom intermediate stream must meet an operating specification, the problem definition previously stated was *the tower cannot meet the bottom stream specification.* The KT methodology suggests refining the problem statement until the question *why* can no longer be answered. This step is similar to the 5-why method, which, as stated earlier, can lead by itself to the root cause or refine the problem definition.
2. *Problem specification,* specify the problem into four dimensions.
   - *What* is the specific object affected by the problem?
   - *Where* is the problem or symptoms manifesting itself on the object?
   - *When* were the symptoms first observed?
   - *Extent* involves other questions related to how many objects are being impacted, what is the size of the deviation, and how many deviations with their associated trends.

   One element unique to the KT methodology is to frame the four dimension questions to an object that could be experiencing the same problem but is not. While reported as confusing by some practitioners, the *Could be, but is not* dimension opens the exploration of differences between objects affected and not affected by the problem leading the troubleshooter closer to the root cause.
3. *Develop a list of probable causes*, based on knowledge and experiences and/or by consideration to what has changed. The CED brainstorming approach could be integrated to this step.

4. *Testing the list of probable causes.* In this step, each probable cause listed must align with the dimensional questions *what, where, when, extent* on the *is and is not.* Misalignment with one or more dimensional questions would disqualify the cause being tested. Note that alignment of some probable causes with the dimensional questions may rely on one or several assumptions.
5. *Determine the most probable cause*, which would typically be the one that requires the least amounts of assumptions or unknowns for aligning with the dimensional questions. Generally speaking, the more assumptions required to make a potential cause fit the dimensional questions, the less likely it is the actual root cause.
6. *Confirmation of the most probable cause* could involve an experiment by either re-creating the undesired behavior or reversing the change that caused the problem. In a chemical plant, this may not always be feasible, and, in some cases, it may require validation through alternate non-destructive methodologies, such as simulations or laboratory experiments.

The problem definition and establishing a list of potential root causes are common to most troubleshooting methodologies. What make KT unique and powerful are the problem specification and potential root causes testing against the dimensional questions steps.

From the 5-Why methodology previously discussed, the case study problem definition could be refined to "Unseal downcomers in the stripping section". However, for the sake of discussion the original problem definition *The tower cannot meet the bottom stream specification* will be used for demonstrating the KT method.

Problem definition: *The tower cannot meet the bottom stream specification.*

The problem specification dimensional questions are described in Table 1.2 along with differences between the *is and is not* dimensions.

Many of the probable causes listed on the CED illustrated in Figure 1.4 can be rejected through field checks and engineering calculations, such as validation of instrumentations, valve alignments, reboiler performance, and material balances. Characteristics/properties of components *A* and *B* do not change when comparing pre and post shutdown, and it is therefore rejected as a probable cause.

The following are potential contributors to the probable cause, but they do not provide any clues as what the probable cause might be:

- process operator experience.
- numerous operating variables and configuration changes.
- potential distraction during startup.

The feed composition change is a result of whether the pre-fractionator is on stream or not. The case study system description omitted to state that the feed from the upstream tower is split between Feed #1 and Feed #2 when the pre-fractionator is off stream. In this configuration, splitting the feed between the two feed points is required due to nozzle and distributor limitations. However, the location of the flowmeters provides measurement of the total flow to the tower but not the split between

TABLE 1.2
Case Study Problem Specification

	Is (Deviation)	Is Not (Logical Comparison)	Differences
What	Tower bottom stream specification	Upstream tower stream specification	• Feed rates • Properties • Process conditions • Composition • Tray type/geometry • Feed configuration
Where	Stripping section	Rectifying section	• Tray type/geometry • Composition • Hydraulic load • Temperature • Properties • Gamma scan
When	Post shutdown	Pre shutdown	• Path to low rate • Gamma scan • Hydraulic load • Feed configuration
Extent	Component *A*: 115–25,240 ppm in bottom stream	N/A	N/A

Feed #1 and Feed #2. At first site, one may speculate on the split fraction, but those differences in feed configuration had been previously experienced without issues, making this probable cause unlikely.

Flooding, fouled or damage trays is ruled out based on the tower gamma scan. However, the low froth height on some of the bottom trays supports the unseal downcomer probable cause. Simulation and process calculations are however predicting the tray load close but above the downcomer seal point. As previously described, the path to low rate from unseal to seal downcomers was stated as a theory that could explain why the downcomer were unsealed. This probable cause is tested below against the dimensional questions.

Relative to ***What***, if the path to low rate requires higher process load for sealing the downcomer, why the upstream tower did not get affected. The upstream tower is also equipped with high-capacity trays that are also sensitive to unsealed downcomers at low rates. This potential problem was recognized during the development of the startup procedure and false load conditions (higher reboiler/reflux ratio) were specified in the procedure. Such specifications were however omitted for the tower experiencing the problem, which suggest to keep the operating procedure on the list of probable causes.

Where, the rectifying section is equipped with conventional two-pass trays, which have downcomer with a positive liquid seal, while the high-capacity trays in the stripping section have dynamic seal requiring a minimum liquid load for sealing the downcomers. This dimension explains why it occurs in the stripping section but not in the rectifying section.

When, the path to low rate before the shutdown occurs from high to low rate with sealed downcomers. On startup, the downcomers were initially unsealed and required sufficient liquid load for breaking the momentum of the vapor velocity flowing through the downcomers.

Extent, unseal downcomers result in little to no fractionation due to vapor bypassing the active area of the tray, therefore resulting in a bottom stream not meeting specification. Once the downcomer gets sealed, the tray regains fractionation efficiency and the bottom stream meets specifications.

Unseal downcomer resulting from the path to low rate (startup to low rate) has been identified as the probable cause. It seems to fit with what was known about the operating range of the high-capacity trays. However, further review of literature around tower stability during startup led to a re-examination of the tray operating limits, more specifically the downcomer critical velocity during startup. This limitation and how to predict it has been discussed by Kister [12, 13] and is characterized by a vapor velocity in the downcomer that prevents liquid from descending freely in the downcomer. This limitation disappears once sufficient liquid head seals the downcomers. This probable cause may also align with the problem specification dimensional questions as stated in Table 1.2. However, relative to *When,* if comparing to previous startup, why now? What was different and what would have caused a high vapor velocity in the downcomers. So far, the analyses of this case study have been done from a static perspective. The process variables' trend illustrated in Figure 1.2 shows significant variability and may require expanding the troubleshooting methodology to a *Sequence of Events Analysis.*

Sequence of events analysis [14] can be based on time (chronological), actions (operational step changes), or outcome behaviors. Process engineers often get overwhelmed when analyzing numerous process variable trends as they are trying to establish relationship among all the variables (cause and effects). Those relationships are already known and are driven by fundamental principles of thermodynamics, transport phenomena, and equipment capabilities. A sequence of event analyses should distinguish process inputs (human actions) from outputs (outcomes from the inputs). The outcome from the inputs should be consistent with fundamental principles and equipment capabilities. Any deviations from that becomes an anomaly that deserve more attention, where the use of the KT dimensional questions *What, Where, When,* and *Extent* will provide more clues relative to the probable cause. For instance, in the case study, numerous input changes have been initiated by process operators with outcomes on the bottom stream specifications. Establishing a sequence of events relies on analyzing process trends, reviewing operating logs, and discussions with process operators. The sequence of events for the case study is summarized below:

- Feed to the unit at time 0.
- Feed to the tower 15 hours after initiating feed to the unit.
- Ramping up steam to the reboilers 10 hours later.

- Tower fed from the upstream tower with feed split between Feed #1 and Feed #2. Through this period, several moves were executed on reflux and steam rates to the reboiler.
- Due to tower bottom stream not meeting specifications, operation decided to commission the pre-fractionator 12 hours later.
- Process engineering support was requested after commissioning of the pre-fractionator, which was unsuccessful in improving the bottom stream specifications (about 30 hours after initiating feed to the unit).
- Tower condition held steady for execution of a gamma scan. Conditions were also used by process engineering for process simulations and tray rating calculations.

The process trend illustrated in Figure 1.2 is extended for covering the entire event, as shown in Figure 1.5. Time 0 hour on the plot corresponds to the time at which the bottom stream analyzer was commissioned. The plot is subdivided into six periods on an outcome-based sequence, where periods 1, 2, 3, 4, and 6 correspond to the weight fraction of component *A* trending down in the bottom stream. Period 5, being the stable period, was requested by process engineering for conducting a gamma scan and gathering process data for simulation purposes.

The first observation from Figure 1.5 is the thermodynamic consistency behavior when observing the effect of the reboiler heat on the weight fraction of A in the bottom stream. When the reboiler duty increases, the weight fraction of A

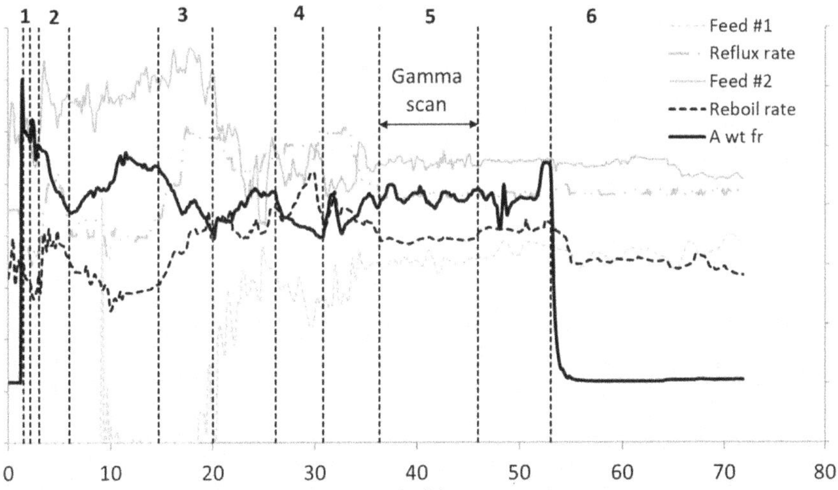

FIGURE 1.5 Case Study Sequence of Events – Bottom Stream Component *A* Outcome.

Reproduced with Permission from figure 1.8, Distillation: Avoid Problems during Tower Startup, André Bernard, Chemical Engineering, June 2012

ramps down. When the reboiler duty goes down, the weight fraction of A rises. This behavior, while not up to fractionation expectations, is thermodynamically consistent. Not meeting the fractionation efficiency might be more the result of poor vapor/liquid contact, which is related to the mass transfer equipment (trays) characteristics. Figure 1.6 shows the expected operating point relative to some of the tray operating limits expressed in terms of liquid and vapor loads. Figure 1.6a illustrates the operating points for the periods where component *A* trends down and Figure 1.6b illustrates the operating points for the periods when component *A* trends up. The location of the operating points in Figure 1.6b is quite revealing as the data is concentrated on each side of a tray limitation. At low vapor and liquid load, most of the operating points are below the downcomer seal point (i.e., minimum liquid load for sealing the downcomer), while at high vapor and liquid load, the operating points are above the downcomer critical vapor rate (i.e. vapor velocity causing liquid entrainment in the downcomer that prevents liquid from settling and sealing the downcomer). Figure 1.6a shows most of the operating points within the stable region of the tray. These observations still support the potential for unseal downcomers. However, while it is well understood that the downcomers were initially unsealed and remain so until the liquid load following the expected operating line rise above the downcomer seal point, it is not clear why the downcomer critical vapor rate was still a factor. In theory, by following the operating line, the downcomer should get sealed once the liquid load crosses the downcomer seal point. A transition zone may exist whether the downcomer seal point is approached from low to high or high to low rates. However, once sealed, the downcomer critical vapor rate limitations should no longer be a factor. In light of these new observations, the KT analyses are reframed with the following problem definition:

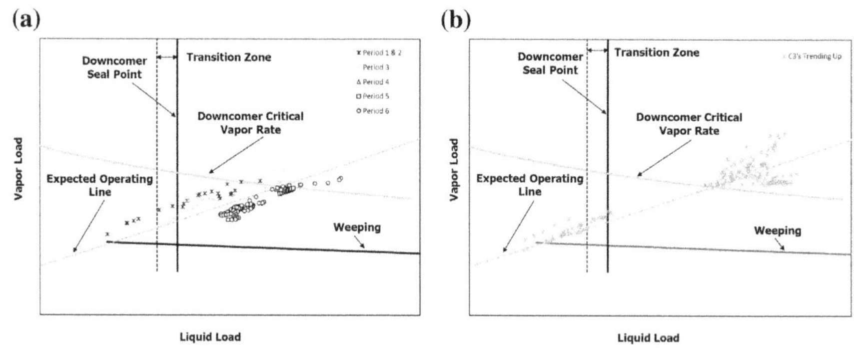

FIGURE 1.6 (a) Component *A* Trending Down versus Tray Operating Limitations; (b) Component *A* Trending Up versus Tray Operating Limitations.

Reproduced with Permission from figure 9, Distillation: Avoid Problems during Tower Startup, André Bernard, Chemical Engineering, June 2012

TABLE 1.3
Case Study Problem Specification Revision

	Is (Deviation)	Is Not (Logical Comparison)	Differences
What	Unseal downcomers	Tray damage	• Gamma scan
Where	Stripping section	Rectifying section	• Tray type/geometry • Hydraulic load • Gamma scan
When	Post shutdown After com. Prefrac.	Pre shutdown After exceeding seal point	• Path to low rate • Feed Configuration • Hydraulic load
Extent	Component *A*: 115–25,240 ppm in bottom stream	N/A	N/A

Problem definition: *Unseal downcomer in the stripping section.*

Two probable causes are being considered:

1. The path to low rates requiring higher liquid load for sealing the downcomer.
2. The feed transition to the pre-fractionator

Relative to **What**, both probable causes could have resulted in unseal downcomers. None of those causes would have resulted in surge events or causing forces that would result in tray damage.

Where, the tray type and gamma scan support both probable causes.

When, Pre and post shut down support the path to low rate probable cause. However, operating data for periods 1, 2, and 3 in Figure 1.6a shows that the downcomer seal point had exceeded well beyond the seal limits and as such the downcomer should have sealed itself and the tower fractionation efficiency re-established. Data points for periods 4 and 5 in Figure 1.6a are sitting on the predicted downcomer critical vapor rate limit. Transition to the pre-fractionator occurred after period 3, from which operating points when component *A* was trending upward were found above the downcomer critical vapor rate. The data suggest that the downcomers were initially unsealed while the operating point was below the downcomer seal point. It is believed that the downcomer sealed itself as the operating point for periods 1–3 moved beyond the seal point. At that point, proper reboiler duty and reflux rate conditions would have probably brought the tower bottom stream within specifications. However, due to numerous operating moves, the tower never had chance to stabilize. Furthermore, the feed transition to the pre-fractionator caused the downcomer to unseal while the vapor velocity was beyond the downcomer critical vapor rate. Based on the above analysis, the feed transition to the pre-fractionator is the likely cause for unsealing the downcomers in the stripping section.

Extent, same as previous evaluation.

A few more questions need answers for fully accepting the feed transition to the pre-fractionator as the root cause to unseal downcomers. Such transition occurred before without a problem, therefore what was different this time? This is where good honest discussion with process operators may fill some blanks. As previously indicated, due to nozzles and distributor limitations, the tower feed is split between Feed #1 and Feed #2 when feeding the tower from the upstream tower with the pre-fractionator offline. It was suspected that in the transition to the pre-fractionator, material from the upstream tower was transition to a single feed point too early, or before the pre-fractionator could offload the upstream tower. As such, the single feed point became a pinch point where significant liquid entrainment occurred, which starved the downcomers of the trays below and caused them to unseal. Observation of the tower sump liquid level trending down and reflux drum trending up around the transition supports those suspicions, which were also discussed with process operators.

Post Event Analyses. The reality of this case study is that it was resolved in two days with a much-abbreviated troubleshooting process. While the corrective action recommended was successful and avoided a costly and unnecessary outage, it did not clearly identify the root cause. While unseal downcomers was effectively diagnosed, the mechanism that caused the downcomer to be unsealed was not properly identified. Using *the path to low rate* as the root cause for unseal downcomer could potentially result in reoccurrence of the same problem. The refined KT analysis, literature review, and sequence of event analysis described earlier were conducted post event and resulted in identifying the true root cause of the problem. By doing so, operating procedures could be revised to prevent re-occurrence of similar events. It is not unusual for troubleshooting activities to occur under time constraints. As such, once the problem gets solved, any post event evaluation is rarely conducted, and the likelihood of re-occurrence is high if the root cause has not been properly identified. Post event evaluations are common when dealing with process incidents, the practice should also extend to operating problems.

Informal or Formal Troubleshooting. The instinct of troubleshooting should become second nature in the mind of process engineers and process operators. Understanding of operational normality and intuitively questioning whatever deviates from normality, why is a system operating a certain way? What, where, when did a deviation occur? Nurturing this thinking process will help resolving numerous problems quickly without the need for formal troubleshooting sessions. In conjunction with good thinking abilities, learning the fundamental principles surrounding the process and associated equipment capabilities is paramount to effective troubleshooting. As discussed earlier, learning on root causes shall be communicated in a way that would prevent re-occurrence. However, there will always be some problems arising, which may require involving experts or where the problem requires multidisciplinary expertise – in which case formal troubleshooting sessions might be required. Those sessions are typically but not always facilitated by someone unbiased and familiar with troubleshooting methodologies. Some methodologies such as KT have formal templates for formalizing and documenting the process. Formal troubleshooting would typically be conducted on more complex problems and may require several sessions. Throughout the process, unanswered questions may require research involving more data analyses, engineering calculations, and/or experiments for validating theories surrounding the problem. It is imperative to understand that the quality of a troubleshooting exercise is not limited to the work process itself but to the knowledge and expertise of those involved.

Nomenclature

AI	artificial intelligence
CED	cause and effect diagram
CPI	Chemical Process Industry
FTA	fault tree analysis
KPI	key performance indicator
KT	Kepner–Tregoe
LP	linear programming
NLP	non-linear programming
P&ID	pipe and instrumentation diagram
PFD	process flow diagram
SIS	safety instrumentation system

REFERENCES

[1] T. C. Arnst, "Optimize Until it Hurts," in *AIChE Spring Meeting, Ethylene Producer Conference*, Orlando, FL, 2006.
[2] S. G. Munoz and J. F. MacGregor, "Big Data – Success Stories in the Process Industries," *Chemical Engineering Progress*, March 2016.
[3] A. Gupta, "Introduction to Deep Learning: Part 1," *Chemical Engineering Progress*, June 2018.
[4] D. Mack and M. E. Naamani, "Practical Use Cases for Artificial Intelligence and Machine Learning," *Chemical Engineering Progress*, August 2022.
[5] A. Bernard, "Distillation: Avoid Problems during Tower Startup," *Chemical Engineering*, June 2012.
[6] A. Young, "Understanding 'The Five Whys': How To Successfully Integrate This Tool into Your Business," *Forbes*, June 2021.
[7] R. Deshotels and R. D. Zimmerman, *Cost-Effective Risk Assessment for Process Design*, McGraw-Hill, 1995.
[8] G. Ramu, *The Certified Six Sigma Yelloe Belt Handbook*, ASQ Quality Press, 2017.
[9] S. Hawksworth, T. Jordan, K. Jeffrey and D. Melideo, "Chapter 5 – Risk Assessment," in *Hydrogen Safety for Energy Applications – Engineering Design, Risk Assessment, and Codes and Standards*, Elsevier, 2022.
[10] I. Nimmo, "It's Time to Consider Human Factors in Alarm Management," *Chemical Engineering Progress*, November 2002.
[11] C. H. Kepner and B. B. Tregoe, *The New Rational Manager*, Princeton Research Press, 1981.
[12] H. Z. Kister, "When Tower Startup has Problems," *Hydrocarbon Processing*, February 1979.
[13] H. Z. Kister, *Distillation Operation*, McGraw-Hill, 1990.
[14] J. A. Klein, "The ChE as Sherlock Holmes: Investigating Process Incidents," *Chemical Engineering Progress*, October 2016.

2 Troubleshooting – The Object

2.1 MAJOR OPERATING UNIT CHARACTERISTICS

In chemical processes, the transformation of raw material is accomplished through several unit operations, which ensures the movement, transformation, and separation of process streams into saleable products. Each unit operation promotes conditions defined by thermodynamic, transport phenomena, and/or chemical reaction required for meeting the intent of the process. This chapter focuses on the main pieces of equipment typically found in the CPI. The discussion covers the fundamental principles, process objectives, equipment characteristics along with typical operating problems symptoms and diagnostics. In troubleshooting, what, where, and why are questions associated with the objects suspected responsible for performance deviation. Meaningful answers to those questions leading to a root cause and remedial actions strongly depends on understanding the parameters influencing the performance of the process equipment at fault.

2.1.1 Fired Equipment

Fired equipment involves the combustion of fuels for generating energy in the form of heat, which can be used as such or converted into power. While the discovery of fire goes back to pre-historical times, its use was limited to heating and lighting [1]. Steam power engines were developed much later in the 1700s and were a key technological breakthrough during the industrial revolution [2]. The main application of fired equipment in the CPI today can be divided into two categories: First, energy in the form of heat, typically accomplished by steam boilers and fired heaters, which could be associated with power generation through steam or gas turbines; second, as chemical reactors for highly endothermic reactions such as pyrolysis and steam reforming. Those applications share two fundamental principles: combustion and heat transfer. Pyrolysis and reformer furnaces involve a third function being chemical reaction.

2.1.1.1 Combustion Fundamentals

Combustion is a rapid chemical reaction of fuel with oxygen. For hydrocarbons, the combustion product is carbon dioxide and water per following stoichiometry:

$$C_nH_m + (2n + m/2)/2O_2 \rightarrow n\,CO_2 + m/2\,H_2O \tag{2.1}$$

If the fuel composition is known, the heat of reaction can be calculated from heat of formation or heat of combustion values tabulated in handbooks such as *Perry*

DOI: 10.1201/9781003449676-2

Chemical Engineering Handbook [3]. Components' heat of combustion tabulated in literature are typically reported in terms of high (HHV) and low (LHV) heating value per unit volume or unit mass. The HHV is based on water being in the liquid phase, while the LHV is established from water being in the vapor phase. The term gross heating value or gross combustion heat is also used in literature when referring to the HHV, while the net heating value or net combustion heat refers to the LHV. Both the HHV and LHV are established on the basis of reactants and products at 25 °C and 1 atm. The LHV can be obtained from the HHV by subtracting the equivalent heat of vaporization of water per unit of fuel as described by Equation 2.2:

$$LHV = HHV - \frac{m_{H_2O}}{m_{fuel}} \lambda_{H_2O} \tag{2.2}$$

where m_{H_2O}, m_{fuel} are the mass of water and fuel, respectively. λ_{H_2O} is the latent heat of water at 25 °C (2,440.4 kJ/kg [3]).

Example 2.1: Table 2.1 provides a natural gas composition and gross heating value on a dry basis provided by a gas supplier. The HHV and LHV values were extracted from *Perry Chemical Engineering Handbook* [3] for validating the heating value stated by the supplier. The combustion reactions of each of the components present in the gas stream are also provided.

The HHV and LHV values can be calculated from the components' individual values using the following expressions. Where the mole factions are first converted into weight fraction for calculating the heating value on a mass basis:

$$m_i = \frac{x_i \ MW_i}{\sum x_i \ MW_i} \tag{2.3}$$

TABLE 2.1
Natural Gas Composition and Associated Combustion Reactions

	MW	Mole Frac	HHV (BTU/lb)	LHV (BTU/lb)	
N_2	28.02	0.005			
H_2	2.016	0.0002	60,957.7	51,571.4	$H_2 + 1/2\ O_2 \rightarrow H_2O$
CO	28.01	0	4,343.6	4,343.6	$CO + 1/2\ O_2 \rightarrow CO_2$
CO_2	44.01	0.003			
CH_4	16	0.947	23,861	21,502	$CH_4 + 2O_2 \rightarrow CO_2 + 2H_2O$
C_2H_4	28.05	0	21,625	20,276	$C_2H_4 + 2O_2 \rightarrow 2CO_2 + 2H_2O$
C_2H_6	30.07	0.042	22,304	20,416	$C_2H_6 + 7/2\ O_2 \rightarrow 2CO_2 + 3H_2O$
PROPANE	44.09	0.002	21,646	19,929	$C_3H_8 + 5\ O_2 \rightarrow 3CO_2 + 4H_2O$
Total		0.9992			

Gross Heating Value, Dry Basis: 38.8 MJ/m^3

The mixture heating values are then obtained by summing up the product of each individual components' mass fraction by their corresponding heating values, with X below substituting H in HHV and L in LHV:

$$XHV = \sum m_i XHV_i \tag{2.4}$$

The HHV and LHV values for the natural gas mixture have been calculated from Table 2.1, Equations 2.3 and 2.4 with the results summarized in Table 2.2.

A validation of Equation 2.2 is done by calculating the LHV from the HHV tabulated for ethane $\left(C_2H_6\right)$ in Table 2.1. As per the combustion reaction of ethane stated in Table 2.1, 1 mole of ethane results in 3 moles of water, or 36 lb of ethane to 54 lb (3 × 18 lb) of water. The latent heat of water stated above is 2,440.4 kJ/kg or 1,050 BTU/lb after unit conversion. The LHV can then be calculated from the HHV:

$$LHV = 22{,}304 - \frac{54 \text{ lb } H_2O}{30 \text{ lb } C_2H_6} \times 1{,}050 \frac{\text{BTU}}{\text{lb}} = 20{,}414 \text{ BTU/lb} \tag{2.5}$$

The value calculated from Equation 2.5 falls within 0.01% from the value provided in Table 2.1 (20,416 BTU/lb).

The natural gas mixture calculated HHV of 23,346 BTU/lb displayed in Table 2.1 corresponds to 53.99 MJ/kg or after calculating the density of the mixture from the ideal gas law to 37.05 MJ/m^3 which falls within 4.5% from the value stated by the natural gas supplier. However, while HHV and LHV data are reported at 25 °C and 1 atm, natural gas volumetric heating value are based on standard conditions, which may vary depending on location. The ISO 13443 [4] standard reference conditions are 15 °C and 1 atm. API standard 2000 [5] define standard conditions at 60 °F and 14.73 psia. Adjusting the above calculation for the ISO 13443 standard condition reduced the gap between the calculated and reported HHV value to 1.2% (38.34 MJ/m^3).

TABLE 2.2
Natural Gas Mixture HHV and LHV Calculations

					HHV	LHV	m_iHHV_i	m_iLHV_i
	MW	Mole Frac	x_iMW_i	m_i	(BTU/lb)	(BTU/lb)	(BTU/lb)	(BTU/lb)
N_2	28.02	0.005	0.1401	0.008351				
H_2	2.016	0.0002	0.000403	2.4E-05	60,957.7	51571.4	1.4651081	1.2395099
CO	28.01	0	0	0	4,343.6	4,343.6	0	0
CO_2	44.01	0.003	0.13203	0.00787				
CH_4	16	0.947	15.152	0.903214	23,861	21,502	21,551.582	19,420.901
C_2H_4	28.05	0	0	0	21,625	20,276	0	0
C_2H_6	30.07	0.042	1.26294	0.075284	22,304	20,416	1,679.1366	1,537.0002
Propane	44.09	0.002	0.08818	0.005256	21,646	19,929	113.78062	104.75534
Total:		0.9992	16.78	1			23,346.0	21,063.9

Why HHV and LHV? The heat of combustion is important from a process engineering perspective as it provides the ultimate heat availability from combusting fuel in fired equipment. The efficiency of fired equipment can be directly calculated by comparing the residual heat contained in the combustion gases exiting the stack to the fuel heat of combustion relative to the reference temperature of 25 °C. The two heating values are known to cause confusion for many. For process engineers, it seems logical that water from a combustion reaction be in the vapor phase. Then why report two different heating values? The answer to this question lies in fuel characterization methodologies. As stated earlier, the heat of combustion can be calculated from tabulated data in literature when the molecular structure and composition of the fuel are known. If data are unavailable, the heat of combustion can always be estimated from molecular group contribution. However, in many cases, numerous and complex components, typically found in liquid fuel, waste, and biomass, make it impossible to calculate the heat of combustion. The heating value of such fuels is then determined experimentally with pressurized oxygen in a bomb calorimeter submerged in a water bath. The heat of combustion is correlated to the temperature rise in the water bath while combusting the fuel in the bomb. The detail procedure, described in ASTM D240–19 [6] or ASTM D4809–18 [7], states a bomb calorimeter and water bath initial temperature of 25 °C. Prior to initiating combustion through a firing wire, the bomb calorimeter is pressured up to 3.0 MPa with oxygen. The high pressure and test temperature of the bomb calorimeter caused the water produced by combustion to remain in the liquid phase, hence providing the HHV value of the fuel being tested.

Determination of the LHV can be done if the molecular hydrogen content of the fuel is determined by analysis. Knowing the atomic fraction of hydrogen contained in the fuel allows calculating the expected water formation from the combustion reaction. As described on Equation 2.1, in the general combustion stoichiometry of hydrocarbons, m moles of hydrogen contained in the fuel will result in $m/2$ moles of water. For example, a weight fraction of 0.2 atomic hydrogen within the fuel is equivalent to 0.2 lb-moles of hydrogen per lb of fuel and would result in 0.2/2 lb-mole of water or 1.8 lb of water/lb fuel. The LHV can then be calculated through Equation 2.5.

Fuel heating values and fuel pricing are typically provided in terms of HHV. In the absence of LHV data, there is nothing wrong in calculating energy efficiency based on HHV as long as the calculation basis is clear and historical data remains consistent.

Reaction characteristics. Mixing fuel and oxygen in the proportion described by Equation 2.1 is not sufficient for initiating a combustion reaction and sustaining a flame. Under ambient conditions, the reaction will still occur but at low rate without the presence of a flame and is defined as an oxidation. Initiating a combustion reaction requires fuel, oxidant, and an ignition source as described by the fire triangle. In a gas mixture, the ignition source provides the energy required for breaking the bonds of the combustible reactants, converting them into highly reactive species. Those species recombine to form the product of the combustion reaction CO_2 and H_2O. For flammable gases under ambient conditions, a minimum ignition energy (MIE) of 0.3 mJ is typically sufficient for initiating a flame. This is a relatively small amount of energy considering that a typical spark plug generates about 50 mJ. Hydrogen has a low MIE value of 0.018 mJ, compared to methane with a value of 0.28 mJ. Increase in fuel/air mixture temperature results in a lower MIE value. It is important to highlight that the minimum ignition energy required for initiating a combustion reaction is much lower than the energy released from the reaction itself. As such, once initiated, a combustion reaction is self-sustained. Removing the ignition side of the fire triangle would therefore be ineffective for terminating a combustion reaction.

Rapid propagation of a combustion reaction in a confined space or in a vapor cloud is defined as an explosion, characterized by high reaction rate typically milliseconds or microseconds [8] and high release of energy. The propagation rate further define explosion as detonation, where the reaction front moves at or faster than the speed of sound or deflagration where it is less than the speed of sound [9].

Flammability Limits define the proportions of fuel/air mixture that can combust. The lower flammability limit (LFL) is the fuel/air composition mixture below which combustion will not occur. The mixture is referred as too lean for sustaining a combustion. On the other hand, a rich fuel mixture above the upper flammability limit (UFL) will not combust either. Combustion can only occur with fuel/air mixture falling between the LFL and UFL. The LFL and UFL are determined experimentally by testing different fuel/air mixtures until flame propagation is visually observed after ignition by methodology, such as ASTM E681 [10]. Numerous references provide LFL and UFL data for pure components/air mixtures under ambient conditions. Table 2.3 provides flammability limits in air for components frequently found in fuels along with other flammability characteristics extracted from literature.

While it is preferable to determine the flammability limits of a mixture experimentally, the data might not always be available when troubleshooting. The Le Chatelier mixing rules can then be applied for providing an estimate of the fuel mixture flammability limits as described by the following equations:

$$LFL_{mix} = \left(\sum_{i=1}^{n} \frac{y_i}{LFL_i} \right)^{-1} \tag{2.6}$$

$$UFL_{mix} = \left(\sum_{i=1}^{n} \frac{y_i}{UFL_i} \right)^{-1} \tag{2.7}$$

TABLE 2.3
Combustion Characteristics of Gases Components [12–15]

Components	LFL (% volume)	UFL (% volume)	Autoignition Temperature (°C)	Maximum Laminar Flame Velocity in Air (cm/s)
Hydrogen	4.0	75	400	265
Carbon monoxide	12.5	74	-	39
Methane	5.0	15.0	540	34
Ethane	3.0	12.4	515	39
Propane	2.1	9.5	450	39
n-Butane	1.8	8.4	405	40

In the absence of data, an approximation developed by Jones (1938) can be used for approximating the flammability limits of components using the following expressions:

$$LFL = 0.55C_{st} \tag{2.8}$$

$$UFL = 3.5C_{st} \tag{2.9}$$

where C_{st} is the stoichiometric concentration of fuel in air:

$$C_{st} = \frac{Moles\ fuel}{Moles\ fuel + Moles\ air} \times 100 \tag{2.10}$$

Another method (Daniel A. Crowl, 2003) [9] correlates the flammability limits to the heat of combustion ΔH_c (10^3 kJ/mol).

$$LFL = \frac{-3.42}{\Delta H_c} + 0.569\Delta H_c + 0.00538\Delta H_c^2 + 1.80 \tag{2.11}$$

$$UFL = 6.30\Delta H_c + 0.567\Delta H_c^2 + 23.5 \tag{2.12}$$

Flammability data are typically reported under ambient conditions. The flammability range however increases as the temperature of the fuel/air mixture rises. The UFL increases and the LFL decreases as temperature increases. Zabetakis et al. (1959) [12] correlated flammability limits adjustments for fuel/air mixture temperature departing from a 25 °C temperature reference. The adjustment correlation is based on the fuel heat of combustion ΔH_c (kcal/mole) as described by the following equations:

$$LFL_T = LFL_{25} - \frac{0.75}{\Delta H_c}(T - 25) \tag{2.13}$$

$$UFL_T = UFL_{25} - \frac{0.75}{\Delta H_c}(T - 25) \tag{2.14}$$

Literature reports little effect of pressure on the LFL, except for pressure below 50 mmHg where combustion is not possible for most gases. The UFL typically trends with pressure, increases with higher pressure, and decreases at lower pressure. Zabetakis (1965) [11] correlated the UFL pressure adjustments based on the following expression where the UFL is the upper flammability limit at 1 atm. (0.101 MPa) and P is the pressure in MPa:

$$UFL_P = = UFL + 20.6(logP + 1) \tag{2.15}$$

Burning velocity and flame speed. The burning velocity of a fuel/air mixture is driven by the rate of the combustion reaction. As for any chemical reaction, it depends on the kinetics characteristics, which varies with fuel/air mixture composition, temperature, and pressure. It can be calculated from the ratio of the volumetric gas flow over the area of the flame front cone or from the gas velocity u and the half angle α of the apex of the flame cone as per Equations 2.16 and 2.17. The effect of pressure is reported as variable while the effect of temperature is more consistent and generally follows Equation 2.18:

$$S_u = \frac{V}{A} \tag{2.16}$$

$$S_u = u \sin \alpha \tag{2.17}$$

$$S_u = A + BT^n \tag{2.18}$$

where S_u is the burning velocity, A and B are constants, T is the temperature, and n denotes a constant for a particular mixture composition. Dugger et al. [12] reported a value of 2 for n for some paraffin hydrocarbons. The burning velocity does not account for turbulence and is also referred to as the laminar flame velocity. The maximum laminar flame velocities for typical fuels in air at 20 °C are tabulated in Table 2.3. The maximum laminar flame velocity generally occurs on the fuel rich side of the combustion stoichiometry.

While the laminar flame velocity depends on the mixture chemical properties, the actual flame velocity is influenced by turbulence created by pressure waves and recirculation of hot gases. The effect of those turbulence contributes to significantly increase the flame velocity at higher values than the laminar flame velocity.

In a chemical reactor, the input rate of reactants affects the reaction in terms of yields. Insufficient reactants may impact reaction rate and residence time resulting in different conversion and potentially total consumption of the reactants. On the other hand, high rates may lead to short residence time and un-converted reactants. A combustion reaction sustaining a flame differs from this behavior as the flame will chase the reactants. If the fuel velocity is less than the laminar flame velocity, the flame will travel backward, a phenomenon called flashback. In the opposite scenario where the fuel velocity exceeds the flame velocity, the flame will move forward away from the burner tip in a phenomenon known as flame liftoff.

Autoignition temperature (AIT) is a temperature at which a flammable mixture will ignite without an external ignition source. It also defines the minimum temperature that would sustain a flame and associated combustion reaction without an external heat source. The AIT of a flammable mixture is determined experimentally by exposing it to different temperature levels until observation of a flame and a sharp rise in temperature as described by the ASTM E659–15 [16] methodology. The AIT is affected by a few factors that could differ from value reported in literature. For instance, the AIT increases with pressures and with lean and rich mixtures. Higher oxygen concentration decreases the AIT. Note also that the AIT is typically measured in clean lab equipment and is not influenced by the catalytic behavior of contaminants such as rust often found in industrial pipes and equipment. The catalytic nature of rust contributes to reducing the AIT. AIT data for pure components typically found in fuel mixtures are provided in Table 2.3.

Adiabatic flame temperature is the temperature at which the enthalpy of the products equals the enthalpy of the reactants plus the heat of combustion. It is the maximum flame temperature theoretically achievable as it assumes no heat losses (i.e., adiabatic conditions). When based on the LHV and assuming complete combustion, it can be calculated using the following heat balance equations:

$$\Delta H_p + \Delta H_r + \Delta H_c = 0 \tag{2.19}$$

$$\sum_{i=1}^{p} n_{pi} \int_{T_0}^{T_2} Cp_{pi} T dT + \sum_{i=1}^{r} n_{ri} \int_{T_1}^{T_0} Cp_{ri} T dT + \Delta H_c = 0 \tag{2.20}$$

where n_{pi}, n_{ri} are the number of moles of product (p) and reactants (r) per moles of fuel, respectively. Cp is the heat capacity of the products (Cp_{pi}) and reactants (Cp_{ri}) and ΔH_c is the heat of combustion (LHV). T_0 is the heat of combustion reference temperature, T_1 is the initial temperature and T_2 is the final temperature or adiabatic temperature. Resolving Equation 2.20 is an iterative process where the adiabatic temperature T_2 is manipulated until closing the equation within tolerance.

While feasible with numerical methods, resolving Equation 2.20 can be extensive. The adiabatic flame temperature can also be easily determined with a process simulator by modeling the combustion in an adiabatic stoichiometric reactor block.

Flame quenching involves mechanisms which reduce the flame temperature. In fired equipment, there is the intended heat loss fulfilling the objective of transferring the heat of combustion to the process (i.e., steam generation or pyrolysis). Some losses are also due to heat leakage, which affects the thermal efficiency of the unit. Composition such as fuel/air mixture is another mechanism affecting the flame temperature. Excess air, for instance, contributes to decreasing the flame temperature due to heat consumed from unreacted oxygen and inert such as nitrogen. Processing waste stream as fuel in firing equipment is a common practice in the industry. Some of those waste streams often contain inert components, which contribute to decreasing the flame temperature. Flame quenching is also done intentionally for reducing the formation of thermal NO_x and is typically accomplished by flue gas recirculation or steam injection. Steam is also used in steam-assisted flares for smokeless operation by increasing turbulence/mixing of the flare gas with air. However, it also contributes to reducing the flame temperature and if not properly controlled could result in poor combustion.

While the adiabatic flame temperature is higher than the actual flame temperature, its calculation provides a methodology for evaluating the compositional impacts of the fuel/air mixture on flame quenching.

Nitrogen oxides NO_x is the result of nitrogen reacting with oxygen to form a variety of NO_x components. In fired equipment, oxygen required for the combustion reaction is supplied from air, which contains about 79% mole diatomic molecular nitrogen (N_2). While diatomic molecular nitrogen is relatively stable, it does react with oxygen at high temperature by the Zeldovich mechanism [13] shown in Equations 2.21a–2.21c and often referred as thermal NO_x. The generation of thermal NO_x is known to increase exponentially with temperature. Little if any NO is being generated at temperature below 760 °C, with generation limited by available oxygen above 1,300 °C:

$$N_2 + O \rightarrow NO + N \tag{2.21a}$$

$$N + O_2 \rightarrow NO + O \tag{2.21b}$$

$$N + OH \rightarrow NO + H \tag{2.21c}$$

Fuel NO_x and Prompt NO_x are two other mechanisms of NO_x formation in combustion. Fuels that contain organo-nitrogen components, such as coals and fuel oil, create fuel NO_x from oxidation and ionization of those components. Fuel NO_x is negligible in gaseous fuels such as natural gas and propane as they have no organically bound nitrogen.

Prompt NO_x is formed from nitrogen in the air, which reacts with oxygen and hydrocarbon radicals. It is referred as a complex mechanism involving hundreds of reactions and tend to occur in fuel rich conditions.

The majority of thermal NO_x in the flue gas of fired equipment is in the form of NO, which can then combine with oxygen once released into the atmosphere and form NO_2 and N_2O. In the presence of air and ultraviolet (UV) light from the sun, NO_2 reacts, and forms ozone and NO. UV lights from the sun creates free radicals from volatile organic compounds (VOCs) which then reacts with NO to form NO_2. This reaction cycle gives the ability of a single molecule of NO to produce ozone several times. The ozone formation is referred as tropospheric (lower atmosphere) ozone and is a primary component of smog, which is known to have adverse health issues. NO_x also contributes to acid rain as it contacts atmospheric moisture.

Many countries have regulations for limiting NO_x emissions from industrial facilities. As described earlier, NO_x emissions increase exponentially with flame temperature and available oxygen. Strategies for reducing NO_x involve low excess oxygen operation and burner design leveraging on flame quenching through flue gas recycle or steam injection. It could also be done post combustion through selective catalytic reduction (SCR) or selective noncatalytic reduction (SNCR). These reduction strategies are discussed in more detail in a later section.

Sulfur oxides (SO_x) result from the oxidation of organically bound and mineral sulfur contained in the fuel. The high-temperature experience in fired equipment favors the formation of SO_2 over SO_3, which tend to form at lower temperatures. Coals and heavy oils are known to contain significant amount of sulfur. Gas fuels such as natural gas have little to no sulfur. SO_2 and SO_3 emissions are a significant contributor to acid rain through reaction with atmospheric moisture resulting in sulfuric acid (H_2SO_4). SCR unit for NO_x removal promotes catalytic oxidation of SO_2 doubling the amount of SO_3.

Many countries are regulating SO_x emissions. The United States expanded their SO_x regulations beyond their borders through the Sarbanes–Oxley Act [17] passed by the Congress in 2002. The act states that foreign companies with listing on a US exchange with 500 or more US-based shareholders must comply with the US act. SO_x reduction can be accomplished by reducing the fuel sulfur content or post combustion through scrubbing.

Soot is the production of carbon particulates promoted by fuel rich mixtures. Often observed as black smoke around the flame, sooth formation enhances flame emissivity and increases radiant heat transfer. Methane does not produce any significant amount of soot, while higher C_2 hydrocarbons tend to generate soot. The formation of soot is known to adhere to walls and equipment surface, such as convective tube banks. The sulfur content of soot contributes to corrosion issues when in contact with moisture. Soot emissions are typically regulated.

Liquid fuels must first be vaporized before burning and sustaining a flame. This is typically accomplished in industrial fired equipment through steam atomization, which creates a liquid spray having a high interfacial area. Heat from the flame vaporizes the small, atomizes liquid droplets, which then combust. The effectiveness of atomization is strongly dependent on the liquid fuel viscosity, which is a function of temperature.

All the combustion characteristics described earlier also apply to liquid fuel. Additionally, the flashpoint temperature characterizes the flammability of liquid mixture. It is determined experimentally using the Cleveland Open Cup Tester as described by the ASTM D92–11 [18] procedure. The methodology involves increasing the temperature of a liquid contained in a small cup, with a test flame passed at specified intervals over the cup until observing ignition of the sample. The flashpoint temperature is more practical from a process safety perspective than for characterizing combustion characteristics of fired equipment.

Solid fuels are not so common in the chemical industry but are significant in the power and pulp and paper industries. Despite downward trends, coal is still a significant fuel in power plants [19]. Pulp and paper are known to burn residual bark and biomass. The characteristics of solid fuels differ from gas and liquid fuels, but the combustion reaction sustaining a flame still occurs in the vapor phase. The combustion of solid particles is characterized by three steps: drying, pyrolysis, and char burn. In small solid particles, drying and pyrolysis occur relatively fast. In large particles, devolatilization caused by pyrolysis of organic matter increases pressure in the solid pores, which push moisture toward the center of the particle, hence reducing the temperature inside the particle. The VOCs produced from pyrolysis ignite as they exit and form a flame in the presence of oxygen at the external surface of the particle. The flame contributes to increase the solid temperature and accelerate the pyrolysis mechanism. The VOCs increase the pore size of the solid particles as they migrate to the surface. As such, oxygen diffuses through those pores once completion of the pyrolysis step and initiates the char combustion step. The char reaction occurs at the gas/solid interface of the pores and generates CO, which then reacts with oxygen as it leaves the particle to form CO_2. Combustion of the solid particles is complete once the mass of combustible material within the particle has been fully consumed, leaving ashes as residue. The ash content depends on the fuel type, with wood residue varying between 0.1% and 1% and coal above 5% on a weight dry basis. Ashes are known to cause problems with slagging, fouling, erosion, and corrosion.

2.1.1.2 Combustion Equipment

The previous section discussed the fundamental principle of combustion. Creating reliable and effective combustion requires equipment which will promote those conditions. The equipment becomes the "what" object in the KT dimensional questions. However, before troubleshooting, normality or stable operating range must be understood, as it becomes the basis for comparison when looking for "What is different". Sustaining a flame requires a few basic operations such as moving the fuel and air to one or several burners. Ensuring a proper air/fuel ratio, operating above autoignition, providing proper flame pattern, and meeting environmental emission regulations are the criteria that are defined by the process. The equipment, although designed for meeting those criteria, have limitations, which defines the reliable operating range.

Fuel storage and transport depend on the nature, origin, and state of the fuel. Natural gas is typically supplied by pipelines and released at fuel gas header pressure. It is not uncommon, depending on composition, to burn waste gas stream from the process. Those streams are often mixed with natural gas at fuel gas header pressure, typically set as a compromise between process and burner-required pressure. Waste gas streams with significant departure from natural gas in terms of combustion properties may cause challenges relative to burner design and heat load controls. The presence of unsaturated hydrocarbons in waste gas streams should be avoided as those components would polymerize when exposed to combustion heat and plug the burners. Mixing waste liquid streams to a fuel gas system shall be avoided as it may cause localize auto refrigeration transient, potentially resulting in embrittlement of pipes and vessels. For example, 1 lb of saturated liquid propane at 70 °F (≈112 psig) mixed with 1 lb of methane at 60 psig and 70 °F result in a mixture of –44 °F. Propane would first cool down to about 36 °F due to the Joule–Thompson effect resulting from the de-pressure to 60 psig. The partial pressure imbalance in the mixture will result in liquid propane migration to the vapor phase, which will further reduce the temperature. Waste liquid streams should be vaporized and superheated prior to their release in a fuel gas system. Mixing of waste gas, while in the vapor phase, may result in condensing of heavy components or moisture when in contact with natural gas. The formation of any kind of droplets is known to clog burners and shall be avoided. Along with proper fuel gas system operating temperature, preventing liquid droplets from reaching the burners is accomplished by the combination of knockout drums and coalescer in proximity, when feasible, to the burners. Individual burner supply piping should take off from the top of the fuel gas header to prevent carryover of potential liquids or solids settled in the header. Besides knock out drums and coalescers, filter units would protect the burners from solid particles such as rust that might be present in the piping system. API 535 [20] recommends gas velocity below 50 ft/s in fuel gas header and below 75 ft/s for the individual burner takeoffs.

Liquid fuel handling depends on viscosity. Storage and transport of high viscosity fuel oil involve continuous circulation between storage and fired equipment's burners with heating set to target recommended viscosity of 20–45 cSt (120–200 SSU) for proper atomization. API 535 recommended practice states a minimum circulation rate equivalent to 1.5 times the total fuel consumed in the burners with pipe velocity not exceeding 6 ft/s. As for fuel gas, takeoff piping to the burners should come from the top of the fuel header to avoid carrying solid particles. Low-viscosity fuel oil may be fed directly to the burners at velocity below 3 ft/s. Care must be exercised relative to pre-heating duties, as vaporization would affect the stability of the burner. The atomization mechanism would also have to prevent vaporization in the burner. Stagnant volumes must be avoided in the handling of fuel oil, as viscosity will increase from low-temperature exposure and chemical reactions. Caution must be taken when blending heavy fuel oil from different origins as it may result in asphaltene precipitation, which will cause pluggage of the fuel oil circulation loop.

Draft refers to the mechanism by which oxygen is supplied to the combustion reaction, with ambient air being the most common source. The mechanisms are categorized in two types: natural and forced draft. Natural draft is one of the most common mechanism and leverage on the small negative pressure created by the chimney effect of the hot flue gas exiting the fire box and/or with premix burners, from the eduction effect created by high-velocity fuel gas discharging from an orifice at the inlet of a venturi. The negative pressure created by the chimney effect is normally less than 1 in w.c. Heat recovery in the convection section of fired equipment while improving energy efficiency increases flue gas pressure drop as it flows across tube banks exchangers. In some cases, the pressure drop would constrain the ability to draw air and may result in incomplete combustion. This problem is avoided by design using induced draft fans (ID fans), which can be seen as extending the natural draft operating envelop. The available negative pressure at the burner, however, still remains about 1 in w.c.

Force draft mechanisms are typically used with air pre-heat and gas turbine/furnace integration. A force draft fan (FD fan) is used with air pre-heat, and would typically increase the available pressure differential at about 4–10 in w.c. The higher-pressure drop is reported to have positive effect on fuel/air mixing which contribute in reducing the flame dimensions. Air supply from FD fans can be controlled through inlet guide vanes or variable speed drive. This provides better excess O_2 control over a wider operating range when compared to natural draft equipment. Operation of both FD and ID fans is not uncommon with this operating mode referred as balance draft. The reliability of fired equipment operating FD and/or ID fans is driven by the fan reliability, which is reported at 98% in API 535.

Burners are devices that provide fuel/air mixing at a desired ratio, and velocity for gas and liquid fuels. They also promote conditions sustaining combustion and flame stability. The design of burners is strongly dependent on fuel, draft, emissions, and furnace configuration. API 535 provides recommended practices relative to burner design and operation for fired heaters in general refinery services.

The fundamental principles previously discussed highlighted differences in combustion characteristics inherent to fuel type and composition. For instance, the AIT depends on fuel composition. Flammability limits and flame velocity are dependent on fuel composition and fuel/air mixing ratios. Gases versus liquids also impact burner design as liquid requires atomization. Environmental consideration, such as NO_x emission, is dependent on flame temperature. The total heat flux required by the process and the furnace geometry are other parameters that influence the design, number, and location of burners. The flame patterns often face constraints by the furnace geometry. The above considerations should typically be understood and addressed during the design of the equipment and burner selection. However, they are often forgotten or not clearly understood over years of trouble-free operation. Deviation from original design is not unusual, with the likelihood of malfunctions increasing with the magnitude of the deviations. It is, therefore, critical to understand the type of burner in operation and its operating range and characteristics. A description of burners commonly found in the industry is provided below.

Raw gas burners segregate the fuel and air until they mix at the ignition point. This configuration eliminates the risk of flashback allowing the burner to handle a wide range of fuel composition. As shown in Figure 2.1, air enters the burner at the inlet, then travels through the plenum or windbox before mixing with fuel released from the gas tip at the burner throat. Figure 2.1 illustrates a burner where fuel can be staged as it includes four-staged fuel gas tips on the circumference of the burner tile. As discussed later, this practice is intended to reduce NO_x formation by reducing the flame temperature. Air can be manually adjusted by the air register handle. The burner tile along with the drilling of the gas tip and aerodynamic of the burner defines the flame shape. Raw gas burners operate over a wide range of pressures, normally varying from 15 to 25 psig. The operating pressure is selected with consideration to the tip drilling and fouling potential. They can be designed for natural or force draft configuration with significant increase in heat release under force draft conditions.

Premix burners entrain partly or fully combustion air by the eduction effect of high-velocity fuel exiting a nozzle at the inlet of a venturi. Complete fuel/air mixing occurs in the venturi throat and through the extended tube length before the ignition point. Figure 2.2 illustrates a typical premix gas burner where it shows the gas inlet, primary air entrance at the spud, and secondary air for partly premix burners. The fuel/air mixture ignites as it enters the combustion zone. The momentum of the flame entrains secondary air, which completes the combustion in partly premix burners. Premix burners are sensitive to flashback and must deliver fuel at velocities above

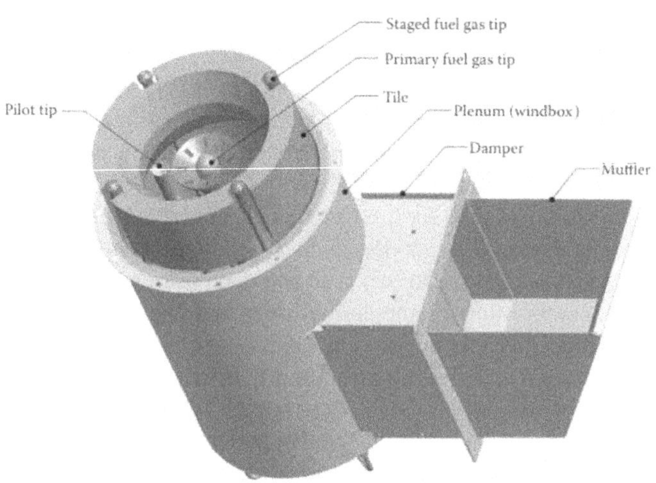

FIGURE 2.1 Raw Gas Burner [21].

Reproduced with Permission from figure 1.1, page 3, *The John Zink Hamworthy Combustion Handbook*, Second Edition, Volume 3 Applications, Edited by Charles E. Baukal, Jr, CRC Press, 2014

FIGURE 2.2 Premix Gas Burner [21].

Reproduced with Permission from figure 1.5, page 4, *The John Zink Hamworthy Combustion Handbook*, Second Edition, Volume 3 Applications, Edited by Charles E. Baukal, Jr, CRC Press, 2014

the flame speed. Reduction in burner pressure could result in fuel velocity below the flame speed, which would result in flashback. Flashback is a phenomenon where the flame chases the fuel backward into the burner and causes significant damage to the burner. This makes the burner sensitive and more specific relative to fuel type and composition. Premix burners are not recommended for fuel mixture containing more than 70% mole hydrogen due to the high flame speed of hydrogen. One advantage of premix burner is the air eduction mechanism, which tends to autoregulate the admission of air with fuel rates, improving overall efficiency. The flame shape is defined by the gas tip and, to a certain extent, the refractory tile. The flame volume is smaller and more defined than raw gas burners. While premix burners can be designed for force draft applications, they are typically used in natural draft fired equipment.

Oil burners require an additional feature when compared to gas burners. As previously indicated, combustion occurs in the vapor phase. Liquid fuel must therefore be vaporized before it combusts. This is accomplished through atomization where liquid fuel oil is broken down into small droplets. Heat from the flame can then propagate through the large oil interfacial area, causing vaporization of the droplets, which then ignite in the presence of air and temperature above autoignition. Atomization is commonly done with dry steam by either mixing fuel oil and steam inside the oil gun or steam assist/port mix atomization. In steam assist, the oil and steam ports join near the exit point, where mixing and atomization occur. Figure 2.3 illustrates a Hamworthy Steam Atomized Residual (SAR) oil gun, which is an internal mix atomization system, where the fuel travels through the inner pipe to the mixing core. From the mixing core, it flows through oil orifice into the outer tip chamber, where it mixes with the atomizing medium (typically steam). The mixture then flows at high velocity to the cone tip and is ejected in a highly atomized state. Figure 2.4 illustrates a Hamworthy Dual Stage (DS) oil gun, where primary atomization occurs by fuel pressure force through a mechanical atomization assembly (fuel body, disk, and tip). The second-stage atomization uses the momentum of the atomization medium where it mixes with the primary atomized oil in the outer tip. The mixture then flows to the

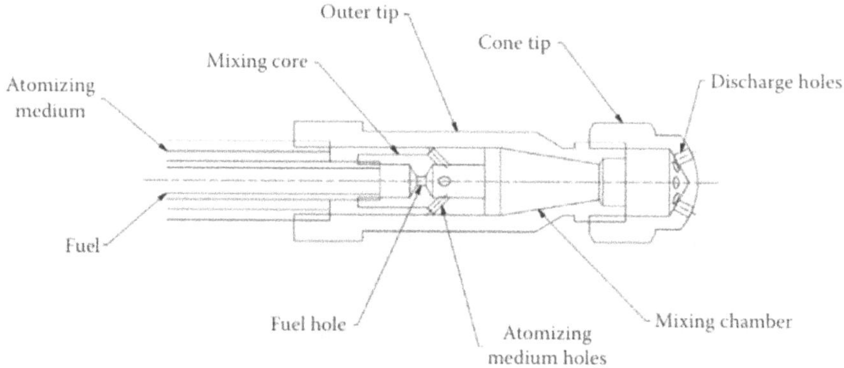

FIGURE 2.3 Hamworthy SAR Oil Gun [21].

Reproduced with Permission from figure 1.52, page 26, *The John Zink Hamworthy Combustion Handbook*, Second Edition, Volume 3 Applications, Edited by Charles E. Baukal, Jr, CRC Press, 2014

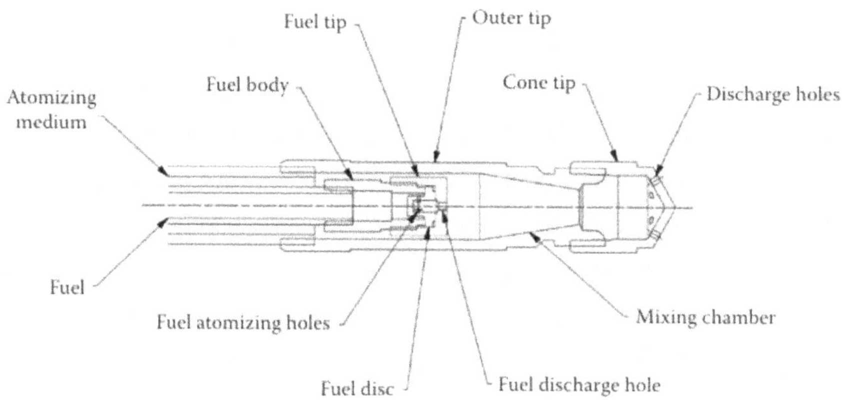

FIGURE 2.4 Hamworthy Dual Stage Oil Gun [21].

Reproduced with Permission from figure 1.53, page 26, *The John Zink Hamworthy Combustion Handbook*, Second Edition, Volume 3 Applications, Edited by Charles E. Baukal, Jr, CRC Press, 2014

cone tip where it is released through the discharge holes. Air or fuel gas can be used for atomizing light fuel oil as steam can cause oil vaporization resulting in vapor lock inside the gun. As previously discussed under handling, fuel oil viscosity is a critical parameter for proper burner operation and must be controlled through a temperature controller.

Low NO_x burners are specific to the reduction of thermal NO_x by either reducing the flame temperature or limiting the oxygen supply [22–24]. One of the early techniques, still used with fuel oil, involves staging the air. The burner is designed such that a primary combustion zone is fuel-rich. The lower flame temperature from an incomplete combustion and oxygen deficiency contributes to reducing NO_x formation. Secondary and sometimes tertiary air, segregated from the primary combustion zone by burner tiles configuration, complete the combustion in the air rich zone.

Most mechanisms with gas burners are based on fuel staging or flue gas recirculation. Fuel staging is done by promoting an air-rich mixture in the primary combustion zone. A fraction of the total fuel gas is supplied through a primary burner tip and mixed with the entire air flow in the primary combustion zone. Secondary burner tips supply the remaining fuel gas outside the primary combustion zone where it can combust as it mixes with excess air supplied to the primary combustion zone. As for air staged burners, the tile configuration ensures proper segregation of the primary and secondary fuel gas supply.

Flue gas recirculation can be done internally or externally. Internal recirculation is accomplished by eduction of flue gas inside the firebox promoted by high-velocity fuel gas at the inlet of a venturi. Internal recirculation is more common in pyrolysis furnaces. External flue gas recirculation, typically used in boilers, is accomplished through recirculation fan, which collects flue gas downstream the convection section. As per other NO_x reduction strategies, the intent is to reduce the flame temperature and oxygen in the combustion zone.

Steam or water injections to the combustion zone are other mechanisms used for reducing NO_x emissions. Low NO_x burners are known to generate longer and wider flames. They are also reported to be more sensitive to fouling due to smaller diameter fuel tips.

Pilots are small burners that provide an ignition source for the main burner. Their purpose is to trigger ignition of the main burner, not to provide flame stability of the main burner. The main burner must be stable on its own and operating at conditions that self-sustain combustion. Pilots are typically designed such that they can be removed for cleaning and maintenance with the main burner remaining in operation. Other than for maintenance, they should be lit at all times and are typically equipped with flame detection devices, such as UV scanners or flame rods. Ignition can be done remotely or locally with electronic ignition devices. It can also be accomplished manually with a torch or portable electronic starter. Pilots are controlled independently from the main burner and are typically supplied with natural gas.

Flame shape of burners are generally classified as round or rectangular [24]. Figures 2.2 and 2.3 illustrate round burners, though they do not show that they could also come in a rectangular shape. A round burner produces a round cylindrical or conical flame. With the tubes on the wall of the firebox, round burners could be located on the floor producing a vertical cylindrical flame or on the wall where they would produce a horizontal conical flame. Rectangular burners are generally installed on the wall of the firebox where they produce a flat flame that heats the refractory wall, which radiate heat to the tubes located in the center of the firebox. Pyrolysis furnaces are generally configured with both wall and floor burners; tube's location allows the entire tube surface exposed to the heat flux. On some fired heaters, especially on water tube boilers, the tubes formed the wall inside the firebox with only a fraction of the tube surface exposed to the heat flux.

Other considerations. Burners operating ranges are defined by a capacity curve [25], which expressed the fuel pressure as a function of the burner heat release (heat/

unit of time – e.g. MMBTU/hr). The curve is generated by the burner vendor and is specific to the burner design and fuel composition. Variation in fuel composition should be communicated with the burner vendor such that operating curves be established for the range of fuel expected to be processed. In practice, the fuel pressure is typically measured downstream a pressure regulator from where individual pipe segment supplies each individual burner. The burner operating curve is however based on fuel pressure at the burner tip. Pressure losses must be accounted for when using the burner performance curves. Operation at pressure above the burner curve shall be avoided unless discussed and approved by the vendor as it could potentially result in flame lift-off.

Excess oxygen is normally measured in a single location typically in the flue gas exiting the radiant section. From an energy efficiency and low NO_x perspective, operating at low excess oxygen is desirable. However, a single measurement does not guarantee that each individual burner was supplied the required volume of air for complete combustion. Maldistribution and air leakage may mislead operation at low excess oxygen, where some burners may operate sub-stoichiometrically. Automating air distribution is an option, which however limited to fired equipment operating with four burners or less due to cost and complexity. Fired equipment, such as pyrolysis furnaces, operates several burners, which require manual adjustments of the air register. Relying solely on the oxygen analyzer does not guarantee that each burner gets the appropriate supply. Balancing of the air supply may require oxygen testing with a portable gas analyzer. Operating at low excess oxygen based on the online analyzer may require additional safeguard such as combustible and CO online analyzers along with safety instrumented systems (SIS).

2.1.1.3 Fired Equipment Heat Transfer

The main objective of combusting fuel is to generate heat that will be transferred to a process. Figure 2.5 provides a simplified schematic of a furnace where a flame, sustained by the supply of a fuel/air mixture to a burner, generates heat. The heat balance on this system can be described by:

Heat inputs:

$$m_{Air}Cp_{Air}\Delta T_{Air} + m_{Fuel}Cp_{Fuel}\Delta T_{Fuel} + m_{Fuel}\Delta H_c \qquad (2.22a)$$

Heat outputs:

$$m_{Process}Cp_{Process}\Delta T_{Process} + m_{Process}V_{frac}\lambda_{Process} + m_{Flue\,Gas}Cp_{Flue\,Gas}\Delta T_{Flue\,Gas} + Losses \qquad (2.22b)$$

The heat inputs include sensible heat to air and fuel relative to the heating value reference temperature (25 °C). The LHV of the fuel is typically used for ΔH_c. However, if not available, the HHV can be used, providing consistency with historical data and comparison basis with other equipment. The heat output includes the process heat gain which involves sensible heat and latent heat when phase change occurs. The sensible heat of the flue gas relative to the stack temperature and the heat of combustion reference temperature (25 °C) represent unused energy from the combustion reaction. The term "Loss" represents heat losses across the wall of the furnace, which are relatively small and often neglected.

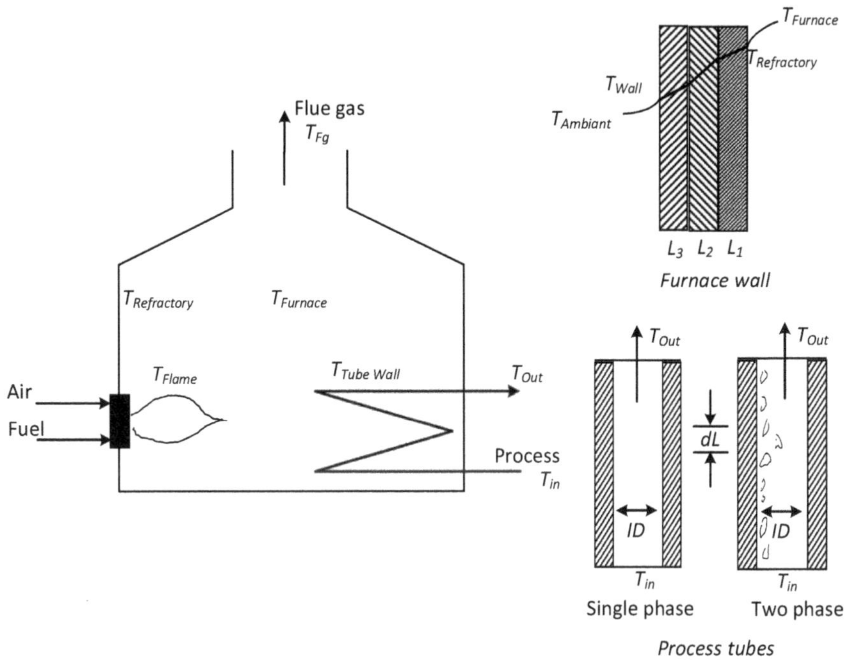

FIGURE 2.5 Simplified Furnace Schematic.

The heat resulting from combustion is transferred to the outer surface of tubes by different mechanisms: convection heat transfer from flue gas, radiant heat from the flame, flue gas, and refractory. Note that flame impingement to the tube surface is an undesired condition, which could cause tube failure due to the high flame temperature.

Convection heat transfer in the radiant section results from the velocity of flue gas at the tube surface. It is highly influenced by the furnace geometry, burner configuration, and rates. The air/fuel flow and combustion reaction at the burners initiates turbulence. The hot flue gas from the flame cools down as it contacts tubes. The cooler flue gas tends to flow downward, which creates flue gas recirculation inside the furnace.

The extent of radiant heat transfer by a flame depends on its luminosity. Oil flames radiate three to four times more than gas due to the high level of soot that promotes luminescence. The luminosity of gas flame is also influenced by the formation of soot, which tends to occur with $C2+$ hydrocarbons. Methane and hydrogen, on the other hand, have relatively low luminosity. Uniformity of radiant heat transfer depends on furnace geometry and burner locations as radiant heat transfer varies inversely to the square of the distance between two surfaces.

The flue gas is primarily composed of CO_2, H_2O, O_2 and N_2. Inert gases of symmetrical composition such as O_2, N_2 do not radiate heat. However, polyatomic gases like CO_2 and H_2O are considered radiant capable, hence radiating heat to a solid surface. While O_2 and N_2 are not radiant-capable, they transfer heat to CO_2 and H_2O as the temperature of those gases trends down due to gas radiant heat losses. The total emissivity of the combined flue gas decreases with temperature.

Refractory is a non-metallic material with composition varying based on application and is often formed of mixtures of silica and alumina. It can withstand high temperatures, has a high heat capacity, and provides good insulation. The high heat capacity causes a significant amount of heat to be stored by the refractory wall and extends the heat-up time on a furnace cold start. Once at steady state and promoted by high emissivity, heat transfer to the refractory wall is radiated back to the process tubes.

Equations 2.22a and 2.22b provide the heat balance between the combustion reaction, flue gas, and process stream. While it validates the thermodynamic consistency of the system, it does not provide any insights relative to heat transfer mechanisms and equipment capability. Without going into fundamental details, heat transfer in the radiant section of fired equipment can be expressed by:

$$Q_R = \underbrace{\sigma_0 \mathfrak{F}_{g,1}(T_g^4 - \theta^4)\,AF}_{\text{Radiant Heat}} + \underbrace{h(T - \theta)\,S_1}_{\text{Convection from Flue Gas}} = \underbrace{m_g\, Cp_g(T_C - T_F)}_{\text{Sensible Heat from Flue Gas}} - Wall\ Loses \tag{2.23}$$

where:

Q_R is the useful heat generated by the furnace radiant section.

σ_0 is the Stefan–Boltzmann constant (5.669×10^{-8} W/m^2-K^4).

T_g or $T_{Furnace}$ shown in Figure 2.5 is the flue gas temperature radiating to the tube surface.

T or $T_{Tube\ Wall}$ shown in Figure 2.5 is the flue gas temperature at the tube wall.

T_C or T_{Flame} shown in Figure 2.5 is the theoretical combustion temperature (adiabatic flame temperature).

θ is the temperature of the tube outside wall or tube metal temperature (TMT).

A is the continuous heat transfer area passing through the center of the tubes.

h is the convection heat transfer coefficient of the flue gas circulation around the tube surface.

S_1 is the exposed surface of the tubes.

F is an overall exchange factor, which is a function of the distance between the tube axes, the tube outside diameter, and the number of tube rows and can be approximated by $F = 0.012\,{}^{E}\!/_{D}{}^{2} - 0.2\,{}^{E}\!/_{D} + 1.2$ where E/D is the distance between tube axes and tube outside diameter.

$\mathfrak{F}_{g,1}$ is an overall exchange factor, which depends on geometry as well as the total emission factor from all the surfaces as expressed by Equations 2.24 and 2.25:

$$\frac{1}{\mathfrak{F}_{g,1}} = \frac{1}{\epsilon_1} + \frac{1}{F_{1,g}} - 1 \tag{2.24}$$

$$F_{1,g} = \epsilon_g \left[1 + \frac{{}^{A_R}\!/_{FA}}{1 + \frac{\epsilon_g}{(1-\epsilon_g)} \times \frac{1}{\mu_{RI}}} \right] \tag{2.25}$$

where:

$\in_1$ is the tube emissivity.

$\in_g$ is the flue gas emissivity.

A_R is the refractory radiating surface.

μ_{RI} is the angle factor between the refractories and the receiving surface and can be approximated by: $\mu_{RI} = -0.16 \ln\left(\frac{A_R}{FA}\right) + 0.49$.

Other than the temperatures and flue gas emissivities, all other variables involved in the radiant heat term of Equation 2.23 are constant and are set by the furnace geometry. The convection term is influenced by temperature and the heat transfer coefficient, which is influenced by flue gas circulation velocity near the tube wall. An increase in heat input is therefore characterized by a higher flue gas temperature, which contributes to increase the flue gas velocity near the tube wall. Other factors influencing the heat transfer rate would be the fuel composition, excess oxygen, and fuel/air inlet temperature. These factors affect the flame temperature and partial pressure of water and CO_2, which, in turn, influence the overall flue gas emissivity.

On the process side, heat transfer is achieved by force convection of the fluid circulating inside the tubes. For single phase, the process sensible heat term from Equation 2.21b can be related to the heat transfer coefficient by the following expression:

$$dQ = m_{Process} Cp_{Process} dT_{Process} = h(2\pi r)\, dL\left(T_W - T_{Process}\right) \tag{2.26}$$

T_W and $T_{Process}$ are the inside wall and bulk process temperature at position L along the process tube. The total heat can be obtained by integrating Equation 2.26 along the full tube length. Or it can be simplified based on average wall and process temperature:

$$Q = m_{Process} Cp_{Process} \Delta T_{Process} = h\, A_{Tube}\left(T_W - T_{Process}\right)_{Average} \tag{2.27}$$

The heat transfer coefficient can be obtained by the relation recommended by Sieder and Tate:

$$h = 0.027 \frac{k}{d_{tube}} Re^{0.8} Pr^{1/3} \left(\frac{\mu}{\mu_W}\right)^{0.14} \tag{2.28}$$

where Re is the Reynolds number $(d_{tube} v\, \rho/\mu)$, which carries the transport properties. Pr is the Prandtl number $(Cp\, \mu/k)$ and carries heat transfer properties.

For two-phase flows (liquid to vapor), the heat transfer coefficient depends on more parameters but can be estimated from the Dengler relation which was established empirically for rising flows in vertical tubes:

$$\frac{h_{TP}}{h_L} = 3.5\left(\frac{1}{X_n}\right)^{0.5} \qquad \text{with } 0.25 < \frac{1}{X_n} < 70 \tag{2.29}$$

X_n is the Lockart–Martinelli parameter

$$X_n = \left(\frac{1-y}{y}\right)^{0.9}\left(\frac{\rho_V}{\rho_L}\right)^{0.5}\left(\frac{\mu_L}{\mu_V}\right)^{0.1} \tag{2.30}$$

where:
h_{TP} is the two-phase flow heat transfer coefficient (W/m^2-K).
h_L is liquid phase heat transfer coefficient per Equation 2.28.
y is the mass vapor fraction.
ρ_L, ρ_V are the liquid and vapor densities (kg/m^3).
μ_L, μ_V are the liquid and vapor viscosities (Pa-s).

Note that for hydrocarbons mixtures, the properties and vapor fraction will vary as the fluid travels up the tubes. For steam generation, while the vapor fraction is expected to vary along the tube, the vapor and liquid properties should remain similar as both water and steam would be expected to remain at saturation.

The entire surface of the tubes is typically exposed to radiant heat from the flame, flue gas, and refractory in fired heater designed for heating hydrocarbons. In water tube boilers, the tubes form the inside wall of the radiant section. The area in refractory is significantly less than in fired heaters and pyrolysis furnaces. As such, heat transfer in the radiant section depends more on radiant heat from the flame, flue gas, and convection from the flue gas circulation. Furthermore, because the water tube forms a wall, only a fraction of the tube surface is exposed to the flue gas. The back of the tube is not exposed and would be cooler. As shown in Figure 2.5, steam bubbles will form on the hot side of the tube.

The furnace configuration illustrated in Figure 2.5 is limited to the radiant section and would be characterized by high stack temperature reflecting low efficiency. Package boilers generating saturated steam would recover some of that heat for preheating boiler feed water and combustion air in a convection section. Modern pyrolysis furnaces and high-pressure boilers are typically designed for maximum heat recovery. Figure 2.6 provides a schematic of a typical D type high-pressure boiler commonly used in the CPI. Residual heat from the flue gas exiting the radiant section is used across a first tube bank, commonly called economizer, for preheating the boiler feed water supply to the steam drum. Two other tube banks are used for superheating saturated steam from the steam drum. Boiler feed water is injected between the two superheating coils at rates required for meeting the superheated steam target temperature. Combustion air pre-heating can be accomplished in another tube bank or as illustrated in Figure 2.5 in an external exchanger. While maximizing heat recovery, it is critical to maintain the flue gas temperature above its dew point. Due to the presence of CO_2 and potential sulfur compounds (fuel dependent), condensation of moisture would create a highly corrosive mixture of sulfurous and carbonic acids.

If no phase change occurs on any of the tube banks, the heat balance can be calculated from $m * CP * \Delta T$ on the flue gas and/or the process side. In such

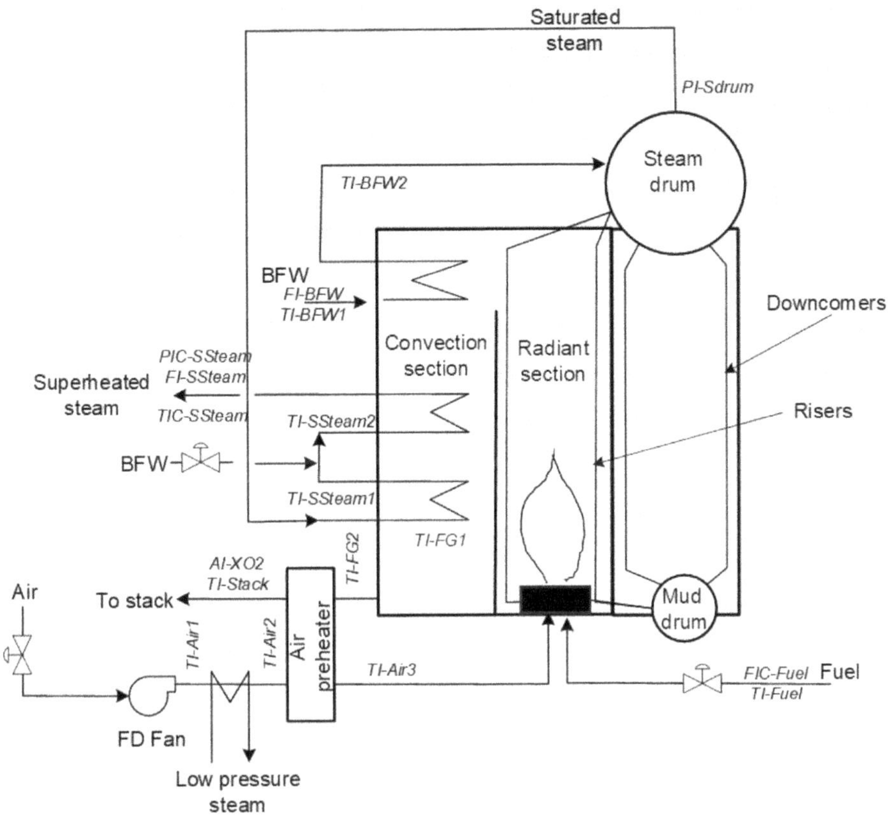

FIGURE 2.6 D-Type High-Pressure Steam Boiler.

case, the heat transfer coefficient can be obtained from the Grimison correlation [26]:

$$h_{Tube\ Bank} = b\frac{k}{OD_{Tube}}Re^{m}Pr^{1/3} \tag{2.31}$$

The coefficient b and m values are provided in Table 2.4. The velocity for the Reynolds number is evaluated at the smallest cross section across the tube bank.

The above correlations might be more useful from a design perspective and may provide little to no practicality for operation process engineers. Their review however helps in understanding the concepts, relationship among operating variables, and expected cause and effects. The high-pressure boiler illustrated in Figure 2.6 is used as an example on how this system should be characterized and monitored from a heat transfer perspective.

Example 2.2: This discusses high-pressure boiler heat transfer performance. While all chemical plants rely on steam for energy, boiler's operation are frequently

TABLE 2.4
Grimison Correlation Coefficient *b* and *m* for Tube Banks of Ten Rows or More [26, 27]

$$X_T = \frac{e_T}{OD_{Tube}} \qquad X_L = \frac{e_L}{OD_{Tube}}$$

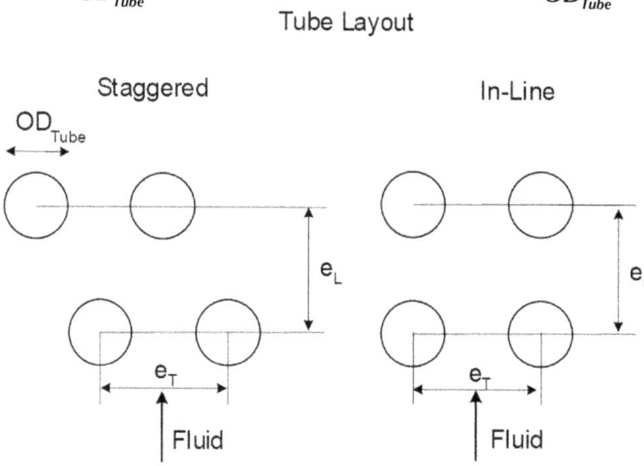

	$X_T = 1.25$		$X_T = 1.5$		$X_T = 2$		$X_T = 2.5$	
X_L	*b*	*m*	*b*	*m*	*b*	*m*	*b*	*m*
In-Line								
1.25	0.383	0.592	0.302	0.608	0.110	0.704	0.0696	0.752
1.5	0.404	0.586	0.275	0.620	0.111	0.702	0.0746	0.744
2	0.460	0.570	0.329	0.602	0.252	0.632	0.218	0.648
3	0.319	0.601	0.393	0.584	0.411	0.581	0.315	0.608
Staggered								
0.6							0.234	0.616
0.9					0.491	0.571	0.441	0.581
1			0.547	0.558				
1.125					0.526	0.565	0.570	0.560
1.25	0.570	0.556	0.556	0.554	0.571	0.556	0.574	0.562
1.5	0.496	0.568	0.506	0.562	0.497	0.568	0.537	0.568
2	0.444	0.572	0.458	0.568	0.530	0.556	0.494	0.570
3	0.341	0.592	0.392	0.580	0.484	0.562	0.463	0.574

overlooked in plant operation. Performance monitoring is often limited to raw variables extracted from plant data historian and trended in the time domain. Figure 2.6 includes process variables in italic that are available from the plant data historian for this specific example. From a load perspective, the steam header pressure (*PIC-Steam*) is controlled by regulating the fuel flow (*FIC-Fuel*). The air flow is controlled for meeting a target excess oxygen as measured by an O_2 analyzer (AI-XO_2) in the flue gas. Process variable monitoring in the time domain from a data historian would likely include the fuel, boiler feed water, superheated steam flow, and the stack excess O_2. Some may also follow the stack temperature as an indirect measure of the boiler efficiency. In this example, it is assumed that the boilers fire natural gas when the plant is down or during startup and shut down. At steady state, it is assumed that a mixture of natural gas and waste gas from the plant is being fired in the boiler. Table 2.5 compares the fuel composition between natural gas and the combined natural/waste gas mixture. The hydrogen content is one of the main differences that draw attention. Natural gas contains very little hydrogen but when mixed with the plant waste gases, the hydrogen mixture composition rises to 68% mol. From the earlier discussion on combustion, this should bring awareness relative to the burner design and flame stability. In this particular case, four wall burners located on each corner of the radiant section are being used. The burners are low NO_x design with raw gas burner contained in a windbox configured for staging combustion air. The raw gas burner design makes it insensitive to flashback when transitioning to a rich hydrogen fuel. Staging combustion air reduces the flame temperature, which however remains warmer on rich hydrogen fuel than on natural gas. The LHV shown in Table 2.5 is calculated from the individual component's heat of combustion, using Equation 2.4. Applying Equation 2.1 (combustion stoichiometric reaction) on all the components contained in the fuel mixture provides the expected stoichiometric quantity (mole and mass) of the combustion product CO_2, H_2O and N_2 per unit mass of fuel. The contribution of NO_x and other potential by-products of combustion is assumed negligible from an energy balance perspective. The excess O_2 does not influence CO_2 and H_2O but does influence N_2. The total air supply per unit mass of fuel along with the

TABLE 2.5
Natural Gas and Natural Gas/Waste Gas Mixture Composition

	Natural Gas		Natural Gas/Waste Gas	
	Mole Fraction	LHV (BTU/lb)	Mole Fraction	LHV (BTU/lb)
N_2	0.0050		0.0020	
H_2	0.0002	51,571.4	0.6770	51,571.4
CO	0	4,343.6	0.0008	4,343.6
CO_2	0.0030		0.0016	
CH_4	0.9470	21,502	0.3144	21,502
C_2H_4	0	20,276	0.0006	20,276
C_2H_6	0.0420	20,416	0.0032	20,416
C_3H_8	0.0020	19,929	0.0004	19,929
Total	0.9992	21,064	1.0	27,168

flue gas composition can be calculated knowing the stack excess O_2 from (AI-XO_2) per following:

$$AIXO_2 = \frac{O_{2Stack}}{FlueGas} \tag{2.32}$$

The mole of flue gas per unit mass of fuel can be expressed by:

$$Flue\ gas = CO_2 + H_2O + N_{2\ Stoich} + N_{2\ Excess} + O_{2\ Excess} \tag{2.33}$$

$O_{2\ Excess} = O_{2\ Stack}$ from which $N_{2\ Excess}$ can be substituted in Equation 2.33 by:

$$N_{2\ Excess} = \frac{0.79}{0.21} O_{2\ Excess} \tag{2.34}$$

After substitution, Equation 2.33 becomes:

$$FlueGas = \frac{CO_2 + H_2O + N_{2Stoich}}{\left(1 - \frac{0.79\ AIXO_2}{100 \times 0.21} - \frac{AIXO_2}{100}\right)} \tag{2.35}$$

Resolving Equations 2.32–2.35 provides the flue gas mole fractions of CO_2, H_2O, and total N_2 per unit mass of fuel. The residual heat from the combustion reaction (i.e., stack heat losses) can be calculated from the stack temperature and the LHV reference temperature of 25 °C using the heat capacities as described by Equation 2.36 with the subscript "*i*" referring to the components CO_2, H_2O, N_2 and residual O_2 in the flue gas:

$$Q_{Stack\ Losses} = \sum_{i=1}^{4} \int_{25}^{T_{Stack}} m_i Cp_i dT \tag{2.36}$$

The individual component's heat capacity formulation is given by Equation 2.37 with the coefficient provided in Table 2.6:

$$Cp = a + bT + cT^2 + dT^3 + e/T^2 \tag{2.37}$$

Equation 2.36 can also be used for calculating the adiabatic flame temperature. In this case, $Q_{StackLosses}$ is substituted by the LHV and T_{Stack} by the adiabatic flame temperature with Equation 2.36 solved iteratively for the adiabatic flame temperature. It is understood that the actual flame temperature is less than the adiabatic flame temperature due to radiant losses and in this case air staging to the burners. The relevance of calculating the adiabatic flame temperature is for estimating the heat duty distribution between the radiant section and across the convection section. It

TABLE 2.6
Heat Capacity Coefficient for Typical Fuel and Flue Gas Components [28]

Component	Formula	*T*	*A*	$b*10^2$	$c*10^5$	$d*10^9$	Temp. Range (in *T*)
Hydrogen	H_2	°C	28.84	0.00765	0.3288	−0.8698	0–1500
Carbon monoxide	CO	°C	28.95	0.4110	0.3548	−2.220	0–1500
Methane	CH_4	°C	34.31	5.469	0.3661	−11.00	0–1200
Ethane	C_2H_6	°C	49.37	13.92	−5.816	7.280	0–1200
Propane	C_3H_8	°C	68.032	22.59	−13.11	31.71	0–1200
n-Butane	C_4H_{10}	°C	92.30	27.88	−15.47	34.98	0–1200
Oxygen	O_2	°C	29.10	1.158	−0.6076	1.311	0–1500
Nitrogen	N_2	°C	29.00	0.2199	0.5723	−2.871	0–1500
Water (vapor)	$H_2O_{(V)}$	°C	33.46	0.6880	0.7604	−3.593	0–1500
Water (liquid)	$H_2O_{(L)}$	°C	75.4				0–100
Air		°C	28.94	0.4147	0.3191	−1.965	0–1500

Cp units are J/(g-mol °C)

also provides clues on the impact of firing waste gas of different compositions. In this example, the plant waste gas is rich in hydrogen, which contributes in elevating the adiabatic flame temperature by about 105 °C, when transitioning fuel from natural gas to the steady-state fuel gas composition (i.e., natural/waste gas mixture). In the scenario where a waste gas stream rich in inert was fired in the boiler for heat recovery rather than burned to flare, a reduction in flame temperature would be expected. The magnitude of the reduction could potentially cause flameout of the burners if the temperature drops below the AIT of the mixture. While the adiabatic flame temperature might still be above the AIT, its calculation would suggest a significant drop in flame temperature, which would raise concerns and support the waste gas composition as the potential root cause for burner flameout.

The heat input to the boilers per Equation 2.22a includes the heat of combustion (LHV) and the sensible heat of air and fuel evaluated per unit mass of fuel using the heat capacity as described by Equations 2.22a and 2.36 applied to the air and fuel composition. An average heat capacity value can be used as a simplification to Equation 2.36 if the temperature departure from the reference temperature of 25 °C is relatively small.

The above calculations are based on the fuel composition, stack excess oxygen measurement, and temperatures. It provides the energy available per unit mass of fuel. While the fuel flow is measured by flowmeter *FIC-Fuel*, the change in fuel composition could likely affect the accuracy of the measurement. A heat balance on the water/steam side will allow back calculation of the fuel flow, which will then be compared to the plant value.

Heat from the radiant section contributes in heating all the boiler feed water entering the steam drum to its saturated temperature as expressed by the *BFW Sensible Heat* term in Equation 2.38. Given an assumed 4% continuous blowdown from the steam drum, only 96% of the supplied boiler feed water gets evaporated. The heat capacity of liquid water is assumed constant under those conditions at a value of 4.2 kJ/kg:

$$Q_{Radiant} = FIBFW\left[\underbrace{Cp_{H_2O}\left(TIBFW2 - T_{Sat}\right)}_{BFW\ Sensible\ Heat} + \underbrace{0.96 * \lambda_{H_2O}}_{Latent\ Heat} \right] \tag{2.38}$$

The heat duty on the economizer can be calculated from the inlet/outlet temperatures using the heat capacity of water and the mass flow of boiler feed water as described by Equation 2.39:

$$Q_{Economizer} = FIBFW\, Cp_{H_2O}\left(TIBFW2 - TIBFW1\right) \tag{2.39}$$

Neglecting pressure drop, the heat duty of the low-temperature superheating coil can be calculated from the enthalpy difference obtained from the steam table between coil inlet and outlet, multiplied by the saturated steam flow as per Equation 2.40:

$$Q_{LTSPHTR} = 0.96\ FIBFW\left(H_{TISSteam1} - H_{SteamDrum}\right) \tag{2.40}$$

The boiler feed water flow injection between the two superheating coils is not measured but can be calculated from a heat balance around the injection station. Since there is no external heat duty, the heat balance can be expressed by:

$$BFW\left(H_{TISSteam2} - H_{BFW}\right) + 0.96\ FIBFW\left(H_{TISSteam2} - H_{TISSteam1}\right) = 0 \tag{2.41}$$

The heat balance can also be expressed after solving for the boiler feed water flow *BFW*:

$$BFW = 0.96\ FIBFW \frac{\left(H_{TISSteam1} - H_{TISSteam2}\right)}{\left(H_{TISSteam2} - H_{BFW}\right)} \tag{2.42}$$

Similar to the low-temperature superheating coil, the heat duty of the high-temperature superheating coil can be obtained from the enthalpy difference and steam flow:

$$Q_{HTSPHTR} = \left(0.96\, FIBFW + BFW\right)\left(H_{TICSSteam} - H_{TISSteam2}\right) \tag{2.43}$$

The total heat transferred from the combustion reaction to boiler feed water and steam can be expressed by:

$$Q_{Total\ Steam} = Q_{Radiant} + Q_{Economizer} + Q_{LTSPHTR} + Q_{HTSPHTR} \tag{2.44}$$

This heat can also be expressed by combining Equations 2.22a and 2.36, in terms of fuel and flue gas:

$$\frac{Q_{Combustion}}{Mass_{Fuel}} = m_{Air}Cp_{Air}\Delta T_{Air} + m_{Fuel}Cp_{Fuel}\Delta T_{Fuel} + m_{Fuel}Q_{Combustion} - \left(\sum_{i=1}^{4}\int_{25}^{T_{TIFG1}} m_i Cp_i dT\right) \tag{2.45}$$

With the mass rate of fuel obtained from the ratio of $Q_{Total\ Steam}/Q_{Combustion}$. The flue gas temperature profile can be calculated from the flue gas temperature TI-FG1 and $Q_{radiant}$, $Q_{Economizer}$, $Q_{LTSPHTR}$ and $Q_{HTSPHTR}$ using Equation 2.36 and the appropriate temperature boundaries. The results of the above evaluation are summarized in Table 2.7.

TABLE 2.7
Example 2.2 Results Summary

	Natural Gas		Natural Gas/Waste Gas	
Heat Duty Profile	**Heat Duty (GJ/hr)**	**Heat Duty Distribution (%)**	**Heat Duty (GJ/hr)**	**Heat Duty Distribution (%)**
Radiant section	123.1	58	160.0	57
Economizer	24.5	12	33.5	12
Low temperature superheater	22.6	11	31.5	11
High-temperature superheater	26.0	12	38.4	14
Air pre-heater	15.1	7	18.4	7
Total	211.4	100	281.7	100
Temperature Profile	**°C**		**°C**	
Adiabatic flame	1,919		2,025	
Bridge wall	1,103		1,206	
Economizer outlet	890		980	
High-temperature superheater outlet	655		709	
Low-temperature superheater outlet	439		474	
Stack	166		184	
Steam generation (% of design)	58		74	
Efficiency (%)	86.8		86.8	
	Calculated (kg/hr)	**Measured (kg/hr)**	**Calculated (kg/hr)**	**Measured (kg/hr)**
Fuel flow	4,325	3,660	4,505	4,755
Difference (%)	18		−5.2	

Despite the change in fuel composition and steam generation output, the two sets of conditions evaluated shows similar heat distribution and overall thermal efficiency. Without the above heat and material balance (H&MB), the higher stack temperature on the natural gas/waste gas mixture could mislead performance evaluation of the boiler by concluding a loss of efficiency. The total heat transfer of 57–58% in the radiant section agrees with values of about 60% reported in literature.

Comparison of the calculated and measured fuel flow suggests a decent match with 5.2% differences for the natural gas/waste gas mixture. A difference of 18% with the boiler operating on natural gas is however observed. As the boiler uses natural gas only during startup or plant shut down, evaluating a different data set from previous startup or shutdown period may provide more insight on this discrepancy. As natural gas does not contain normal operating fuel composition, the flow element becomes suspicious. Reviewing the datasheet for the type of device, design specifications and whether it provides any form of compensation may provide clarification to the above discrepancy. Flow measurement deviations are not unusual on streams exhibiting variability in composition and operating under different pressure and temperature conditions. Nevertheless, boiler feed water and superheated steam flow measurements are believed to be more accurate given the absence of operational variability. Conversely, it is less common to observe significant deviations in temperature, pressures transmitters, and online analyzers if properly calibrated. The fuel composition in this evaluation was based on the natural gas supplier specifications and the natural gas/waste gas mix stream composition based on plant material balances. Those become assumptions in the evaluation that could be challenged when dealing with large discrepancies in calculations. Getting laboratory analyses of the stream would be a way to firm up those assumptions, but given the dynamic nature of plant operation, there are no guarantees that variability in composition may not have occurred during the period from which the plant operating variables data set was sampled.

The raw data collected from the plant and results provided in Table 2.7 can be used for calculating the *UA* value of the different tube banks in the convection section. Trending of the coil *UA* value over time will show sign of heat transfer performance losses over time, often attributed to fouling mechanisms.

From process engineering and troubleshooting perspectives, the above boiler characterization can easily be programmed on a spreadsheet for daily calculations. Whether the fuel flow discrepancy gets resolved or not, historical trends of the discrepancy itself, thermal efficiency, heat distribution, and temperature profile would provide a definition of normality and show early signs of transition toward abnormality and potential failure.

The above heat and material balance characterization provides most of the data required for assessing the likelihood of numerous malfunctions associated with combustion and heat transfer in the boiler operation. Additionally, from Equation 2.23, a deeper analysis can be done around the variables influencing the heat transfer mechanism in the radiant section. The radiant heat term, A, defined as the continuous area passing through the center of the tubes, can be approximated by the surface of the cube delimiting the boundaries of the radiant section of the boiler. In the radiant section of boilers, the distance between the tube axes and the tube outside diameter is close to unity because the tubes are close to each other. The overall exchange factor F can then be approximated to 1. As there is no refractory surface emitting in the radiant section, $F_{1,g}$, in Equation 2.25, is reduced to. $F_{1,g} = \epsilon_g$ Equation 2.24, which provides the overall exchange factor $\mathfrak{F}_{g,1}$ relative to geometry and total emission factors can be simplified to the following expression:

$$\frac{1}{\mathfrak{F}_{g,1}} = \frac{1}{\epsilon_1} + \frac{1}{\epsilon_g} - 1 \quad (2.46)$$

With ϵ_1 the tube metal surface emissivity assumed at 0.79 in this example. ϵ_g, the gas emissivity is the summation of the water and CO_2 emissivity subtracted by a correction factor. The water and CO_2 emissivity are obtained from the Hottel and Egbert [21, 26, 27] charts, which were developed by correlating multiple sources of experimental data. As shown in Figure 2.7, these charts correlate the emissivity as a function of the average gas temperature in the radiant section and the product of the partial pressure by the equivalent thickness e (sometimes referred as the mean beam length). The equivalent thickness can be determined from tabulated values in literature or approximated by:

$$e = 3.6\frac{V}{A} \quad (2.47)$$

where V is the total volume and A the total surface of the cube defined earlier as the boundary of the radiant section. Assuming a pressure of 1 atm., the partial pressure of water and CO_2 is equivalent to their mole fraction. The average gas temperature is initially assumed equivalent to the bridge wall temperature and then iterated until the heat from the radiant term of Equation 2.23 matches the radiant heat value reported in Table 2.7. θ in Equation 2.23 is the outside TMT, which is unknown in this case and approximated by the temperature of the fluid circulating inside the tubes (steam/water mixture) plus 50 °C.

For the convection heat term of Equation 2.23, S_l represents the exposed surface of the tubes. Only half the external surfaces of the tubes forming the wall of the radiant section are exposed to the flue gas. The exposed area is therefore equivalent to $S_l = \pi \times Outside$ diameter/2 x Tube length x Tube count. Some studies [26, 30] suggest a relatively small contribution by the convection heat transfer term in Equation 2.23, with typical heat transfer coefficient h value of about 10–12 W/m^2 °C.

The low flue gas velocity in the radiant section primarily explains the low heat transfer coefficient. T, the temperature of the flue gas in contact with the tube, is unknown and approximated to the bridge wall temperature.

Table 2.8 summarizes the parameters used in Equation 2.23 with the water and CO_2 emissivity estimations summarized in Figure 2.7. The total heat duty reported in Table 2.8 reconciles within 1–2% with the radiant section heat reported in Table 2.7. The results given in Table 2.8 suggest that convection heat contributes for only 5–8% of the total heat transferred in the radiant section. For the purpose of understanding how the change in fuel composition affects the heat transfer mechanism, it would not

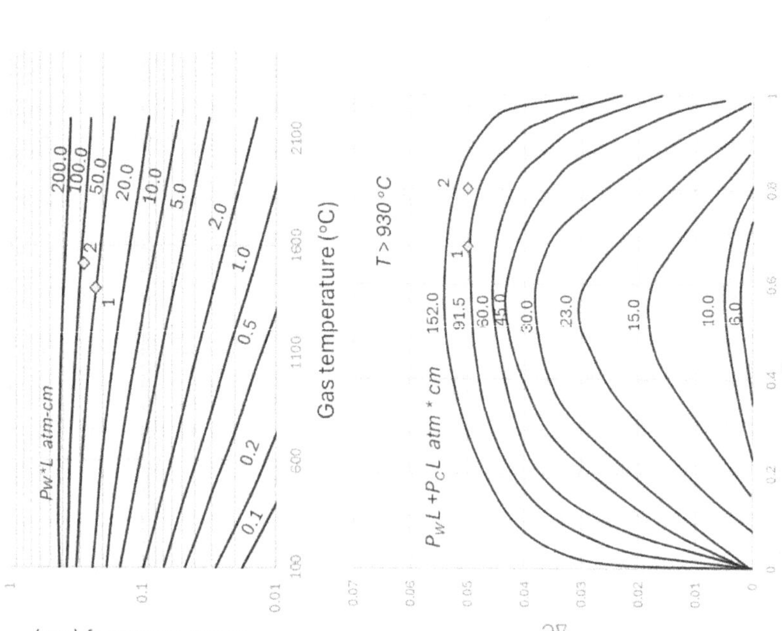

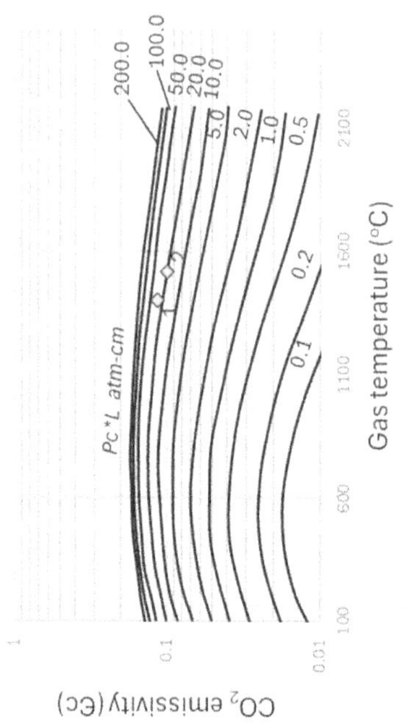

Flue Gas Partial Pressure		Natural Gas	Natural Gas/ Waste Gas
CO_2	atm	0.081	0.0569
H_2O	atm	0.1719	0.2303
O_2	atm	0.0181	0.0141
N_2	atm	0.7219	0.6987
$P_w/(P_w+P_c)$		0.6797	0.8019
P_c *L	atm-m	0.3106	0.2182
P_w *L	atm-m	0.2009	0.2692
P_c *L + P_w *L	atm-m	0.5116	0.4874
Emissivity			
CO_2		0.11	0.094
H_2O		0.19	0.215
ΔЄ		0.053	0.052

FIGURE 2.7 Steam and CO_2 Emissivity [29].

Source: Reproduced with Permission from figures 7.20, 7.21, and 7.22, pages 188–189, *The John Zink Hamworthy Combustion Handbook*, Second Edition, Volume 1 Fundamentals, Edited by Charles E. Baukal, Jr, CRC Press, 2013

TABLE 2.8
Radiant Section Heat Transfer Mechanism Characterization

Parameters	Units	Natural Gas	Natural Gas/Waste Gas
Geometry			
Boiler tube OD	mm	89	89
Boiler tube pitch	mm	102	102
Gas heat transfer surface – *A*	m^2	306.5	306.5
Gas heat transfer volume – *V*	m^3	326	326
Tube outside surface – S_1	m^2	350	350
Equivalent thickness – *e*	m	3.84	3.84
Temperatures			
Average flue gas temperature – T_g	°C	1,400	1,515
Assumed outside TMT – Θ	°C	367	367
Assumed gas temperature at tube wall – *T*	°C	1,103	1,206
Equation 2.23 Parameters			
Overall exchange factor – $\mathfrak{F}_{g,1}$		0.2318	0.2406
Overall exchange factor – *F*		1	1
Stefan–Boltzmann constant – σ_0	W/m^2 K^4	5.68 X 10^{-8}	5.68 × 10^{-8}
Convection heat transfer coefficient – *h*	W/m^2 °C	10	10
Heat Transfer Mechanism			
Radiant heat	GJ/hr	111.1	151.3
Convection heat	GJ/hr	9.2	7.1
Total	GJ/hr	120.3	158.4

be inappropriate to neglect the convection heat term of Equation 2.23. By doing so, all constants cancel each other out, leaving the ratio of the overall exchange factors $\mathfrak{F}_{g,1}$ along with the average flue gas and outside TMT as variables:

$$\frac{Q_{R1}}{Q_{R2}} \propto \frac{\sigma_0 \mathfrak{F}_{g,11}\left(T_{g1}^4 - \Theta^4{}_1\right) AF_1}{\sigma_0 \mathfrak{F}_{g,12}\left(T_{g2}^4 - \Theta^4{}_2\right) AF_2} \& \frac{h_1(T_1 - \theta_1)\, S_1}{h_2(T_2 - \theta_2)\, S_1} \propto \frac{\mathfrak{F}_{g,11}\left(T_{g1}^4 - \Theta^4{}_1\right)}{\mathfrak{F}_{g,12}\left(T_{g2}^4 - \Theta^4{}_2\right)} \quad (2.48)$$

Besides the change in fuel composition, the radiant section heat duty is 30% higher on the case where the boiler fires a mix of natural gas/waste gas. The corresponding ratio of the overall exchange factor $\mathfrak{F}_{g,1}$ increases by 4% and the ratio of the fourth power temperature difference increases by 31%. While the numbers do not align perfectly, they provide an indication on the parameters influencing heat transfer. The value of the overall exchange factor $\mathfrak{F}_{g,1}$ is driven by a change in emissivity of the flue gas. The natural gas/waste gas mixture is a hydrogen-rich fuel, which results in a higher average flue gas temperature in the radiant section. As illustrated in Figure 2.7, the emissivity value goes down as the temperature increases. When transitioning to a hydrogen-rich fuel, this effect is offset by the incremental partial pressure of water. However, the emissivity of CO_2 drops by the combined effect of higher temperature and lower partial pressure. Nonetheless, the overall impact of the fuel composition on emissivity contributes in increasing the heat duty by 4%. The higher heat duty in the radiant section is therefore driven mostly by the higher average flue gas temperature. As a relevant observation, an increase in steam generation on the same fuel composition would result in a reduction of the overall exchange factor $F_{g,1}$. The partial pressure of water and CO_2 would remain similar, but the increase in average flue gas temperature required for the higher steam generation would cause a reduction in water and CO_2 emissivity.

Performance of the economizer, high- and low-temperature superheating coils can easily be evaluated by calculating the *UA* value from the heat duty and temperature profile, as reported in Table 2.7, using the following expression:

$$UA = \frac{Q}{LMTD} \tag{2.49}$$

where Q is the heat duty of the coil. The LMTD (log mean temperature difference) can be evaluated by:

$$LMTD = \frac{\left(T_{hot\ in} - T_{cold\ out}\right) - \left(T_{hot\ out} - T_{cold\ in}\right)}{Ln\left(\dfrac{T_{hot\ in} - T_{cold\ out}}{T_{hot\ out} - T_{cold\ in}}\right)} \tag{2.50}$$

Combining the overall heat transfer coefficient U and coil area A together is for simplification as it does not require any coil geometry data. Data specific to equipment design may not always be available due to proprietary technology or plant documentation control effectiveness. As such the above *UA* estimates would provide some indication on the thermal effectiveness of the coil. It must be noted however that the overall heat transfer coefficient is flow-dependent. As such, good judgment must be exercised in analyzing the data before deciding on loss of effectiveness. Trending the *UA* value over an extended period is an effective practice in assessing loss of performance where the data could also be filtered relative to rates. Departure in *UA* value can be further assessed through Equation 2.31. The ratio of Equation 2.31 for *UA* values that shows significant deviation from each other could provide more insight on what would explain the difference while canceling most the terms related to the coil geometry. Equation 2.31 is based on flue gas properties and velocity through the Reynolds number. Expressing the Reynolds number in term of volumetric flue gas rate would eliminate the cross-section term as described below:

$$Re = \frac{dv_{\infty}\rho}{\mu} = \frac{dq\rho}{A\mu} \tag{2.51}$$

where q denotes the volumetric rate of flue gas and A indicates the smallest cross section across the tube bank. The coefficient m shown in Equation 2.31 becomes the only value that must be estimated or assumed. This can be done by estimating from drawings if available or by guessing a value and conducting sensitivity based on the range of values provided in Table 2.4, where the value of the coefficient m varies from 0.556 to 0.752. While the value of the coefficient m could significantly influence the absolute value of the heat transfer coefficient, it seems neutral relative to the ratio of two sets of conditions. The simplified ratio of Equation 2.31 is provided below:

$$\frac{h_1}{h_2} = \frac{K_1 (q_1 \rho_1)^m \left(\frac{Cp_1 \mu_1}{K_1} \right)^{1/3}}{K_2 (q_2 \rho_2)^m \left(\frac{Cp_2 \mu_2}{K_2} \right)^{1/3}} \tag{2.52}$$

While Example 2.2 may not be a situation requiring troubleshooting, the difference in operating conditions makes it a good case for looking at the parameters that drives differences. Table 2.9 summarizes the heat transfer calculations results for the convection section tube banks. The ratio of *UA* values suggests a 22% increase on the

TABLE 2.9
Heat Transfer Coefficient by Section

Parameters	Units	Natural Gas	Natural Gas/Waste Gas
Economizer			
Flow	m^3/hr	412,950	462,557
Heat duty	GJ/hr	24.5	33.5
Density	kg/m^3	0.27	0.24
Heat capacity	J/kg K	1,364	1,436
Thermal conductivity	W/m K	0.0999	0.1099
Heat transfer coefficient *h*	W/m^2 C	1,331	1,436
UA	kJ/hr C	30,582	37,337
High-Temperature Superheater			
Flow	m^3/hr	412,950	462,557
Heat duty	GJ/hr	26.0	38.4
Density	kg/m^3	0.33	0.30
Heat capacity	J/kg K	1,311	1,377
Thermal conductivity	W/m K	0.0817	0.0893
Heat transfer coefficient *h*	W/m^2 C	1,248	1,372
UA	kJ/hr C	72,063	88,335
Low-Temperature Superheater Duty			
Flow	m^3/hr	412,950	462,557
Heat duty	GJ/hr	22.6	31.5
Density	kg/m^3	0.42	0.39
Heat capacity	J/kg K	1,247	1,304
Thermal conductivity	W/m K	0.0644	0.0694
Heat transfer coefficient *h*	W/m^2 C	1,158	1,264
UA	kJ/hr C	120,386	136,816

natural gas/waste gas fuel mixture for the economizer and high temperature superheater. The overall heat transfer coefficient is the combination of the heat transfer coefficient on the flue gas, steam side, and conduction heat transfer across the tube thickness per following expression:

$$\frac{1}{U}=\frac{1}{h_{Flue\,Gas}}+\frac{\Delta x}{k}+\frac{1}{h_{Steam}} \tag{2.53}$$

On the flue gas side, the heat transfer coefficient ratio from Equation 2.52 increases by about 10%. The heat transfer coefficient on the water/steam side can be estimated by the correlation given by Sieder and Tate, as described by Equation 2.28. The ratio of the Sieder and Tate correlation can be reduced to the ratio of the velocity at the 0.8 power, which can also be approximated by the ratio of the steam rates between the two cases at the 0.8 power as described by Equation 2.54 with m being the mass rate of water or steam. Using the steam generation expressed as percentage of design stated in Table 2.7, a heat transfer coefficient ratio of 1.22 (i.e., $75^{0.8}/58^{0.8}$) can be derived on the water/steam side of those coils. This represents a 22% increase in heat transfer coefficient for the natural gas/waste gas fuel mixture case:

$$\frac{h_{Steam1}}{h_{Steam2}}=\frac{\left[0.027\dfrac{k}{d_{tube}}Re^{0.8}Pr^{1/3}\left(\dfrac{\mu}{\mu_W}\right)^{0.14}\right]_1}{\left[0.027\dfrac{k}{d_{tube}}Re^{0.8}Pr^{1/3}\left(\dfrac{\mu}{\mu_W}\right)^{0.14}\right]_2}=\frac{Re_1^{0.8}}{Re_2^{0.8}}=\frac{v_1^{0.8}}{v_2^{0.8}}=\frac{m_1^{0.8}}{m_2^{0.8}} \tag{2.54}$$

By just using heat transfer coefficient ratio and without more data on the coils geometry it would be difficult to correlate directly to the U value. The above method has some inaccuracies but does provide key insights into the parameters responsible for change in heat transfer performance. On the flue gas side, the increase in heat transfer coefficient would have been expected to be driven by the higher flow. However, as shown in Table 2.9, the increase in heat transfer coefficient is primarily driven by the fluid thermal conductivity constant K. Though the flue gas volumetric flow increases by about 10% on the natural gas/waste gas fuel mixture case, the density drops by 10% and offset the impact of the flow. On the tube side of the coil (water/steam), the increase in heat transfer coefficient is directly driven by flow rates.

Liquid fuels would make the above assessment more difficult due the complexity of the mixture. Liquid fuel's mixture typically contains several complex hydrocarbons molecules, which may involve more than carbon and hydrogen chemical bonds. The characteristics of liquid fuel are often limited to viscosities, specific gravity, boiling curves, and heating value typically expressed as HHV. For proper reconciliation of the heat balance, liquid fuel characterization should be based on the LHV, which is done by measuring the fuel hydrogen molecular content as stipulated in ASTM D240–19 standard.

In the fuel gas example, the flue gas composition was established from the combustion reaction stoichiometry. For liquid fuels, a similar approach can be followed by using flue gas analysis rather than reaction stoichiometry for the major combustion products CO, CO_2, SO_2, NO_x, H_2O, and excess O_2. In many cases, the contribution of SO_2 and NO_x is small relative to the heat balance and can be neglected. By doing so, stoichiometric oxygen can be correlated to CO_2 and H_2O using Equation 2.1, which would then provide the stoichiometric value of nitrogen in Equation 2.33. As for fuel gas, this approach has the advantage of not requiring fuel and air flow metering. As previously discussed, liquid fuel normally circulates in a loop between storage and the burners with only a fraction of the circulation rate supplying the burners. The fuel flow to the burners then either relies on the difference between the loop supply and return flow elements or is calculated form the storage tank draw down. The air flow is not always measured, especially on natural draft fired equipment.

Analysis of flue gas was conventionally done by Orsat [31] analysis, which provided volumetric fraction of O_2, CO_2, and CO content of flue gas on a dry basis. The method is based on the flue gas sample circulating into different absorption chambers containing solvent which selectively absorb O_2, CO_2, and CO. The decrease in gas volume as the sample travels through the different chambers is correlated to the volumetric composition of the flue gas components. Nitrogen is determined by difference. SO_2 is absorbed with CO_2 and would require a more extensive method for its determination. However, while SO_2 content has value from an environmental perspective, it has a minor impact on the heat balance. Calculation of the flue gas water content can be done through wet and dry bulb temperature measurements and appropriate humidity charts.

The use of online analyzers such as Fourier transform infrared (FTIR) [32] analyzer has the advantage of providing historical data, allowing performance evaluation under different set of circumstances. Flue gas composition can be established by other type of instruments, such as chromatography or combination of chromatography with spectrometry.

AlAbbad et al. [33] have proposed a methodology for characterizing heavy fuel oil mixture to a surrogate formulation or single component emulating the stoichiometric characteristics and properties of heavy fuel oil. While the methodology is intended to facilitate characterization and simulation of gasification processes, it could also be used for characterizing combustion reaction of fuel oil. The characterization involves a fractionation step, splitting the mixture into an enriched heavy fraction and a volatile distillate, which can be analyzed by a different methodology that is more suited to their characteristics. Both fractions undergo characterization through nuclear magnetic resonance (NMR) spectroscopy, then Fourier transform-ion cyclotron resonance mass spectrometry (FT-ICR-MS) for the heavy fraction and gas chromatography-mass spectrometry (GC-MS) for the volatile fraction. The end result provides an average molecular weight and CHNSO (organic compounds containing carbon [C], hydrogen [H], nitrogen [N], sulfur [S], and oxygen [O]) molecular structure. As intended, the method is more suited for providing valuable design information than its use for troubleshooting existing fired equipment.

It is worth noting that, without flue gas online analysis, troubleshooting the effectiveness of fired equipment burning liquid fuel would be limited to trending stack temperatures. Depending on variability of liquid fuel characteristics, this could become misleading. Historical flue gas composition, on the other hand, would provide valuable information relative to potential variation in fuel characteristics. This would also require to frequently evaluate the fuel LHV.

Pyrolysis furnaces, steam reformers, and fired heaters differ from steam boilers by their radiant section configuration. While steam boiler tube risers form the walls of the radiant section, pyrolysis furnaces, steam reformers, and fired heaters have the radiant tube confined between different refractory walls. In the boiler example, Equation 2.25 was simplified to the gas emissivity $€_g$ since there are no refractory wall or A_R equal to zero. For the sake of comparison, let us assume that the refractory wall area is equal to the continuous surface area passing through the center of the tubes (A). This corresponds to an A_R/A ratio of 1, for which the angle factor μ_{RI} is estimated at about 0.5. Recalculating the value of the overall factor $F_{g,1}$ would result in an increase of about 55% when compared to the value reported in Table 2.8. In other words, the radiant heat term of Equation 2.23 would increase by about 55% when assuming refractory walls for similar flue gas and outside tube wall temperatures. Radiant heat from refractory surfaces to the radiant tubes is known to contribute to a significant fraction of the heat transfer mechanism in pyrolysis furnaces, steam reformers, and fired heaters.

Distribution of the heat flux is another key difference between steam boilers and pyrolysis/steam reforming furnaces where uniform heat flux across the radiant tubes is critical from the perspective of reaction, run lengths, and longevity of radiant coils. As such, while steam boilers are characterized by a small number of burners, pyrolysis furnaces in ethylene plants may be equipped with over 100 burners strategically located for promoting an optimum heat flux.

In the boiler example, the heat duty of the radiant section was correlated to the saturated steam generation. This is relatively simple as it involves phase change of a single component without chemical reaction. This approach may also apply to fired heaters supplying heat to streams of known composition without experiencing any form of chemical reaction. However, it would not be applicable to the case of pyrolysis furnaces, which are highly endothermic processes involving hundreds to thousands of sides and parallel reactions. In this case, good evaluation of heat transfer would require temperature measurements from the bridge wall, on each side of the convection tube banks and stack temperature. In addition, knowledge of the fuel composition and excess oxygen would be required for performing an evaluation similar as the one demonstrated in Example 2.2.

2.1.1.4 Water Side of Steam Boilers

As shown in Table 2.8, the average flue gas temperature for Example 2.2 was estimated between 1,400 and 1,515 °C. This is relatively close to the melting point of carbon steel (1,425–1,520 °C) and significantly exceeds maximum recommended operational temperatures of carbon steel and other boiler tube metallurgies. Boiler tubes operating at temperature above recommended maximum can result in graphitization, which is a mechanism causing microstructural change occurring in carbon and low allowed steels from long-term exposure to temperature above 425–550 °C [34]. The tube material becomes brittle and prone to failure after long-term exposure to those high temperatures. The above discussion points to two operating practices that may result in boiler tube failure, if not followed. First practice is to ensure a constant boiler water supply to the tubes by maintaining a sufficient boiler water inventory in the steam drum. Running out of water inventory in the steam drum will likely result in catastrophic failure where the TMT will exceed the melting point of the tube material. The extent of damage may range from tube damage to destruction of the boiler. This risk is well known, and most steam boilers are typically equipped with a mechanism that stops fuel supply to the burners on a low steam drum level excursion.

The second consideration involves TMT operating above graphitization temperatures, while mostly design-related consideration can be associated with improper deviation from design capabilities of the equipment. Example 2.2 demonstrates the impact of fuel composition on the average flue gas temperature in the radiant section. The TMT was not rigorously calculated but rather assumed based on rule of thumb. It is not unusual that fuel composition changes may occur as waste stream is being recovered for fuel values or opportunities arising from the availability of cheap external fuels. Other changes such as burner replacement without consideration to the full boiler operation may also impact the TMT. Those changes may not always follow a rigorous design process and, in many instances, be addressed through a trial run first. In such case, improper balance between firing duty, heat transfer area, and water circulation could cause TMT exceedance from design and may result in tube failure.

Water quality is an important element in steam generation and varies based on steam pressure and utilization. API RP 538 [35] provides boiler water quality in relation to steam drum pressure. Poor water quality is the primary cause for corrosion and deposition of residue on heat transfer surfaces of boilers and steam generators. The primary mechanisms responsible for corrosion in steam generation equipment are driven by the presence of dissolved gases, such as oxygen and carbon dioxide, under-deposit attack, and improper pH. Boiler water normally undergoes mechanical deaeration prior to supplying steam generation equipment. Mechanical deaeration is accomplished by contacting water with steam in a process vessel commonly called deaerator, where internals (i.e., nozzles, trays) promote vapor/liquid contact for maximizing the transfer of dissolved gases to the vapor phase, which is then released to atmosphere. The transfer of dissolved gases to the vapor phase is promoted by temperature and partial pressure imbalance between the liquid and vapor phase. Contacting boiler water with steam in a deaerator first raises water temperature close to its boiling point, thus reducing solubility of the dissolved gases. Second, steam free of those gases promotes latter's transfer from liquid to the vapor phase as the system aims at establishing partial pressure equilibrium. In many cases, mechanical deaeration is insufficient and must bc complemented with chemical oxygen scavengers. Sodium sulfite, sodium bisulfite, hydrazine, and organic oxygen scavenger are commonly used. The selection depends on reaction rate, residence time, and operating temperature and pressures. Oxygen scavengers are typically injected in the deaerator exit pipe connected to a boiler water storage vessel, which then supplies boiler feed water pumps. Sodium sulfite reacts with oxygen to form sodium sulfate, hence contributing to the level of dissolved solids in boiler water, which is handled through blowdown rates. For steam system operating above 1,000 psig, hydrazine and organic scavengers are preferred as their products of reaction release no dissolved solids. For instance, the product of reaction from hydrazine and oxygen is water and nitrogen. In high-pressure boilers, sulfite could decompose into SO_2 and H_2S, which would promote corrosion.

Under deposit, corrosion can be the result of caustic where salts may accumulate on tubes surface under steam blanketing conditions. Steam blanketing refers to a vapor film inside the tube metal surface. This is typically localized in the hotter section of tubes, with horizontal tube being more sensitive. Steam blanketing is also rate-related; exceeding the design heat flux would cause higher circulation in the boiler tubes with higher vapor fraction as the steam/water mixture rises. Higher circulation also means higher velocities, where a combination of erosion and corrosion will lead to tube failure faster. Caustic corrosion also occurs when caustic substances (highly corrosive alkaline substances like sodium hydroxide [NaOH, caustic soda]) get concentrated at the tube surface and dissolves the magnetite (Fe_3O_4) protective layer of the tube. Without eliminating the caustic concentration, the attack on the magnetite layer continues until the loss of metal result in tube failure. Protection against caustic corrosion involves coordinating phosphate dosage with pH control. Phosphate provides a buffer that prevents large pH excursion caused by elevated caustic concentration. Disodium phosphate reacts with excess caustic substance to form trisodium phosphate and prevents caustic accumulation. Addition of excess trisodium phosphate can, however, reverse the reaction to caustic. Monosodium phosphate (NaH_2PO_4) converts caustic soda $(NaOH)$ into disodium phosphate (Na_2HPO_4), which, as stated earlier, can react with residual caustic soda. Reaction control is achieved by adding the appropriate phosphate formulation for reducing or increasing pH. Literature [36] reports suitability of pH range between 9.2 and 9.6 for minimizing corrosion of carbon steel at boiler feed water temperature. pH online monitoring and frequent alkalinity analysis (once per shift) are required for ensuring proper phosphate dosage.

Galvanic corrosion is a mechanism that occurs when conductive materials of different redox potentials are coupled together. A noble material, less prone to corrosion, is characterized by high redox potential. Coupling two different metals or alloys in a conductive solution or electrolyte can cause the one with the lower redox potential (anode) to lose electron to the material of higher redox potential (cathode). This is also associated with a loss of metal mass from the anode. The corrosion rate depends on the redox potential difference between the two materials. A common type of galvanic corrosion in boiler system involves iron and copper. Change in metallurgies, which can promote galvanic corrosion typically, occurs at joints or welds. Accumulation of conductive deposit can also create localized galvanic corrosion.

Deposition of residue on the tube surface is another problem that could lead to failure. Deposits involve the formation of scale adhering to the tube surface. Scale is characterized by salts of limited solubility in water. As water transitions from liquid to steam, the salts start precipitating on the tube surface. They are known to form a crystal structure that grows on the tube surface by either bonding or capturing other particles. A sludge is another form of deposit and is the result of solids that precipitate in the boiler water or travels through the boiler tube as suspended solids. Both scale and sludge can form hard surface inside the tube that are difficult to remove chemically or mechanically. Deposition of residue on the tube surface restricts the water flows as it increases the hydraulic resistance of the thermosiphon circuit between the steam drum and the radiant section. This results in a higher degree of vaporization in the tube risers and can promote steam blanketing (i.e., dry vapor film on the tube surface), which negatively impacts heat transfer. Both steam blanketing and the accumulation of residue add an additional resistance to heat transfer, thereby impacting the capacity and effectiveness of the boiler. Most of all, it inhibits the metal surface from the cooling provided by water circulating in the thermosiphon circuit, resulting in high tube metal temperatures which would eventually lead to tube failure. Scale and sludge deposition is prevented through chemistry and proper blowdown operation. The strategies aim at preventing adherence and/or solubilizing residues. Phosphate treatment on its own or in combination with polymers prevents residues from adhering on the tube surface. While the resulting particles remain insoluble in water, their accumulation is prevented through appropriate blowdown rate. Use of chelant additives solubilizes the residue in the water, hence preventing adherence to the tube surface. As for phosphate treatment, proper blowdown rates prevent accumulation.

The above problems extend beyond the steam generation equipment. Figure 2.8 shows a simplified high-level schematic of a steam system where, steam generated from high-pressure boilers supplies heat exchangers and turbines. Heat exchangers transfer heat to process streams by condensing steam; mechanical energy is supplied to rotating equipment, such as pumps, compressors, and/or power generators, by expanding steam in turbines. Steam turbines can exhaust to lower steam header pressures, where steam is being consumed by lower pressure level users or by condensing turbines where exhaust steam is condensed at low pressure by cooling water. Maximizing turbine power per unit mass of steam involves designing and operating turbines at low exhaust pressure to level where the residual energy has no value from a process perspective. Turbine configuration (back pressure vs. condensing) is

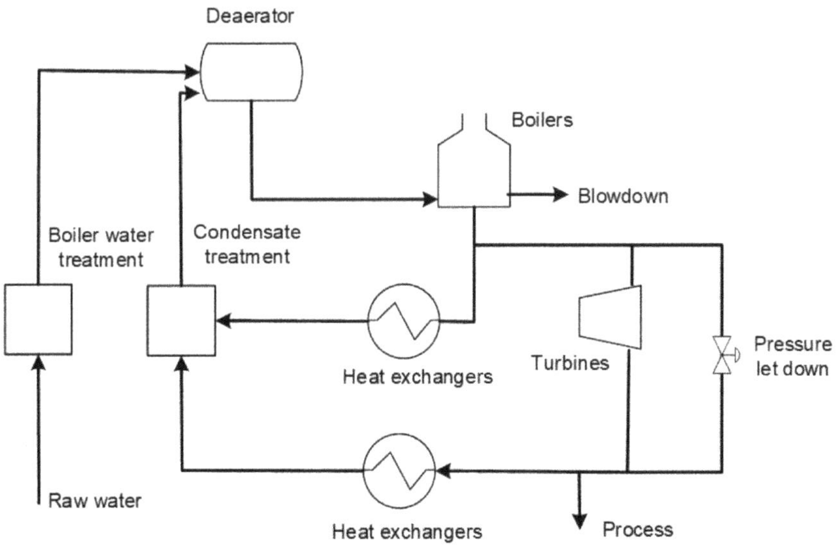

FIGURE 2.8 Steam Balance.

something that is set by design based on steam balance optimization. While a back pressure turbine may operate at a higher steam/power ratio, the value of the exhaust steam used by process users energetically outperforms isenthalpic expansion from steam let-down valves. Condensing turbines makes more sense when exhaust steam exceeds process demands, requiring steam venting or condensing for balancing the system. Condensate from heat exchangers and condensing turbines is normally collected, treated, and recycled as boiler feed water to a deaerator, which supplies boiler water to boilers and steam generators. Raw water, after treatment, supplements losses resulting from blowdown and direct steam injection in process units.

Raw water quality is dependent on geography relative to available water supply. Many plants are located nearby waterways such as river and lakes and supplied with fresh water while others have only access to seawater. The source of raw water and its associated contaminants along with the expected operating steam pressure dictate the treatment technology. Some common pre-treatments, regardless of sources, would involve screening, which removes gross particles and protects raw water supply pumps. Chlorination through gas or sodium hypochlorite injection protects and prevents microbiological growth on process equipment. Prevention of microbiological growth is more impactful on cooling water systems but also protects boiler water treatment equipment. Clarification and filtration are used for removing suspended solids. Addition of chemicals such as polymers and flocculants is frequently use to enhance settling in clarifiers.

Desalination of seawater can be accomplished by distillation or two-pass reverse osmosis. Sodium is the primary constituent responsible for seawater conductivity. Desalination of seawater can be sufficient as treatment for boilers or steam generation equipment operating below 900 psig. Mixed-bed polishing is recommended for operation above 900 psig. Mixed-bed polishing involves layers of anion and cation ion exchange resins contained inside a single vessel.

Depending on water quality, raw water treatment can be limited to sodium zeolite softening or in combination with upstream lime (hot or cold) treatment for steam generation below 600 psig. At higher pressure, the total dissolved solids contained in softened water are known to cause foaming issues.

Demineralized water is typically recommended for high-pressure steam generation above 1,200 psig. Conventional processes use ion exchangers with strong acid cation (SAC) resin bed, which could also be preceded by a weak acid cation (WAC) bed. A degasifier unit can be used downstream the WAC and SAC for reducing the load on the downstream strong base anion (SBA) resin bed. As for cation resin, a weak base anion (WBA) can be used upstream the SBA bed. Multiple ion exchanger trains are normally in operation for allowing regeneration of the ion exchanger resin once saturated. Regeneration of ion exchanger train can be based on volume or breakthrough of some quality indicators, such as conductivity or silica.

Condensate recovery is driven by water conservation and heat recovery. As steam travels and condenses in process units, the resulting condensate collects contaminants from the process units; it must be treated prior to supply to the boiler water system. Contaminants include iron and copper from piping system and steam traps. Exchangers can leak hydrocarbons and minerals from cooling water. Condensate treatment could involve activated carbon filters for removing organic material. Some plants operate activated carbon filters on an as-needed basis, as they would monitor and troubleshoot occurrence of organic excursion. Removal of minerals, such as iron and copper, is accomplished through operation of mixed-bed ion exchangers.

The water circulating in a steam system – raw water and condensate – initially originates from the raw water supply. Condensate recycles increases the residence time of a molecule of water within the system. Infinite residence time would be equivalent to no system losses and 100% condensate recycles. In reality, losses are compensated by raw water supply (after proper treatment) to the system. Residence time is typically associated to the concept of cycles. The number of cycles is directly proportional to water losses, which – other than blowdowns – cannot be controlled as they are defined by process needs. Water blowdown then defines the number of cycles undertaken by boiler water and becomes a control point relative to boiler water quality. The boiler water cycle is defined as the ratio of boiler water supply over blowdown plus losses. The boiler water supply is the summation of the flow from the boiler water treatment unit and condensate from the condensate treatment unit for a steam system, like the one illustrated in Figure 2.8. The cycle of concentration of a specific component can be calculated from the ratio of the component concentration in the water supplied from the raw water treatment unit to the concentration of the component in the combined treated condensate/water supply. Monitoring of the controlling components would normally dictate the blowdown rates for maintaining the cycle of concentration of the component within control limits. The component monitored can be a contaminant or the residual concentration of chemicals used for protecting the system against corrosion and/or deposition.

2.1.1.5 Process Side

Fired equipment is used for supplying heat to highly endothermal chemical reactions occurring at temperatures beyond the capability of steam with steam cracking and steam reforming being the two main applications. Figure 2.9 provides a simplified schematic of a steam cracking furnace where hydrocarbon feed is first preheated by flue gas in the convection section. It then mixes with superheated dilution steam and undergoes further preheating in the convection section prior to its entry into radiant coil where the cracking reaction takes place. The mixture entering the radiant coil must be in vapor phase. As such, in liquid cracking such as naphtha, partial vaporization occurs through the feed preheating coil followed by full vaporization and superheating after mixing with dilution steam. The hydrocarbon/dilution steam mixture is further superheated as it travels through additional tube banks, which cross exchange heat with flue gas. The temperature at the inlet of the radiant coil is referred as the crossover temperature. High crossover temperature reduces the required residence time for the cracking reaction, hence the length of the radiant coil. However, depending on feedstocks, the crossover temperature is limited to a maximum range of 500–700 °C for preventing initiation of cracking in the superheating coils. As the hydrocarbon/dilution steam mixture enters the radiant section, it can split into two or more inlet radiant tubes depending on the radiant coil configuration. In such arrangement, venturis are typically used for ensuring equal distribution among the tubes. This is accomplished by operating the radiant coil inlet/outlet pressure above the critical pressure ratio, causing each venturi to operate at sonic velocity. This limits

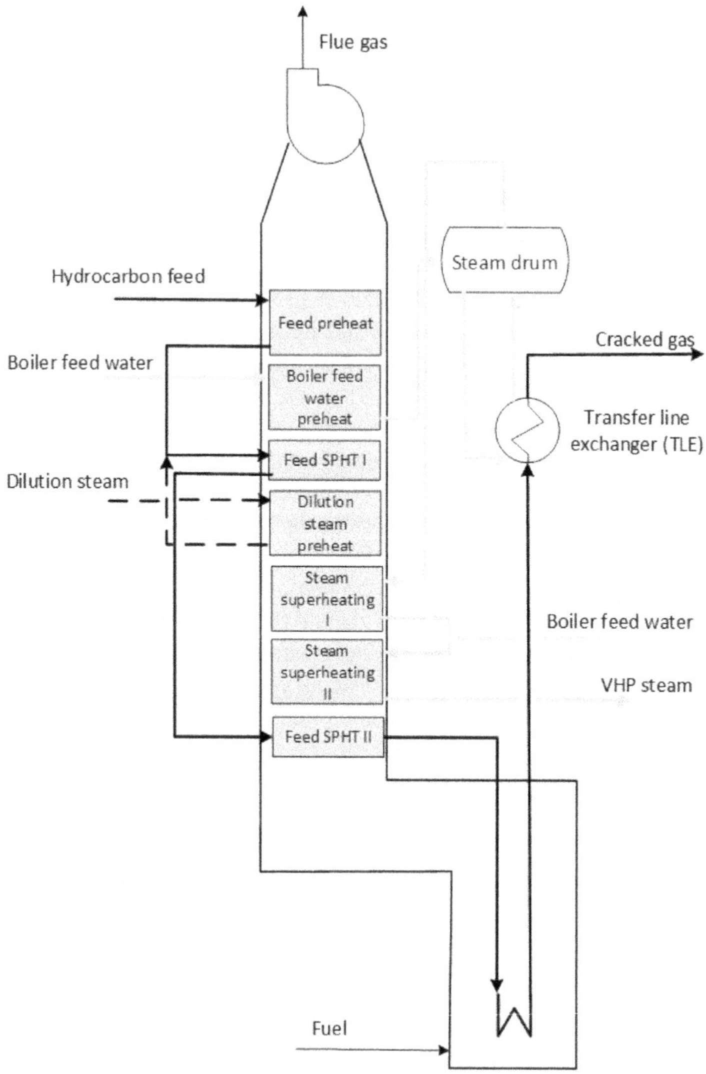

FIGURE 2.9 Typical Steam Cracking Furnace.

turndown of the radiant coil as it defines minimum critical flows for conditions to remain above the critical pressure ratio of the venturis.

In the radiant coil, the chemical bonds of the reactants are broken down when exposed to high temperatures, promoting free radical chain reactions. At elevated temperatures, the C-C bond gets broken near the end of the molecules, which, in the case of high-molecular-weight paraffin, results in the formation of a light alkane and a heavier component, which undergo further radical decomposition. The propagation of free radicals is followed by termination through rapid quenching resulting in a mixture rich in olefins (ethylene and propylene), C4's, raw pyrolysis gasoline (RPG), and pyrolysis fuel oil (PFO). The existence of olefins and aromatics in feedstocks is not desired, as it typically promotes coking and fouling reactions. In naphthenes and aromatics, the C-H bond being the weakest gets broken first, resulting in dehydrogenation and condensation reactions. Feedstocks that are free or lean in olefin and aromatic content may get contaminated by recycle streams when co-cracked.

Directionally speaking, higher temperatures are required with small paraffin hydrocarbons. The reaction mechanisms and kinetics of steam cracking are complex and involve hundreds to thousands of sequential and parallel reactions. The cracking reaction is driven by temperature, hydrocarbon partial pressure, and residence time. The combination of those variables defines the severity, conversion, and selectivity of the reaction. Severity is defined as a relative measure of the conversion of feedstock to products. While the concept of conversion might be straightforward when cracking a specific chemical component, such as ethane, it is not as clear when cracking the wide mixture of components typically encountered with liquid feedstocks such as naphtha. The concept of severity indirectly relates overall conversion to furnace effluent component ratio. Popular choices are methane/propylene and ethylene/propylene ratios [37]. When equipped with furnace effluent online analyzers, the severity can be used as a control variable for regulating the intensity of firing by manipulating the fuel flow. Severity can also be indirectly controlled by the radiant coil outlet temperature, as it correlates to severity.

Dilution steam is mixed with the hydrocarbon feed for providing the partial pressure required for suppressing coke formation. It also helps achieving the desired selectivity by lowering hydrocarbon partial pressure. Water has little interaction with hydrocarbons but reduces coke accumulation by converting it to carbon monoxide and hydrogen per the following reaction:

$$C_{(S)} + H_2O \rightarrow CO + H_2 \tag{2.55}$$

(S) is for solid. Coke is basically solid carbon. The ratio of dilution steam to hydrocarbon feed varies with feedstock. Directionally, heavier feedstock requires higher dilution steam ratios than light feedstock. For instance, ethane feed typically uses dilution steam ratio of 0.3, naphtha 0.5 [38], and heavier feedstock can be as high as 1.0. Additionally, dilution steam contributes in mitigating coke formation promoted by pyrolysis fuel oil and tar in exchangers downstream the radiant coil.

The following expression relates radiant coil residence time to the coil geometry and operating parameters [39]:

$$\textit{Radiant coil residence time} = \frac{\textit{Coil ID}}{4} \times \frac{H}{Q} \times \rho \tag{2.56}$$

Selectivity and yields are known to increase at short residence time. As described by Equation 2.56, the coil ID is set by design and the total heat transfer per unit mass of gas H is the total energy required for a given feed conversion, set by thermodynamic and cannot be manipulated. Reducing the radiant coil residence time can, therefore, be accomplished by increasing the heat flux Q and/or reducing the density of the gas ρ. The density is influenced by the radiant coil outlet pressure and explains why lower coil outlet pressure improves yields. Increasing the heat flux is the parameter that exerts the most influence on reducing residence time. High heat flux should then result in lower residence time and increase selectivity and yields. However, high heat flux also increases TMT, which increases the rate of coke formation, hence reducing the furnace run length and its availability in making products. There is therefore an optimum operating heat flux, which maximizes production and run length.

Although designed for specific feedstocks, feed flexibility can be accommodated by adjusting operating conditions. However, alternative feedstocks may require compromising on the optimum furnace operating conditions. Constraints such as venturis critical pressure ratios may require operating at rates deviating from the optimum residence time. This, to some extent, can be mitigated by manipulating the heat flux. Nonetheless, throughput range limitations and inefficiencies would be expected. For instance, changes in heat load may translate into more or less heat transfer in the convection section, hence impacting the overall energy balance of the furnace. Table 2.10 shows comparison of the yield profiles of ethane and naphtha, with emphasis on the differences in molecular weight. As a feed, ethane has a molecular weight that is about 60% less than naphtha [38]. The molecular weight of the gas exiting the radiant coil is about 30% less on ethane cracking when accounting for dilution steam. Using the ideal gas law, for same pressure and temperature, the density of the ethane feed and resulting cracked gas is then driven by the molecular weight and would be lower than naphtha by the same magnitude. Lower density means higher volumetric rates, hence higher-pressure drops in the convection section resulting in lower pressure at the inlet of the venturis. The lighter gas in the radiant coil would further reduce the residence time. As previously stated, light paraffins such as ethane required higher cracking temperature, which will further reduce the residence time from a hydraulic perspective. Operation outside the design characteristics could result in suboptimal performance and possibly shorter furnace run length.

TABLE 2.10
Ethane and Naphtha Cracking Yields Comparison [38, 40]

Product Yields	Ethane (% Mass)	Naphtha (% Mass)
Hydrogen	5.9	1.0
Methane	7.0	16.9
Ethylene	80.9	38.7
Propylene	1.9	15.5
Butadiene	1.8	4.8
Butylenes/butanes	0.8	5.1
Pyrolysis gasoline:	1.6	15.5
Fuel oil	0.1	2.5
Total	100	100
Approx. molecular weight	19	30

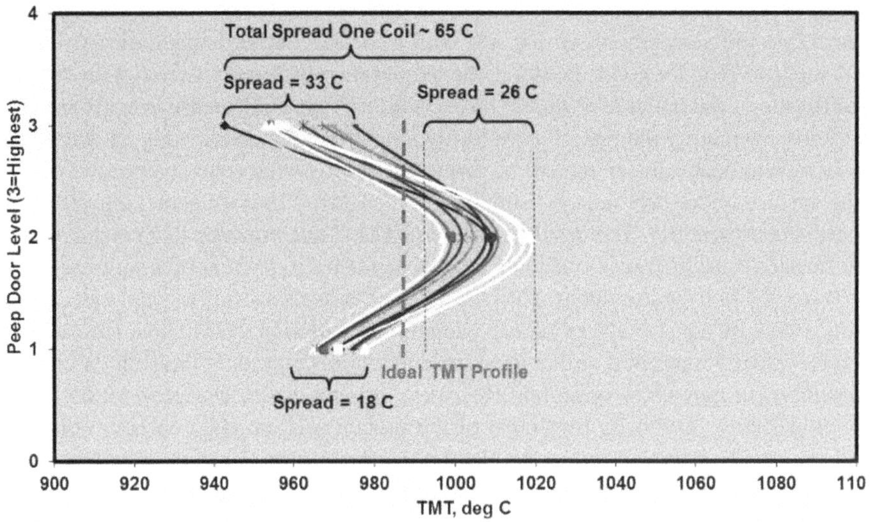

FIGURE 2.10 Tube Metal Temperatures [38].

Reproduced with Permission from figure 2.2, Ethylene Cracking Heater Troubleshooting Tutorial, Brian K. Sullivan, 2015 AIChE Spring Meeting – Ethylene Producer Conference

Furnace coil configuration provides plug flow hydrodynamic characteristics designed for short residence time (0.15–0.6 seconds). The reaction ideally requires uniform temperature across the length of the radiant coil. Furnaces are designed using sophisticated software for configuring the geometry of their components for optimum performance from yields to run length. This includes position and number of burners for ensuring optimum heat flux. In practice, there is no way to measure the process temperature profile inside the coils. The temperature profile is inferred through external TMT measurements using a pyrometer at peep door locations on each deck and on both sides of the furnace. Figure 2.10 [41] illustrates an example of TMT's measurement on a furnace designed with heart and wall burners. The ideal TMT profile is represented by the vertical dash line. While the ideal profile would be difficult to meet by design, the magnitude of deviation provides valuable information on the furnace operation. Deviations from the ideal profile may be caused by design, where depending on the height of the radiant section and with only hearth burners, a lower temperature at the base would occur as the flame might not be fully developed. The heat flux may also decrease at the top of the radiant section. This deviation can be minimized with side wall burners. From an operation perspective, non-optimal heat fluxes are caused by excess air due to improper burner air register settings, with infiltration occurring through leaks or improperly closed peep doors. Poor burner operation could occur due to fouling, damage, or improper burner tip installation. Fuel distribution imbalance could also affect the heat flux distribution.

The TMTs of different coils illustrated in Figure 2.10 suggest very consistent coil-to-coil measurements with temperature range varying between 18 and 33 °C. This suggests a good gas distribution across the different coils and would point toward a heat flux imbalance between the side wall to hearth burner firing duties. Sullivan [41] suggests that, for good operation, the temperature difference between the bottom and the top of the outlet tube should be of 30 °C or less. TMT measurements are also useful in evaluating the rate of coke formations. Coke formation adds an additional resistance to heat transfer, as will be discussed later. As the coke layer grows on the tube surface, more heat must be supplied to the radiant coil for achieving the same conversion or severity. This results in a rise in TMTs over time, up to a point at which the furnace must be transitioned to a decoke sequence to prevent tube failure.

Beyond TMT measurement, Sullivan recommends measuring oxygen and carbon monoxide with a gas analyzer at each peep door location in the fire box. Excess oxygen is typically measured online in a single location, commonly the arch. While the measurement may show optimum 2% excess oxygen at the measured location, the air profile may vary along the height of the radiant section. High carbon monoxide and low oxygen levels are symptomatic of poor combustion due to oxygen deficiency. Conversely, high oxygen content contributes to cooling the flame temperature, hence reducing the heat flux to the radiant coil. The above deficiencies cause a localized heat flux deficiency that is compensated by the coil outlet temperature (COT) or conversion analyzer to increase firing. The peak heat flux then becomes concentrated in localized area, which promotes coke formation and reduces run length. Localized oxygen deficiencies could be caused after burning of accumulated non-combusted fuel near the radiant coil or carried over to the convection section. This may also promote flame impingement on the radiant coil. Another methodology promoted by Sullivan for assessing the heat flux distribution consists of operating each coil at identical COT and comparing the spread of flow rates between the coils.

Coke formation is a by-product of reaction, which affects the run length of the furnace. There has been significant work done surrounding coke formation mechanisms given the impact it has on furnace availability, capacity, and radiant coil life. The formation of coke is described by three different mechanisms: catalytic, pyrolytic, and condensation coking. Alloy used for radiant coil is rich in nickel and iron, which along with their oxidated molecular structures catalyzes coking reaction. The formation of coke on tube surface is initially described by the formation of coke filaments. Catalytic coking is reported as the primary coking mechanism. As previously stated, the presence of olefins and aromatics in the feed will contribute to the formation of coke by radical or pyrolysis coking. Condensation of heavy components results in the deposition of tar on the tube surface or on equipment downstream of the radiant coil.

Coke layer growing on the inside surface of tubes creates an additional resistance to heat transfer. This affects the heat flux required for achieving the desired reaction yields and, like deposits in boiler tubes, increases the TMT, which, if not rectified, can result in tube failure. Observation of tube bulging is an indication of a hot spot at the tube location created by localized coke deposition. Coke layers adhering on the tube surface are characterized by a different expansion coefficient than the tube metallurgy. As such, the coke layer induces significant shear forces on the tube surface when experiencing temperature transients during decoking operation, trip, and/or shut down. As the coke layer grows, the pressure drops across the tubes increase, thereby impacting capacity. In reformers, coke formation within the catalyst restrains access to reactive site and affects reaction yields. Carburization and creep are the main mechanisms affecting the life expectancy of a radiant coil. Coking is promoting carburization, which is defined by carbon absorption through the alloy. This contributes to weakening the tube alloy and is one of the main causes of tube failure. As stated earlier, carburization is often noticed by tube bulging. Creep is defined as tube elongation over the life span of the radiant coil. It is driven by the number of thermal cycles that the radiant coil experiences. Coking indirectly affects creep as it exposes the radiant coil to higher TMTs and more frequent temperature transients due to shorter run length required for decoke sequences.

Coke formation can be minimized by first providing a protective catalytically inactive layer. This can be accomplished by pre-sulfidation or pre-oxidation when starting up a furnace. In pre-oxidation, a protective layer such as chromium and silicon oxide (Cr_2O_3/SiO_2) covers the surface of the tube, hence providing a barrier for Ni and Fe to catalyze coking reaction. The protective layer also protects against carburization. That protective layer is expected to degrade over the on stream/decoke cycles that the furnace will experience. This can be alleviated by continuously adding coke formation inhibitors. Sulfur compounds such as dimethyl sulfide (DMS) and dimethyl disulfide (DMDS) are commonly used in ethylene steam cracking furnaces. The dosage is however critical, as over-dosage could result in the opposite outcome. Tube coating during the manufacturing process is another strategy used for minimizing the impact of coking. ANK 400, jointly developed by NOVA Chemicals and Kubota [42], is an alloy with a stable inert oxide on the inside surface of the tube created during the manufacturing process. Kubota has developed another design that further reduces coke formation and enhances heat transfer using mixing element configured inside the tube ("Mixing Element Radiant Tube", MERT).

Decoke of furnaces is a necessary operational activity and is driven by TMTs or schedules depending on plants' operating strategies and circumstances. While scheduling decokes may not be optimal, it might be required depending on the expected run length of the furnaces, and/or limitations of utility systems, such as dilution steam and plant air. The objective of a decoke cycle is to burn the coke contained inside the furnace with a steam/air mixture. A decoke operation first requires transitioning the furnace from on stream to decoke site while minimizing thermal gradient for avoiding tube damage. As stated earlier, the coefficient of thermal expansion of coke differs from that of the radiant coil tube metallurgy. Dilution steam rates are increased as the hydrocarbon feed is withdrawn from the furnace. Dilution steam rates are increased, such that the venturis still operate above the critical pressure ratio. Firing adjustments are required during the transition for avoiding large thermal transients. This often requires shutting down some side wall burners and balancing the heat flux. The furnace effluent is then switched from the normal process stream to a decoke destination. Air is then mixed with dilution steam and preheated to target temperature for combusting the coke accumulated in the furnace process flow path. Air addition is done by increments to avoid/minimize thermal gradients. The decay of CO_2 in the decoke effluent provides indication on progress, with completion when CO_2 reaches a minimum threshold and stabilizes. The furnace is then transitioned back in service by following the reverse sequence. Improper decoke operation could reduce run length and potentially damage the radiant coil. Note that reduced run length also affects the life expectancy of the radiant coil as it will operate more frequently at high TMTs and experience more transition cycles between normal operation and decoke sequence.

Furnace yields, though discussed earlier from the perspective of operating conditions' cause and effect, may not always be directly measured or the spectrum of component analyses might be limited. The primary objective of the furnace is to convert raw material to valued product. Proper process engineering should back calculate the overall yields from a plant material balance and correlate it to a global furnace (i.e., all furnaces together) yield. This would confirm conversion and severity. Current yields should be compared to design and deviations understood and reconciled. Keeping historical records of furnace performance would first help in recognizing changes in behavior and second speed up troubleshooting when deviation occurs. Note that changes in furnace yields will affect the process loads of downstream equipment and in some cases may exceed their capability, hence constraining plant capacity.

2.1.1.6 Fired Equipment Safety System

Operation of steam boilers and furnaces is associated with risk of fire, explosion, overpressure, and major asset damages, if not operated properly. Besides causing financial impacts, this could also result in injuries and/or fatalities. Such risks are mitigated through SIS, operating procedures, and training. In some countries, burner management systems (BMSs) are regulated by codes. Focus of these codes is to have safety systems in place for preventing accumulation of combustibles in the fire box, which could then result in an explosion. In most instances, the codes are not retroactive, and existing facilities operate under a grandfather clause until they undergo modifications, at which point they must comply with the applicable codes. Codes are based on standards such as NFPA 85 and NFPA 86 in the United States and EN-746–2 in Europe. Codes could also include elements of recommended practices such as API-538 and others. The standards require to have flame monitoring devices on each burner and/or pilots, depending on configuration and combustion characteristics, unless robust safety systems are put in place to prevent operation of the burners at temperature less than 760 °C (NFPA 86) or 750 °C (EN-746–2).

Steam boilers are characterized by a limited number of boilers and can easily supply each burner and/or pilots with flame scanners. Whether scanners are required on both pilot and the main flame depends on whether the pilot can ensure temperatures above autoignition, in which case scanners can be limited to the pilot.

Steam cracking furnaces are equipped with numerous burners. It can be argued that requirement for flame-monitoring devices do not apply as the fire box (i.e., radiant section) operates well above 760 °C. However, furnaces must still operate through transient conditions such as startup, shutdown, and decokes. The number of burners requiring flame-monitoring devices can be minimized in different ways. As with boilers, if it can be demonstrated that the pilots can ensure a temperature above the autoignition point, then scanners become limited to the pilots. Another strategy is to have a limited number of burners identified as startup burners equipped with flame scanners and demonstrating that operation of those burners would raise the fire box temperature above 760 °C, at which point all remaining burners can be lit. This strategy requires an interlock system, which prevents fuel supply to the burners operating without flame monitoring devices if the fire box temperature is below 760 °C, as typically measured at the arch.

Before igniting any burners, NFPA 85 and NFPA 86 standards require a purge cycle with air or inert gases equivalent to at least four times the system volume, which includes the fire box and other passages handling the recirculation and exhaust of combustion products. The purge cycle must be completed before any startup and re-initiated in the event of a failed attempt in lighting burners.

The startup burner strategy requires ensuring that no fuel is supplying the other burners. Leak testing by pressuring the volume downstream the trip valve of those burners and validating that pressure has been holding will validate the readiness of the system for startup. This must be implemented through operating procedures.

As discussed in earlier sections, burner capacities are characterized by burner curves, which express the expected heat release as a function of burner pressure. Operating beyond the burner capacity or burner pressure beyond the curve would potentially result in un-combusted fuel. While low pressure may result in flashback, it would damage the burner. As such, high and low fuel pressure trips of burners are not uncommon. It is important to understand that burner curves are fuel-specific. Burner vendors shall be contacted when operating on fuel mixtures that have not been evaluated.

The loss of hydrocarbon feed to a cracking furnace may not have any direct impacts on combustion, but it does create a significant thermal transient on the convection banks and radiant coil. As previously discussed, the accumulation of coke will induce significant stress on the radiant coil under those circumstances. As such, it is common practice to perform a partial trip on the furnace. In other words, a partial trip logic program in a programmable logic controller (PLC) would bring the furnace to conditions equivalent to the initial state of a decoke procedure by readjusting dilution steam rates and firing. The furnace would operate in a hot steam standby mode until feed can be reestablished.

The boiler feed water thermosiphon circulation between the steam drum and the radiant section of a steam boiler, if interrupted, would result in significant tube damage due to high TMT. As such, boilers are typically protected from shortage of boiler feed water supply through low steam drum level interlock, which stops the fuel supply to the boiler. Similarly, for the protection of the primary quench exchangers in a cracking furnace, the fuel supply to the furnace is stopped by an interlock that gets energized by a low steam drum level.

The above section discussed the most common SIS found in fired equipment. Other safety systems intended to protect people and assets can also be put in place and are typically specific to the operation and process units associated with fired equipment. Some of those systems have a safety integrity level (SIL), which is typically driven by the consequences of an undesired event. SIL rating is ranked from 1 to 4, with 4 being the most stringent. The SIL level is driven by the required risk level reduction and the reliability of the system, hence reducing consequences and likelihood to acceptable levels. High SIL rated systems involves multiple instruments voting (e.g., 2 out of 3 energized) for recognizing an undesired event and triggering the protective actions. Given the many components involved in interlocks and SIS, upfront development of "Fault Tree Diagram" similar to that illustrated in Figure 2.11, would, in addition to serving as an effective training tool, provide a good head start if troubleshooting of those systems becomes necessary.

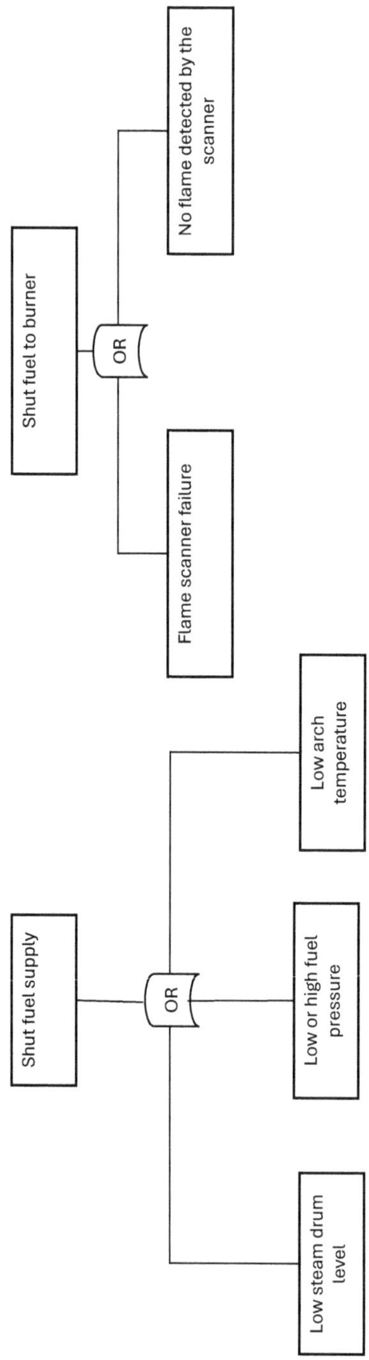

FIGURE 2.11 Fired Equipment Safety Logic FTA.

2.1.1.7 Flare

A flare is a recognized and accepted technology for destroying hydrocarbons released in refineries and petrochemical facilities. Under normal conditions, released hydrocarbons originate from purges or vent streams intended to prevent accumulation of non-desired chemicals or impurities, which may affect the quality of the product streams. Breathing loss from storage tanks is another release, occurring under normal conditions. Under abnormal circumstances, a flare provides destruction of excess hydrocarbon inventory from relief devices, which would otherwise be released to atmosphere. Destruction of hydrocarbon during de-inventory of process units during plant shut down is also accomplished through the flare system. Plant startup is another contributor of elevated release to the flare system until intermediate and product streams meet specifications. A flare system is associated with one or more headers collecting gases from multiple locations. A flow of a non-condensable gas, typically nitrogen, is circulating in all the headers for providing a positive pressure, which is intended to prevent air ingress in the headers. The above circumstances provide a wide range of chemical composition and rates causing significant variability in combustion characteristics. This brings design and operational challenge for ensuring safe and reliable operation of the flare system.

2.1.2 Reactors

The main purpose of chemical reactors is to convert raw materials into a mixture of products of higher value. They are also used for converting undesired components in feedstock or process streams when separation alone is not feasible, with the most common applications being hydrotreating and hydrogenation.

The previous section partially covered the topic of chemical reactions with steam cracking and reforming, in which the radiant coil of fired equipment fulfills the role of a chemical reactor. While it was stated that fired equipment was a good fit in those processes due to the highly endothermic reactions, there are several applications in the chemical industry that involve reaction mechanisms requiring different conditions. Operations process engineers are not expected to develop reaction mechanisms or design chemical reactors but should understand the underlying chemistry and the conditions that influence conversion, selectivity, and yield. They should also understand the design range and associated operating parameters. For illustration, poor reactor yields result in higher unconverted feedstocks, which are recycled, hence increasing process load on separation equipment. It also causes higher energy usage and lower production/unit mass of feedstocks.

Stoichiometry defines the proportion of chemical species involved in chemical reactions. When in balance, the molar contribution of each atomic element in the reactants equals that of the products. Imbalance can only work one way, with excess reactants. In other words, you cannot create products without reactants, but you can have excess of one or more unconverted reactants. Although the atomic elements are in molar balance, their re-arrangement into different molecular structure may create molecular molar imbalance between reactants and products; mass conservation is, however, maintained. For example, in the synthesis of methanol, a total of 3 moles of reactants results in 1 mole of product. However, from a mass standpoint, 32 g of reactants equal 32 g of products:

$$CO + 2H_2 \rightarrow CH_3OH \tag{2.57}$$

To illustrate stoichiometric balance, assuming complete reaction, if 28 kg of CO (1 kg-mole) is supplied with only 2 kg of hydrogen (1 kg-mole), all the hydrogen would be consumed, but only 1/2 of the CO would react due to hydrogen deficiency. This would result in 16 kg of methanol and 14 kg of unreacted CO. As basic and simple as this may appear, it can often be missed as many see reactors as black boxes and forget about the fundamentals. Material and stoichiometric balances are probably one of the first things to validate when troubleshooting reactor performance.

The feasibility of chemical reactions is thermodynamically defined by the difference in Gibbs free energy between the products and the reactants. Literature [43] suggests that chemical reactions are expected to occur when the difference in Gibbs free energy is negative, doubtful but possible for Gibbs free energy between 0 and 10,000 cal/g-mole, and highly unfavorable above 10,000 cal/g-mole. This is not an absolute rule as some chemical reactions are known to occur in the unfavorable range. There is still a fair amount of empiricism in chemical reactions mechanisms, especially when catalysts are involved. Nonetheless, understanding thermodynamic limitations may help in diagnosing potential problems associated with reactor performance.

The difference in Gibbs free energy can be calculated from the heat of formation (ΔH^o) and entropy (ΔS^o) differences as described by Equation 2.58:

$$\Delta G^o = \Delta H^o + T\Delta S^o \tag{2.58}$$

Values of ΔG°, ΔH°, and in some cases the entropy ΔS° at 25 °C are documented in literature [44] for many known components. Equation 2.58 can be adjusted for temperature using the following expressions:

$$\Delta H^T = \int Cp\, dT \tag{2.59}$$

The heat capacity (Cp) term can be integrated using Equation 2.37, with the integral solved as an indefinite integration for providing the heat of formation as a function of temperature. The heat capacity coefficients (Δa to Δe) are calculated based on the difference between the products and reactants normalized by their stoichiometric value:

$$\Delta H^T = I_H + \Delta aT + \frac{1}{2}\Delta bT^2 + \frac{1}{3}\Delta cT^3 + \frac{1}{4}\Delta dT^4 - \Delta e/T \tag{2.60}$$

I_H, integration constant that is determined from known conditions, is calculated from ΔH^{25}, which is calculated from individual component's heat of formation at 25

°C extracted from literature. At constant pressure, $dS = Cp\ dT/T$, then after solving the indefinite integral:

$$\Delta S^T = I_S + \Delta a \ln T + \Delta bT + \frac{1}{2}\Delta cT^2 + \frac{1}{3}\Delta dT^3 - \frac{\Delta e}{2T^2} \quad (2.61)$$

I_s, the integration constant, is calculated from ΔS^{25} similarly as for Equation 2.60. If components entropy ΔS^{25} values are not available, ΔG^{25} can be used by combining Equations 2.58, 2.60, and 2.61:

$$\frac{\Delta G^T}{T} = \frac{I_H}{T} + (\Delta a - I_S) - \Delta a \ln T - \frac{1}{2}\Delta bT - \frac{1}{6}\Delta cT^2 - \frac{1}{12}\Delta dT^3 - \frac{\Delta e}{2T^2} \quad (2.62)$$

The synthesis of ethylene oxide is used to illustrate the relevance of the above concept. Ethylene oxide is produced by partial oxidation of ethylene as described by the stoichiometric reaction Equation 2.63a. However, contacting ethylene with oxygen can be described by other balanced stoichiometric reactions as expressed by Equations 2.63b–2.63d. From the previous section on combustion, given the right conditions, ethylene in the presence of oxygen is known to oxidize or combust. The likelihood of the other reaction occurring is not so clear, but evaluating the difference in Gibbs free energy would provide the thermodynamic preference of those reactions relative to each other:

$$C_2H_4 + \tfrac{1}{2}\,O_2 \rightarrow C_2H_4O \text{ (ethylene oxide)} \quad (2.63a)$$

$$C_2H_4 + 3\,O_2 \rightarrow 2CO_2 + 2H_2O \quad (2.63b)$$

$$C_2H_4 + O_2 \rightarrow 2CH_2O \text{ (formaldehyde)} \quad (2.63c)$$

$$C_2H_4 + \tfrac{1}{2}\,O_2 \rightarrow C_2H_4O \text{ (acetaldehyde)} \quad (2.63d)$$

The entropy, heat of formation, and heat capacity coefficients for the components involved in the reactions described by Equation 2.63a–2.63d were gathered from literature with the value summarized in Table 2.11. The difference in Gibbs free energy for those reactions was calculated using Equations 2.58–2.62, with the results plotted in Figure 2.12. The plot on the left side shows little dependence of the Gibbs free energy on temperature, which is caused by the magnitude of the *x* axes. The *y* axes scale was magnified on the plot shown on the right to illustrate the temperature dependency.

The plot given on the left side shows a significant difference between the Gibbs free energy associated with the formation of CO_2 and water when compared to the formation of formaldehyde, acetaldehyde, and ethylene oxide. It also suggests the thermodynamic order of favorability starting with CO_2 and water followed by formaldehyde, acetaldehyde, and ethylene oxide.

Estimating the thermodynamic equilibrium of the combined reactions 2.63a–2.63d can be done through the relation between the equilibrium constant and the Gibbs free

TABLE 2.11
Gibb's Free Energy, Heat of Formation and Heat Capacity Coefficient for the Synthesis of Ethylene Oxide, Formaldehyde, and CO_2

Component	Entropy @ 25 °C (J/gmol K)	Heat of Formation @ 25 °C (kJ/gmol)	Heat Capacity Eq. 2.37 Coefficient				
			a	*b*E*[3]	*c*E*[6]	*d*E*[9]	*e*E*[6]
Ethylene	219.32	52.47	−6.39	184.40	−112.97	28.50	0.316
O_2	205.15	0	31.32	−20.24	57.87	−36.51	−0.007
Ethylene oxide	243.00	−52.64	−23.26	275.70	−188.97	51.03	0.387
Formaldehyde	218.95	−115.90	5.19	93.23	−44.85	7.88	0.551
Acetaldehyde[3]		−166.19	34.14	4.00	0.156	−0.016	
CO_2	213.78	−393.51	25.00	55.19	−33.69	7.95	−0.137
$H_2O_{(g)}$	188.84	−241.83	30.09	6.83	6.79	−2.53	0.082

Sources: NIST Chemistry WebBook, SRD 69. Acetaldehyde data from Perry and Chilton *Chemical Engineering Handbook,* 5th Edition and Ludwig's *Applied Process Design for Chemical and Petrochemical Plants*

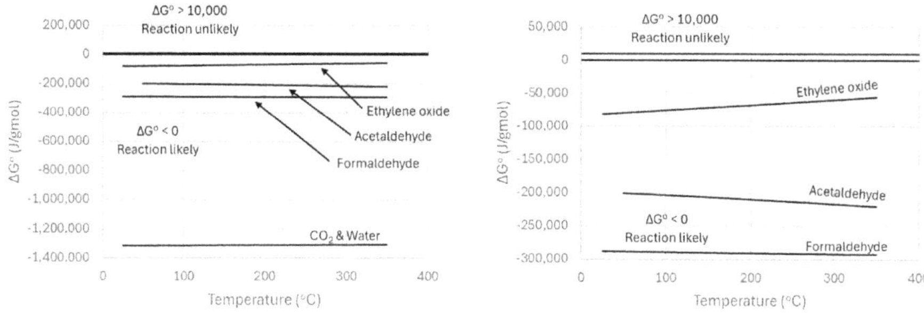

FIGURE 2.12 Gibbs Free Energy – Oxidation of Ethylene.

energy difference of the global reaction. The equilibrium constant of a chemical reaction is thermodynamically defined by multiplying the equilibrium activities of the products to the power of their stoichiometric coefficient over the product of the equilibrium activities of the reactants to the power of their stoichiometric coefficient as described by Equation 2.64. The equilibrium activity coefficients are determined from the ratio of the fugacity of the component over their standard state $\left(a_i = \frac{f_i}{f_i^o}\right)$.

$$K_{eq} = \frac{\prod a_P^{np}}{\prod a_R^{nr}} = \frac{\prod f_P^{np}}{\prod f_R^{nr}} = exp\left(\frac{\Delta G^0}{RT}\right) = exp\left(-\frac{\Delta H^0}{RT} + \frac{\Delta S^0}{RT}\right) \tag{2.64}$$

For an ideal gas, the fugacity can be substituted by the partial pressure or molar fraction. For liquids, the activity can be approximated by the molar concentration or fraction. Exponents np and nr are the stoichiometric coefficients of the components in the product and reactants, respectively.

The selectivity of Equations 2.63a–2.63d can be evaluated by combining them as a single reaction. It can be determined by manipulating the stoichiometric coefficients by trial and error until reaching the minimum Gibbs free energy. For simplification, this calculation can be accomplished with a process simulator using a Gibbs reactor block, which uses Gibbs free energy minimization for establishing the overall reaction product equilibrium. Calculating the expected selectivity of the overall reactions described by Equations 2.63a–2.63d at a temperature of 200 °C provides results in alignment with the plots illustrated in Figure 2.12, whereas, with excess oxygen, the entire ethylene reacts to CO_2 and water. By suppressing the formation of CO_2, the product of reaction is composed of formaldehyde and excess oxygen. The simulation must be configured with suppression of CO_2, formaldehyde, and acetaldehyde for allowing formation of ethylene oxide. The significant departure in Gibbs free energy explains why from a thermodynamic standpoint the products of those reactions cannot coexist at equilibrium.

Ethylene oxide is produced by partial oxidation of ethylene, as described by Equation 2.63a with the most competing reaction being the full oxidation of ethylene described by Equation 2.63b. The departure from the thermodynamic equilibrium is primarily explained by differences in reaction kinetics, operating conditions, and catalysts described in more detail later.

Equilibrium and reversible reactions limit the ability to fully consume the reactants. This concept can be illustrated by comparing a typical combustion reaction, where full depletion of the reactants can be accomplished by providing excess oxygen. For example, ethylene can be fully consumed by the combustion reaction described by Equation 2.63b in the presence of excess oxygen when above the autoignition temperature. On the other hand, the reaction of methanol with acetic acid to form methyl acetate and water as described by Equation 2.65 is a reversible reaction. If above the equilibrium concentration, methyl acetate would react with water to form methanol and acetic acid:

$$CH_3OH + CH_3CO_2H \rightleftarrows CH_3COOCH_3 + H_2O \tag{2.65}$$

The concentration of the reactants and products can be estimated by simulation using a Gibbs reactor block as previously described or from equilibrium constant data published in literature. The equilibrium constant for the reaction described by Equation 2.65 is reported at a value of 30 for a temperature of 30 °C and 17 for a temperature of 120 °C [45, 46]. Using molar fractions, Equation 2.64 can be simplified to:

$$K_{eq} = \frac{X_{CH_3COOCH_3} * X_{H_2O}}{X_{CH_3OH} * X_{CH_3CO_2H}} \tag{2.66}$$

If stoichiometrically balanced, $X_{CH_3COOH_3}$ equals X_{H_2O} and X_{CH_3OH} equals $X_{CH_3CO_2H}$. The summation of the mole fractions being unity, the equilibrium constant K_{eq} can be expressed in terms of a single mole fraction:

$$K_{eq} = \frac{X^2_{CH_3COOCH_3}}{\left(\frac{1}{2} - X_{CH_3COOCH_3}\right)^2} \tag{2.67}$$

Resolving the above equations suggests that if this reaction proceeds at 120 °C and reaches equilibrium, a residual molar fraction of about 9.8% of methanol and acetic acid is expected in the product stream. If the reaction proceeds at 30 °C, the residual molar fraction drops to 7.7%. Because of thermodynamic equilibrium, total depletion of the reactants is not possible, which has implications relative to downstream separation units. The equilibrium constant is a thermodynamic entity that is influenced by temperature and not by kinetics or catalyst. It is worth noting that the magnitude of the equilibrium constant reflects the extent of reactants' consumption. Full consumption of the reactants makes the denominator in Equation 2.66 equal to zero or near zero, making the equilibrium constant approaching infinity. On the contrary, a small equilibrium constant is equivalent to higher reactant residual in the product stream. This principle applies whether the reaction is reversible or not.

Le Châtelier principle states that a mixture will aim at re-establishing equilibrium in response to changes affecting that equilibrium. Excess reactants will move the reaction forward to the products, whereas accumulation of products beyond the equilibrium will reverse the reaction toward the reactants. Leveraging upon this principle, introducing an inert component would shift the equilibrium, which may benefit the reaction. In a gas system, the introduction of an inert component increases the total number of moles in the system, hence reducing the partial pressure of the product and contributing to displacing the equilibrium from reactants to product. This strategy has previously been discussed for steam cracking, where injection of dilution steam improved selectivity by reducing partial pressure of the products. Increasing the reactor pressure in gas system would shift the reaction direction toward the side of fewer moles.

Kinetics refers to the rate of a chemical reaction. A chemical reaction might be thermodynamically feasible, but it has no commercial practicality if conversion of the reactants to products occurs over years instead of minutes or seconds. Oxidation reactions at ambient temperature, while not necessarily desired, are examples of reactions occurring over extended periods when compared to steam cracking, which occurs over less than a second. Kinetics is key in the determination of reactor volume as it relates directly to residence time required for achieving the desire reaction and conversion. It therefore dictates the operating range of a chemical reactor, where higher rates would limit conversion of the reaction. Contrarily, low rates would increase residence time and may result in undesired competing reactions.

Kinetics is strongly associated with reaction mechanisms where it depends on the controlling step. In many cases, it can be relatively complex and beyond the scope of this discussion. For simplification and discussion, the reaction of *A* with *B* to form product *C* is considered as described by Equation 2.68:

$$A + B \rightarrow C \tag{2.68}$$

Figure 2.13 provides the results of laboratory experiments, where a mixture of A, B, and water was incorporated in different test tubes, which were exposed to the same temperature but for different time durations. The test tubes were then analyzed for the residual concentration of component A. Two kinetic models are being evaluated to describe the rate of reaction, being a first-order model as described below:

$$\frac{dC_A}{dt} = k_1 C_A \tag{2.69}$$

With k_1 being the rate constant, Equation 2.69 can be integrated

$$\int \frac{dC_A}{C_A} = k_1 \int dt \tag{2.70}$$

After resolving the indefinite integration residual constant using known initial conditions $C_A = 0.1$ at time = 0, Equation 2.70 becomes:

$$C_A = exp\left(k_1 t + \ln\left(C_0\right)\right) \tag{2.71}$$

The data points calculated from Equation 2.71 are illustrated in Figure 2.13 by the square identified as first order. As shown, the data points provide a good fit to experimental data for reaction times below 30 minutes. The model starts to diverge from experimental data beyond 30 minutes.

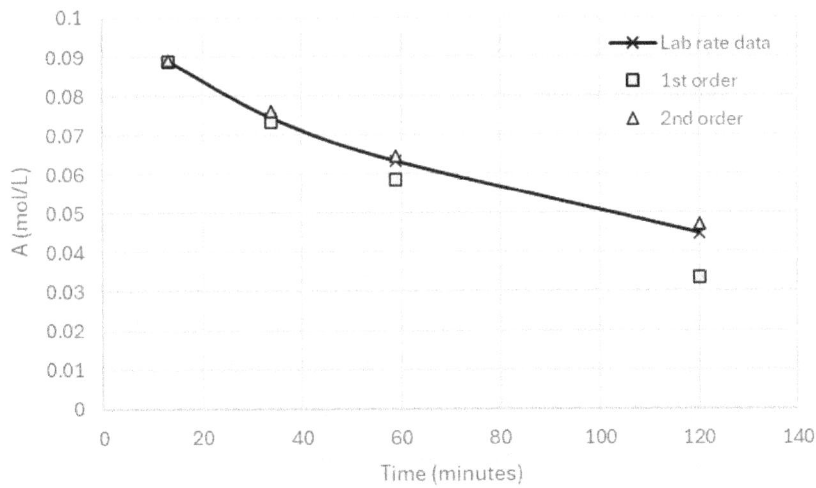

FIGURE 2.13 Rate of Consumption of A – Experimental Data versus First- and Second-Order Kinetics.

A second-order Type I kinetics described by Equation 2.72 is also evaluated against experimental data:

$$\frac{dC_A}{dt} = k_2 C_A^2 \tag{2.72}$$

With k_2 the rate constant, as previously done, Equation 2.72 can be integrated as shown in Equation 2.73:

$$\int \frac{dC_A}{C_A^2} = k_2 \int dt \tag{2.73}$$

Resolution of the indefinite integrals with the same known initial conditions as above provides a relationship between the residual concentration C_A and time as described by Equation 2.74:

$$C_A = 1/\left(k_2 t + 1/C_{A0}\right) \tag{2.74}$$

As shown in Figure 2.13, this second-order kinetic model provides a better fit with experimental data. Understanding of reaction mechanisms and kinetics is critical in reactor design. It provides the basis for reactor scale-up, which also involves optimization with the recovery section. From an operational perspective, the volume of the reactor is fixed. Understanding the kinetics provides insight related to operation range and impacts on downstream separation equipment.

Example 2.3: The second-order kinetics model developed earlier is used for comparing the operational range of a reactor having a useful volume of 15 m^3. Aqueous mixtures of *A* and *B* are fed to a continuous stirred tank reactor (CSTR) in the same molar proportion as the laboratory test data. The expected production of component *C* and its molar concentration in the reactor effluent is calculated as a function of feed rates. As illustrated in Figure 2.14, at throughput below 30,000 kg/hr, the production of component C is below 50 kg/hr. It remains above 50 kg/hr between throughputs of 30,000 and 200,000 kg/hr and drops below 50 kg/hr beyond 200,000 kg/hr. This would define the operating range if maximization of product *C* is the objective. However, due to the dilution effect of water, which is considered an inert in this reaction, the concentration of component *C* rapidly decreases as feed rates increase. This may restrict the capacity of the downstream recovery units and may limit the reactor operating range.

Operation of the reactor in batch mode provides an interesting comparison. The production rate of component *C* from a batch reactor was estimated assuming a period of 60 minutes total for charging and unloading a reactor having the same useful volume (15 m^3). As illustrated in Figure 2.14, the production rate seems to reach a plateau after 60 minutes of reaction time, which is equivalent to a total batch time of 120 minutes. Further increase in reaction time does increase the concentration of component *C*, but not its total mass per batch time due to a decrease in reaction rate. While the continuous reactor provides a higher mass production of component *C*, it delivers the product at a much lower concentration by a factor of 2–10 depending

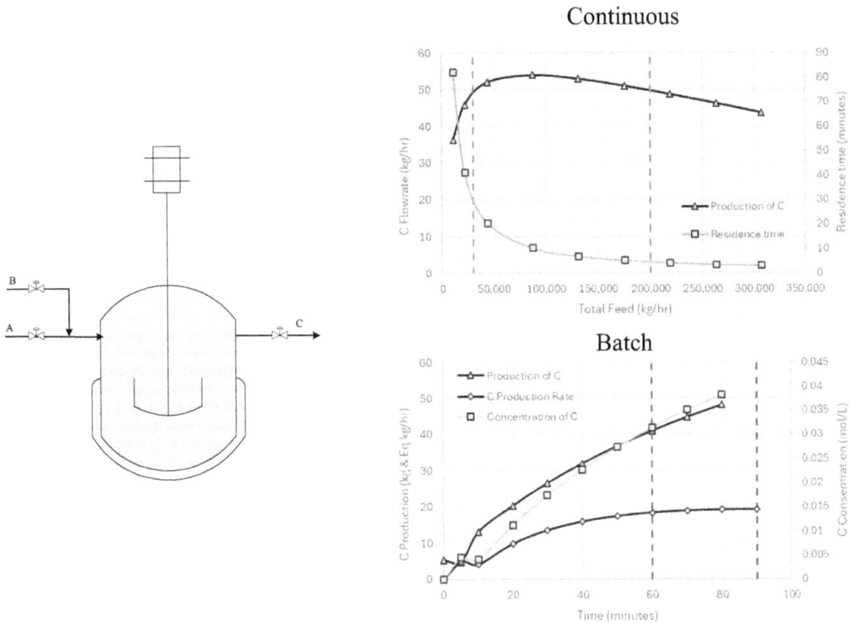

FIGURE 2.14 Continuous versus Batch Reactor Performance.

on feed rates or reaction time. The decision between one operation mode over the other is a design decision based on the recovery unit and production volume targeted. Batch reactors are often selected in specialty chemicals where small volumes of different product formulations are manufactured. It is also worth noting that in batch system, the reactor vessel is frequently used to perform tasks other than chemical reaction such as blending and separation. This reduces capital cost of the units and is a good fit for small volumes involving varieties in product slates.

The rate constant k is a function of temperature typically following the Arrhenius equation:

$$k = k_0 e^{-E/RT} \tag{2.75}$$

where k_0 is a pre-exponential factor that is catalyst dependent, E is the activation energy, R the gas constant, and T the temperature.

Catalysts reduce the activation energy, speed up the reaction rate, and improve selectivity toward the desired reaction. They can be in a homogeneous form, such as sulfuric acid used in the synthesis of methyl acetate as described by Equation 2.65, or heterogeneous like the alumina-supported silver catalyst used in the partial oxidation of ethylene to ethylene oxide. On a macro level, the catalyst remains unchanged by the reaction, that is, they return to their original form at the end of the reaction. On a micro level, they are actively involved in the reaction by creating an unstable intermediate, which reduces the activation energy, speeds up the overall reaction, and drives it toward the desire products. In the methyl acetate reaction, sulfuric acid transfers a proton to acetic acid, which creates an active site that reacts with methanol to form methyl acetate as described by Equations 2.76, 2.78, and 2.79. A hydrated proton (H_3O^+) is created along with the formation of methyl acetate. This hydrated proton can combine with the HSO_4^- anion generated in Equation 2.76, releasing water and the sulfuric acid catalyst back to its original molecular configuration. It is worth noting that the proton released from sulfuric acid can also bond with methanol as described by Equation 2.77, thus reducing the reactivity of methanol with the acetic acid cation that follows Equation 2.78. Literature [47] suggests that under small sulfuric acid concentration, the protonation of acetic acid is more favorable and more than compensate the potential protonation of methanol. Dosage of sulfuric acid must therefore be within a certain range to avoid promoting the protonation of methanol.

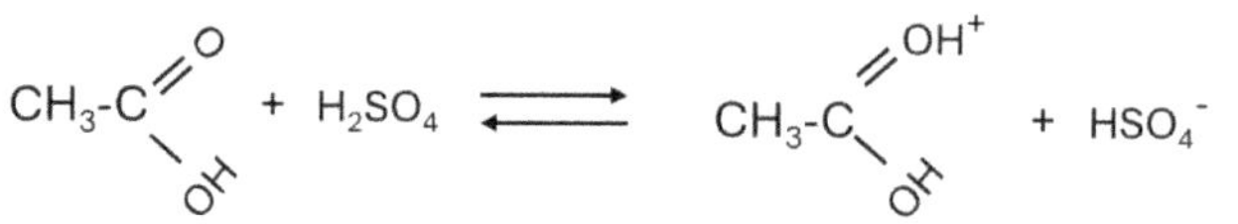

Acetic Acid Protonation by Sulfuric Acid (2.76)

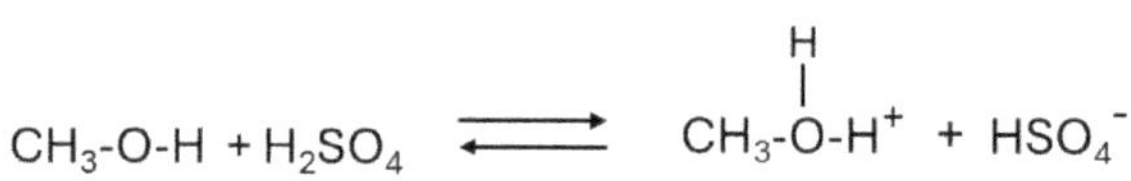

Methanol Protonation by Sulfuric Acid (2.77)

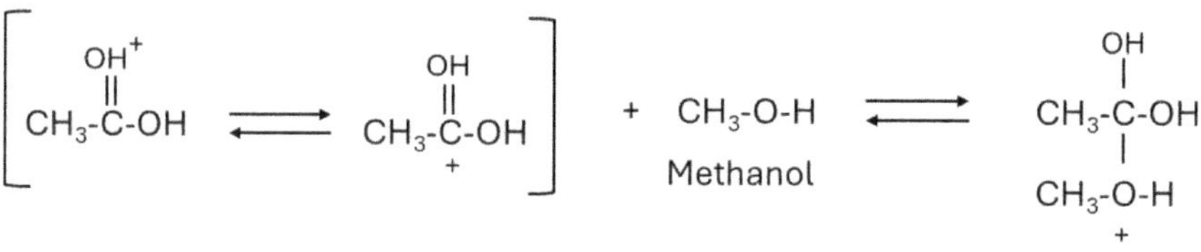

(2.78)

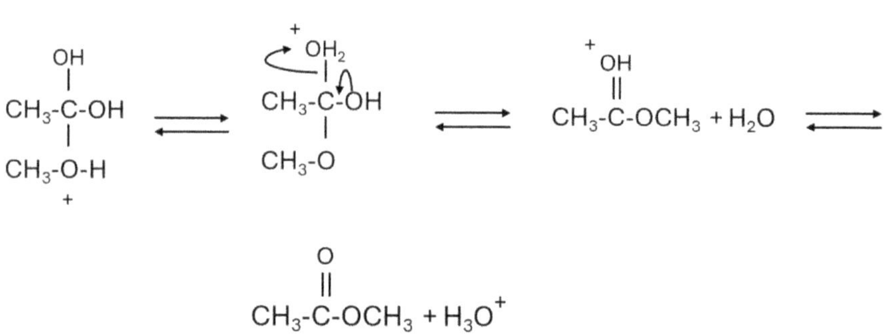

Methyl Acetate (2.79)

Homogeneous catalysts are typically involved in liquid phase reactors as the catalyst mixes with the reactants in a homogeneous mixture. Heterogeneous catalysts, on the other hand, are characterized by active sites contained within a solid support. They can be used in liquid or vapor phase reactors. The support is typically designed to restrict access to the active site for specific molecules based on molecular configuration, size, and geometry. Access to the active site is driven by adsorption through the pores of the catalyst support. Two types of adsorption mechanisms typically occur. Physical adsorption is characterized by weak forces attracting the molecules, and it is associated with the release of heat comparable to the heat of condensation. Physical adsorption is rarely associated with chemical reactions and does not impact the activation energy or the reaction rate. It is a relatively fast process, which is easily reversible. Chemisorption is the second adsorption mechanism associated with solid catalyst. The forces involved are much stronger than those in physical adsorption, where the molecules are held to the active site by valence forces. The heat release is also greater and typically equivalent to the heat of reaction. Unlike physical adsorption, the process is irreversible. Chemisorption can be classified into two types: activated and nonactivated. Activated chemisorption follows the Arrhenius equation, in which the activation energy and rate constant are dependent on temperature, as previously described by Equation 2.75. Nonactivated chemisorption is less frequent and does not depend on temperature. It is characterized by a low or near zero activation energy and rapid reaction rate.

The capacity of a catalyst bed is characterized by the volume of reactants that can be adsorbed, which is also influenced by temperature as illustrated in Figure 2.15

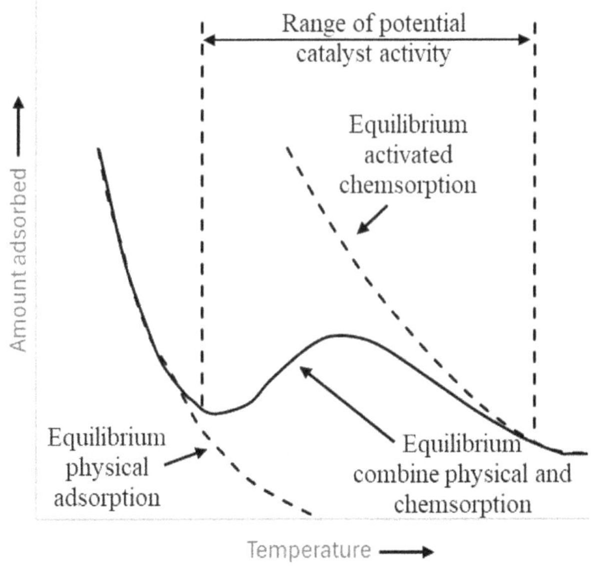

FIGURE 2.15 Effect of Temperature on Adsorption Capacity [44].

Reproduced with Permission from figure 2.2, page 19, *Multiphase Catalytic Reactors Theory, Design, Manufacturing, and Applications*, John Wiley & Sons Limited

[48]. While further increase in temperature might be beneficial from a chemical reaction perspective, the combined equilibrium adsorption curve is characterized by a maximum followed by a decrease with temperature. Beyond a certain point, increasing the temperature provides no additional capacity gains. At lower temperature, adsorption is entirely done by physical adsorption, which, as stated earlier, rarely involves chemical reaction. The catalyst bed is therefore characterized by an operating range identified as *Range of potential catalyst activity* shown in Figure 2.15.

The product of reaction is released from the catalyst active site by desorption as it must travel out through the support pores.

Heterogeneous catalysts are characterized by selectivity, stability, and activity. Selectivity is defined by the fraction of the desired product over the total secondary products. It can also be defined as the fraction of the desired product over the controlling reactant consumed. Stability is defined by the catalyst ability to retain its activity when exposed to physical and chemical changes. From a kinetics perspective, the catalyst activity can be defined by the reaction rate of the desired reaction. Heterogeneous catalysts are expected to lose activity over time due to accumulation of impurities such as coke on the active site. Sinter occurs due to long-term operation or exposure to undesired conditions, such as high temperature. Poisons which permanently bond with the active site cause a permanent loss of catalyst activity. Heavy metals such as lead, mercury, cadmium, and others are known catalyst poisons. Other components, such as CO in acetylene hydrogenation, are referred to as temporary poisons and are frequently used to temper the catalyst activity.

Catalyst regeneration is required, periodically, as a result of decline activity. Depending on catalyst type, regeneration can be executed in situ (i.e., on site) or offsite. In some cases, deactivation is permanent, for example when poisoned by heavy metals. On-site regeneration typically involves burning coke and oligomers on active sites using a dilute stream of oxygen or air. The burn is typically followed by a reduction step for the active metal using a dilute hydrogen stream. Improper regeneration/reduction will result in lower catalyst activity and shorter run length. Parallel reactors, one on stream and one on standby, are typical for onsite regeneration. This avoids unnecessary plant outages for regenerating the catalyst. Reactor run length should be established based on the cost associated with the loss of catalyst activity against the cost of regeneration. For offsite regeneration and single reactor operation, a full plant outage is required. In such case, the financial impact of a reduction in catalyst activity should be balanced against a total plant outage. Furthermore, storage of spare catalyst would minimize production impact in case of unexpected loss of catalyst activity.

In fluid catalytic cracking (FCC) used for converting heavy hydrocarbon fractions to gasoline, the catalyst is continuously regenerated. The catalyst is rapidly deactivated due to coke deposition in the fluidized bed section. It is then regenerated in the regenerator section as it is contacted by air before being recycled to the fluidized bed reactor.

The reaction mechanisms define the stoichiometry of all the parallel or sequential reactions expected to occur in a reactor. Some of those reactions are desired, whereas others negatively impact the value of the reactor effluent. As previously discussed, ethylene oxide is produced by partial oxidation of ethylene as described by

Equation 2.63a, with the most competing reaction being the full oxidation of ethylene described by Equation 2.63b. The departure from the thermodynamic equilibrium is primarily explained by differences in reaction kinetics, operating conditions, and catalyst. The process requires the air or oxygen/ethylene mixture supplying a reactor to be above the upper flammability limit (UFL). This limits the rate of Equation 2.63b to a straight oxidation rather than combustion, hence a much slower reaction rate. Operating above the UFL is also required from a process safety perspective to prevent a potential deflagration inside the reactor. Commercial processes use an alumina-supported silver catalyst, sometimes in combination with additives such as 1,2-dichloroethane for improving selectivity by suppressing total oxidation to CO_2 and water. Operating pressures vary between 30 and 100 bar depending on using air or oxygen as reactant, with operating temperatures in the range of 220–250 °C. Higher temperatures tend to promote the selectivity toward full oxidation of ethylene.

Heat of reaction is the heat generated or required by a chemical reaction. A reaction is classified as exothermic when it generates heat, and endothermic when it requires energy. For example, combustion reactions generate heat and are classified as exothermic. Steam cracking and reforming, on the other hand, require heat and are classified as endothermic. The heat of reaction is calculated by doing the difference between the products' and reactants' heats of formation or heat of combustion prorated by the stoichiometric coefficient, as described by Equation 2.80. Components' heat of formation or heat of combustion can be obtained from literature or estimated from group contribution:

$$\Delta H^{T25}_{Reaction} = \sum_{Product}\left(n_p\ \Delta H^{25}\right) - \sum_{Reactant}\left(n_r\ \Delta H^{25}\right) \tag{2.80}$$

The heat of reaction is based on the reference temperature of 25 °C at 1 atm. Similar to the flame adiabatic temperature discussed in combustion, the adiabatic temperature of a chemical reaction can be estimated from the heat of combustion using Equations 2.36 and 2.37, in which $Q_{Stack\ losses}$ is replaced by $\Delta H^{T25}_{Reaction}$. Unlike combustion, high reaction temperatures may negatively impact the expected selectivity, damage the catalyst, and could also lead to runaway reactions. Hence, in many cases, reaction conversions are limited by controlling or limiting the temperature rise for preventing those undesired outcomes.

Calculating the heat of reaction is another tool for process engineers along with stoichiometry for monitoring the performance of chemical reactors. The presence of inlet/outlet analyzers is highly recommended, and with flow and temperature indication it should provide sufficient process data, beyond the degree of freedom required for assessing the performance of the reactor. Sudden gaps trending away from the expected H&MB are signs that something deserves more attention.

Reactor design depends on the characteristics of the chemical reaction, endothermic or exothermic, catalytic or not, vapor or liquid phase. As previously discussed, steam cracking occurs in an endothermic tubular reactor at high temperature, which makes this suitable for fired heaters. At moderate temperatures, conventional reactors with heated jackets or heating coils are frequently used for liquid phase reactors. Similarly, those reactors can be used for mild exothermic reaction with a cooling medium circulating in the coil or reactor jacket. Adiabatic and isothermal reactors are typically used for heterogeneous catalytic reactors. Going over the different reactor designs is beyond the scope of this book and the discussion will be limited to liquid phase reactor vessel, adiabatic and isothermal packed bed reactors, which are the most common design used in the chemical industry.

Liquid phase reactor vessels are typically equipped with mixers, and the content is assumed to be completely mixed. Such reactors operating in continuous mode are referred to as CSTRs. Depending on kinetics, several reactors can be operated in series to achieve the desired conversion and yields. The assumption of a completely mixed vessel means that the reactor content composition and temperature are identical to the reactor effluent. Deviations from ideal mixing may influence the material balance and kinetics. For instance, the residence time of a completely mixed vessel equals the reactor volume divided by the volumetric flow rates. Deviations from ideal mixing influence the true residence time of the reactor. Back mixing, for instance, is like an internal recycle and contributes to increasing the residence time. Contrarily, dead or stagnant volume would reduce the residence time. These reactors by design have some level of non-ideality, which can be evaluated through computational fluid dynamics (CFD) or through residence time distribution experiments with tracers. Reactor internals anomalies, such as damage or loose baffles, accumulation of sludge would alter the mixing behavior of the reactor, hence affecting its performance. Stagnant volume or non-uniform accumulation of fouling material on heat transfer surfaces could cause inconsistent temperature distribution and is another example of reactor malfunction. Damaged mixer or deviations from design rotational speed is another deviation affecting reactor performance.

Adiabatic heterogeneous catalytic reactors are characterized by a packed bed of catalysts contained in a process vessel. The chemical reaction proceeds as reactants travel through the bed. The extent of the reaction rises as the process fluids move away from the inlet, which can be observed by the temperature profile across the height of the reactor. The height of the reactor or catalyst volumes is limited by the outlet temperature, which, if too high, may affect the selectivity of the catalyst, or, worse, result in a runaway reaction. Reactors in series would typically be provided with inter-bed cooling in the design, for mitigating against the temperature rise going beyond an acceptable level for the desired conversion. The ideal mixing pattern of packed bed reactors would follow a plug flow hydraulic. The mixing pattern is strongly dependent upon the feed distributor and volumetric rates. Some levels of radial dispersion are expected at design rate, but this should be relatively minor relative to the reactor performance. Poor distributor design or operation below turndown limits will likely create maldistribution and the potential for channeling. Extended operation below turndown may cause irreversible preferential flow paths and channeling, which will impact reactor performance. Improper distribution may also result in hot spots that will negatively impact catalyst activity in localized areas. It must be understood that thermowell location along the height of a reactor does not provide any indication about the radial temperature profile. It is not unusual to hydraulically false load pack bed reactors when operating below turndown, which prevents channeling. However, caution must be applied, as recycles introduce dilution and lower heat of reaction. Excessive dilution may cause the mixture to enter the reactor below the activation energy required for initiating the reaction or the heat of reaction may not be sufficient for sustaining the reaction inside the reactor.

Isothermal heterogeneous catalytic reactors are configured like shell and tube exchangers, with the process fluid circulating through the tubes filled with catalyst. A heat transfer fluid is circulating through the shell side at rates required for removing or supplying heat released or required by the reaction. The distribution of the heat transfer fluid is configured such that the process side temperature profile along the reactor tubes remains relatively constant. It is not unusual for those kinds of reactors to have a large number of tubes, which makes them more complex in design and catalyst loading/unloading activities when compared to packed bed reactors. However, significantly less reactors would be required when compared to pack bed design, as they are not limited by the temperature rise. Although not immune to runaway reactions, they provide more robustness than packed bed reactors due to uniform temperature control. As for packed bed, they follow a plug flow hydraulic, with radial dispersion initiated by the distributor. Radial dispersion is less likely at the tube level given the small size. Temperature profile measurement is a challenge and limited to fairly small number of tubes. Therefore, hot spot caused by maldistribution is harder to detect. They are equally affected, as with packed bed reactors, from hydraulic false load due to dilution, which may result in unsustainable reaction.

Monitoring of H&MBs around reactors should provide early warning around loss of catalyst activity. Inconsistency between H&MBs and temperature profiles may suggest the potential for hot spots and localized catalyst deactivation.

Example 2.4: Due to the proprietary nature of catalysts, heterogeneous catalytic reactors are often perceived as black boxes by operation process engineers. This example is based on C_2 hydrogenation reactors in ethylene plants and is intended to demonstrate how the theory described earlier can be used for assessing the performance of a reactor. The primary objective of those reactors is to eliminate acetylene from the C_2 stream [49]. Acetylene is a poison to catalysts used by polyethylene plants. As such, polymer-grade ethylene has a typical product specification of less than 5 ppm acetylene [50]. Some ethylene plants operate their C_2 hydrogenation reactors to full conversion of acetylene. Acetylene contamination of the ethylene fractionator is associated with prolonged flaring periods due to the binary vapor/liquid equilibrium characteristics between ethylene and acetylene. The C_2 hydrogenation reactors operate upstream the ethylene fractionator; as such, a reactor acetylene off-specification excursion requires the process operator to re-direct the reactor outlet to flare so to prevent contamination of the ethylene fractionator. Getting the reactor back within specification is an exercise that is significantly faster than freeing acetylene from the ethylene fractionator. Therefore, it can be stated that the primary function of the C_2 hydrogenation reactors is to eliminate the acetylene from the C_2 stream. The three primary chemical reactions involved in the C_2 hydrogenation reactor are described by Equations 2.81–2.83:

$$C_2H_2 + H_2 \rightarrow C_2H_4 \tag{2.81}$$

$$C_2H_2 + 2H_2 \rightarrow C_2H_6 \tag{2.82}$$

$$C_2H_4 + H_2 \rightarrow C_2H_6 \tag{2.83}$$

Given that ethylene is the value product, there are incentives in maximizing the conversion of acetylene to ethylene. However, as described earlier, this reaction competes with two other reactions, which lead to the formation of ethane. From a thermodynamic perspective, the conversion of acetylene to ethane described by Equation 2.82 is more favorable with a difference in Gibbs free energy of about –74,000 cal/g-mol when compared to Equations 2.81 and 2.83 for which the difference in Gibbs free energy was estimated at –40,710 cal/gmol and –31,235 cal/gmol, respectively, for a reactor operating pressure of 25 bar.

Not shown earlier are other reactions which are less significant but detrimental to the catalyst performance. Those reactions involve the conversion of acetylene to intermediate C4s, which can then further react to oligomers and coke. Although the formation of oligomers and coke is expected over extended periods, their formation can be accelerated by operation at high temperature beyond what the catalyst has been designed for. The catalyst manufacturer would normally provide a recommended temperature range. Other than operating temperature, the formation of oligomers and coke can be promoted by the presence of impurities such as water or methanol. The rate of formation of oligomers and coke would typically define the run length of the catalyst as it impacts catalyst activity and selectivity.

The reactor evaluated in this example is an adiabatic packed bed reactor as illustrated in Figure 2.16, where flow measurements are done on the C_2 and hydrogen streams. Analyzers at the inlet and outlet of the reactor are providing stream composition of acetylene, ethylene, and ethane. The hydrogen residual is measured at the outlet of the reactor. The methane content of the hydrogen supplied to the reactor is measured by an analyzer on the hydrogen stream. The reactor inlet and outlet stream temperatures are measured; temperature probes installed throughout the catalyst bed provide a temperature profile.

The hydrogen flow is regulated by a flow controller for which the set point (SP) is set by a stoichiometric ratio calculated from the C_2 stream flow and acetylene analyzer.

Evaluating the material balance is the first step for evaluating the performance of the reactor. This involves converting mass flows into molar flow rates for alignment with the analyzers and reaction stoichiometry. To do so, the C_2 stream molecular weight can be estimated from the analyzer values and components molecular weights as described below:

$$MW_{C_2In} = \left(C_2H_2*26 + C_2H_4*28 + C_2H_6*30\right) / \left(C_2H_2 + C_2H_4 + C_2H_6\right) \quad (2.84)$$

where C_2H_2, C_2H_4, and C_2H_6 are the mole fractions of acetylene, ethylene, and ethane, respectively, obtained from the reactor inlet stream analyzer. The hydrogen stream may contain between 4% and 6% mol methane, with the value provided from an online analyzer on the hydrogen stream. The stream molecular weight can be calculated following the same methodology as for the C_2 stream:

$$MW_{H2In} = \left(CH4*16 + \left(1 - CH4\right)*2\right) / \left(CH4 + H2\right) \quad (2.85)$$

The reactor outlet molecular weight can be calculated in a similar fashion from the analyzer outlet stream values. Although relatively accurate for the determination of the molecular weight, the mole fractions at the reactor outlet may require

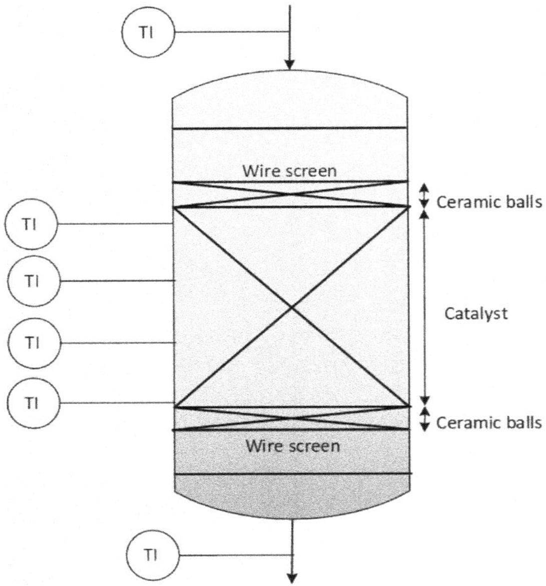

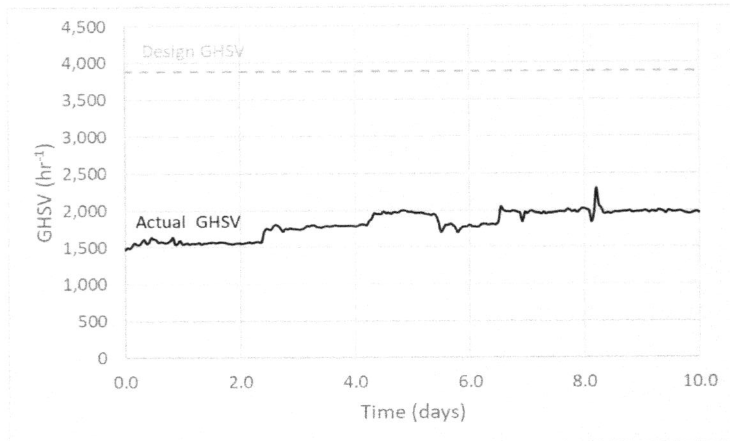

FIGURE 2.16 C_2 Hydrogenation Reactor Example 2.4.

normalization due to the small hydrogen residual composition that could impact the methane balance. Methane is an inert in this reaction; its mass and molar flow rate should be based on the inlet rather than calculated by differences from the analyzer.

From the calculated inlet and outlet stream molecular weight along with the normalized molar composition, a mole balance can be calculated:

$$MolarFlow\_In = C2Flow * MW_{C2In} + H2Flow * MW_{H2In} \tag{2.86}$$

$$MolarFlow\_Out = \left(C2Flow + H2Flow\right) * MW_{C2Out} \tag{2.87}$$

$$C2H4 = MolarFlow\_In * C2H4_{Out} - MolarFlow\_Out * C2H4_{In} \tag{2.88}$$

$$C2H6 = MolarFlow\_In * C2H6_{Out} - MolarFlow\_Out * C2H6_{In} \tag{2.89}$$

The reactor selectivity toward ethylene can then be calculated from:

$$\text{Selectivity}_{C_2H_4} = C_2H_4 / C_2H_2 * 100 \tag{2.90}$$

where C_2H_2 is the total mole of acetylene fed to the reactor. The selectivity toward ethane can also be calculated by doing the ratio of ethane mole generated per mole of acetylene supplied to the reactor. The margin from 100% on the summation of the ethylene and ethane selectivity is an indication of other reactions such as intermediate C4s which can result in the formation of oligomers and coke.

The above selectivity can be validated by comparing the measured reactor outlet temperature to the calculated temperature obtained from a heat balance around the reactor. This being an adiabatic reactor, the heat of reaction is the only source of heat and is calculated from the heat of formation of the reactants and products:

$$Q_{\text{Reaction}} = \left(\Delta H_{fC_2H_4} - \Delta H_{fC_2H_2}\right) * C_2H_4 + \left(\Delta H_{fC_2H_6} - \Delta H_{fC_2H_2}\right) * C_2H_6 \tag{2.91}$$

The outlet temperature can then be calculated from

$$Q_{Reaction} = mCp\,\Delta T = m \int Cp\,dT \tag{2.92}$$

With the reaction occurring at a pressure of 25 bar, the heat capacity coefficients were adjusted from a regression using a process simulator with the resulting expression shown below on a molar basis:

$$\begin{aligned} Q_{Reaction} - \Bigg[& \Sigma n_i\, a_i \left(T_{Out} - T_{In}\right) + \Sigma \frac{n_i\, b_i}{2}\left(T_{Out}^2 - T_{In}^2\right) + \Sigma \frac{n_i\, c_i}{3}\left(T_{Out}^3 - T_{In}^3\right) \\ & + \Sigma \frac{n_i\, d_i}{4}\left(T_{Out}^4 - T_{In}^4\right) + \Sigma \frac{n_i\, e_i}{5}\left(T_{Out}^5 - T_{In}^5\right) + \Sigma \frac{n_i\, f_i}{6}\left(T_{Out}^6 - T_{In}^6\right)\Bigg] = 0 \end{aligned} \tag{2.93}$$

where n_i are the components molar flow and e_i a_i, b_i, c_i, d_i, e_i and f_i are the regressed heat capacity coefficients for each component. Equation 2.93 is resolved numerically for the outlet temperature T_{out}.

TABLE 2.12
C_2 Hydrogenation Reactor Heat and Material Balance

Reactor Feed			
C_2 feed (kg/hr)	54,480		
Hydrogen feed (kg/hr)	51.6		
Analyzers			
	C_2 inlet stream	H_2 inlet stream	Outlet stream
H_2 (mole %)	N/A	Not measured	0
CH_4 (mole %)	N/A	6.15	Not measured
C_2H_2 (mole %)	0.56	N/A	0
C_2H_4 (mole %)	60.00	N/A	60.16
C_2H_6 (mole %)	39.08	N/A	39.42
Calculated Mole Balance			
	Reactor Inlet		**Reactor Outlet**
H_2 (kmol/hr)	18.25		0
CH_4 (kmol/hr)	1.18		1.18
C_2H_2 (kmol/hr)	10.44		0
C_2H_4 (kmol/hr)	1,128.2		1,134.5
C_2H_6 (kmol/hr)	732.3		736.8
Reactor Performance			
Selectivity			
Ethylene (%)	60.3		
Ethane (%)	41.8		
Gas hourly space velocity (hr^{-1})	1,573		
Heat of reaction (GJ/hr)	2.5		
Measured outlet temperature (°C)	52.2		
Calculated outlet temperature (°C)	49.2		

Table 2.12 summarizes a snapshot reactor H&MB, and Figure 2.17 illustrates the selectivity of the reactors over a ten-day period. The differences in measured and calculated temperature are also plotted for the same period, with the upper and lower expected temperature based on 100% ethylene or ethane selectivity plotted. The catalyst design gas hourly space velocity (GHSV) is 4,000 hr^{-1} with an expected ethylene selectivity of 50% at the beginning of run and 30% at the end of run.

At first sight, the reactor performance data displayed in Table 2.12 suggest an ethylene selectivity higher than expected at 60.3%. The ethane selectivity is, however, calculated at 41.8%, suggesting a small error in the mole balance as the combined ethylene and ethane selectivity should equal or be less than 100% if accounting for

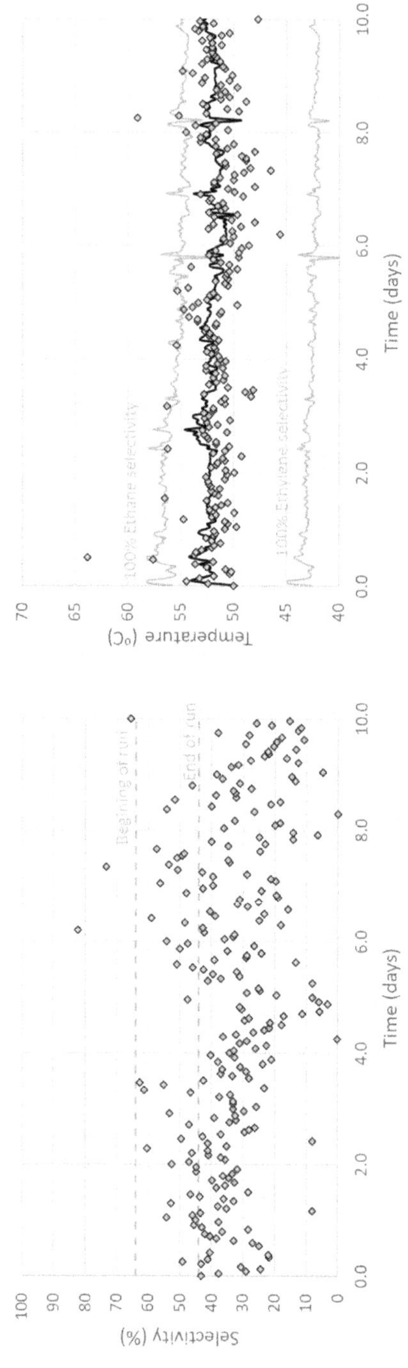

FIGURE 2.17 C_2 Hydrogenation Example – Selectivity and Outlet Temperature.

secondary reactions. Regardless, a 2% error is reasonable for assessing the reactor performance. The calculated temperature is underpredicted by 3 °C, which could be an indication of other reactions taking place inside the reactor or accuracy of the temperature transmitter or analyzers. It could also be an indication of secondary reactions taking place in the reactor, which would however contradict the calculated selectivity slightly exceeding 100%. Regardless, overall, those results do not raise any alarming concerns. However, performing the calculations on an hourly basis for a period of 10 days and trending the ethylene selectivity, the predicted outlet temperature, and the GHSV provide a different perspective on the reactor performance. As illustrated in Figure 2.17, the ethylene selectivity is averaging 32%, varying from negative values to about 60%. The reactor outlet temperature predictions are for the most part relatively close to the actual temperature with an average difference of 0.6 °C. There are, however, some significant differences at times that could be as high as 14 °C. The actual GHSV, illustrated in Figure 2.16, rises by step increment over the ten-day period evaluated, from 1,500 to about 2,000 hr^{-1}. This corresponds to 37–50% of the design GHSV. Operation at such low rates relative to design can be associated with several problems. First, distribution would be impacted if operating below the turndown of the reactor inlet distributor. This could likely result in channeling, stagnant volumes, and potential back mixing, which would all impact the reactor performance. Lower rates increase residence time and may negatively affect the expected selectivity, also leading to secondary reactions. In this example, there are no correlations of any operating parameters that can explain the inconsistency in ethylene selectivity other than operating the reactor at significantly less GHSV than design. This example is a classic case of a plant revamp, which minimizes capital expenditure by retrofitting existing equipment. While the primary objective of the reactor in consuming the acetylenes is met, there is a loss of valuable ethylene. In other applications, where the reactor process objective is to convert feedstocks into valued chemicals such as partial oxidation of ethylene to ethylene oxide, a similar reactor performance as described in the above example would be more problematic as it will significantly impact downstream units as well as unit overall performance.

2.1.3 Separation

The combined stream from the reaction section of a chemical plant contains one or several chemicals of value. By itself, the stream has no practicality and must undergo further processing for removing impurities, contaminants, and separating value chemicals into streams meeting product quality specifications. As discussed in the fired equipment and reactor section, chemical reactions rarely proceed through completion due to equilibrium and/or yield optimization. Reactor effluents then hold unreacted reactants that must also be separated and recycled to the reaction section. Separation processes leverage on differences in physical and thermodynamic properties between the components contained within a stream. For instance, separation of liquid/solids, and vapor/solids streams are driven by density differences and particle size distribution. Liquid/liquid separations are based on fluids' immiscibility, density differences, droplet size of the disperse liquid as well as interfacial tensions. Vapor/liquid separations are characterized by differences in boiling points and relative volatilities between the components contained in a mixture. This section reviews the fundamental concept involved in separation units, the design criteria, and potential problems associated with operation of those units.

2.1.3.1 Distillation

Distillation is the most common unit operation used in the CPI for purifying process streams to sales specifications. It is a vapor/liquid application that leverages on difference in boiling points/relative volatilities between the components being separated. By adding energy to a feed stream, a distillation tower will simply make product streams that have improved purities. Distillation towers are designed for promoting multiple vapor/liquid equilibriums where bi-directional mass transfer between the vapor and liquid phase occurs. Understanding and evaluating the performance and capacity of distillation towers can be divided into two elements: vapor/liquid equilibriums and hydraulics.

Vapor/liquid equilibrium concepts are discussed here through the separation of ethanol from water, which is probably one of the oldest applications, initially used for producing alcohol beverages. As for any other binary systems, the temperature, vapor/liquid composition characteristics of water/ethanol mixtures can be determined by boiling the mixture in a closed container and collecting liquid and vapor samples once equilibrium is reached at a given composition and pressure. The Othmer still, illustrated in Figure 2.18, is one of the early devices that were used for the determination of the vapor/liquid equilibrium. The device consists of a still followed by a condenser, which recycles the condensed vapor back into the still. Sampling and analyses of the condensed vapor from the condenser and liquid from the still, once reaching a stable temperature, provide vapor/liquid equilibrium at the measured temperature. Repeating the experience for different mixtures provides the vapor/liquid/temperature equilibrium data at a given pressure. Figure 2.19 illustrates the vapor/liquid/temperature equilibrium for a water/ethanol mixture at atmospheric pressure [51]. The upper curve defines the boundary between a vapor/liquid mixture (below the curve) and a single phase vapor (above the curve). The upper boundary limit is known as the dew point or saturated vapor mixture. Similarly, the lower curve defines the boundary between a vapor/liquid mixture (above the curve) and single-phase liquid (below the curve). The lower boundary limit is known as the bubble point or saturated liquid mixture.

Figure 2.19 can be used for calculating the vapor/liquid distribution and composition resulting from a flash. The plot shows that if a saturated liquid containing 40% mol ethanol was heated to a temperature of 82.3 °C (point A on Figure 2.19) and supply at a constant rate of 100 mol/hr to a vessel, the vapor leaving the vessel would have an ethanol concentration of 56% mol (point C on Figure 2.19) and the liquid stream 26% mol (point B on Figure 2.19). The total material balance and ethanol balance are described by Equations 2.94 and 2.95:

$$F = V + L \tag{2.94}$$

$$Fx_F = Lx_L + Vy_v \tag{2.95}$$

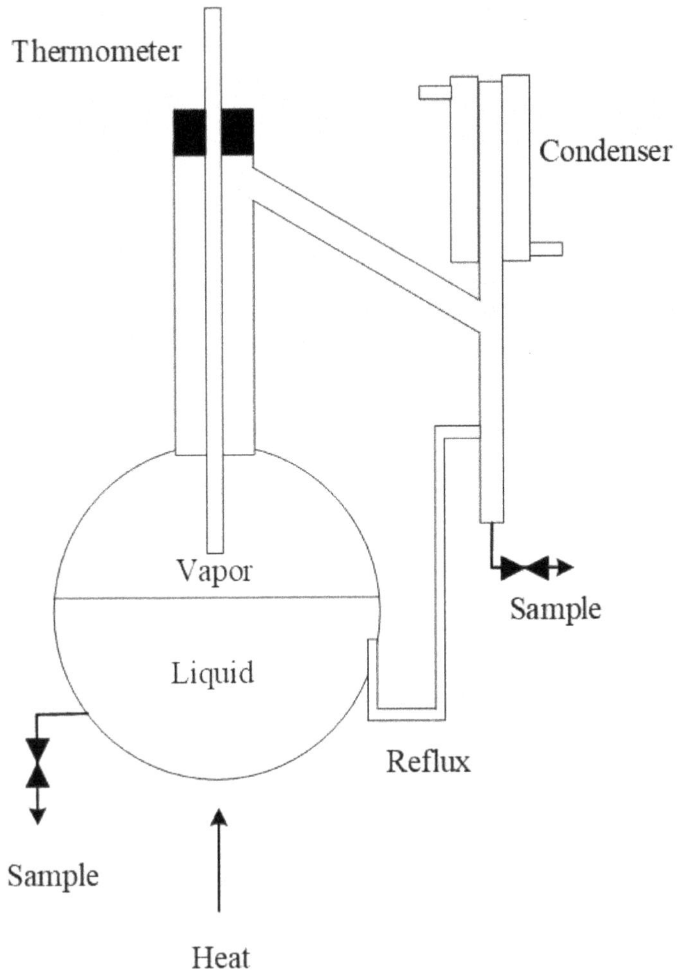

FIGURE 2.18 Othmer Still.

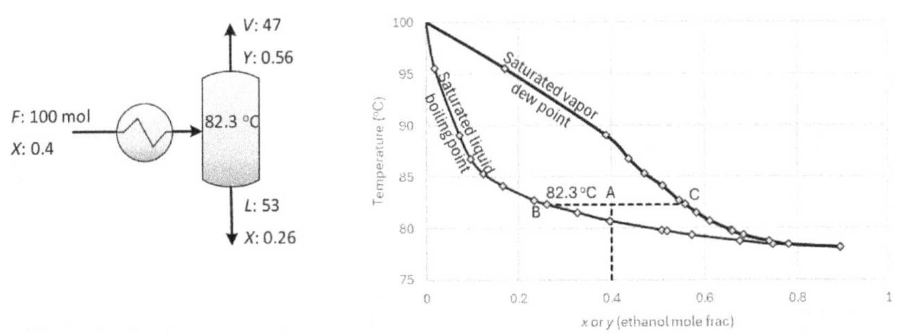

FIGURE 2.19 Ethanol/Water T–x–y Diagram and Flash Calculations.

Equations 2.94 and 2.95 can be combined into:

$$\frac{L}{V} = \frac{(y_V - x_F)}{(x_F - x_L)} = \frac{\overline{AC}}{\overline{AB}} \tag{2.96}$$

where Equation 2.96 correlates L/V to the ratio of the lengths AC/AB on Figure 2.19, which is known as the lever rule. The lever rule implies that the ratio AC/CB corresponds to the fraction of the feed F that is released from the vessel as liquid (47 mol/hr) and AB/CB the fraction of the feed released as vapor (53 mol/hr).

The above single equilibrium is obviously insufficient for separating those two components. Repeating the process by contacting the vapor stream in another vessel with a liquid stream of composition richer than the equilibrium will create another equilibrium level contributing to enriching ethanol in the vapor leaving that second vessel. Similarly, the liquid stream leaving the feed vessel can be contacted with a vapor stream, creating a different equilibrium, which would reduce the ethanol content of the liquid stream leaving the bottom vessel. As shown in Figure 2.20, the liquid supply to this system can be accommodated by condensing the vapor exiting the top vessel and recycling a fraction of it back to the top vessel. This recycled liquid stream is commonly called reflux in distillation. In a similar fashion, the vapor supply to the bottom vessel is done by vaporizing a fraction of the liquid leaving the bottom vessel.

The points x_D and x_B have been arbitrarily set on the T–x–y diagram of Figure 2.20 and by doing so satisfy the degrees of freedom needed for solving the system material balance. Table 2.13 summarizes the material balance equations describing this system. The molar compositions required for resolving the material balance are extracted from the T–x–y diagram. Stage $n+1$ is represented by the vapor/liquid equilibrium relating x_D (i.e., y_{n+1}) to x_{n+1} and Stage $n+3$ to x_B (i.e., y_{n+4}) and y_{n+3}.

McCabe–Thiele's [52] graphical methodology (Figure 2.21) was developed in 1925 and was used successfully by process design engineers until the emergence of computers. The methodology assumes constant molar overflow, meaning that the molar heat of vaporization is assumed constant while neglecting the effect of sensible heat and heat of mixing. It also assumes constant liquid and vapor molar flow in each section separated by a transition such as feed point. Figure 2.21 illustrates the use of the method on the ethanol/water separation example discussed earlier. The method uses an x–y equilibrium diagram rather than a T–x–y diagram. The feed entry is represented by the q-line. The q represents the thermal state of the feed, defined as the number of moles of liquid generated inside the tower per 1 mole of feed. In the case of a saturated liquid feed, $q=1$, while for a saturated vapor feed, $q=0$. Table 2.14 shows how the q-line is calculated for different feed conditions. The q-line would be vertical and intersect the equilibrium curve if the feed is a saturated liquid. For a saturated vapor, the q-line travels horizontally, moving left from the diagonal until meeting the equilibrium curve. Figure 2.21 also shows operating lines on each side of the feed point. The rectifying section operating line is on the right side, where fractionation is characterized by the down-flowing liquid condensing the heavy component, hence enriching the light component as vapor moves upward from stages to

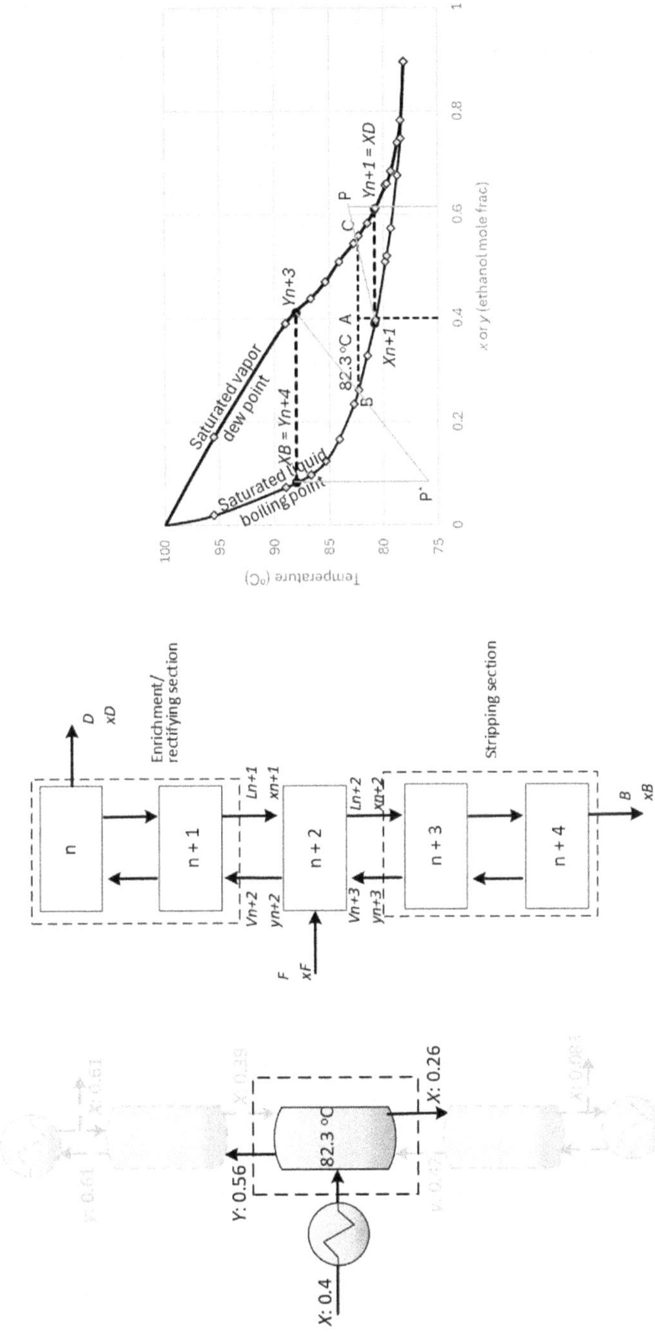

FIGURE 2.20 Ethanol/Water Multi-Stage Equilibrium Calculations.

TABLE 2.13
Ethanol/Water Multi-Stage System Material Balance Summary

Equation	Value
Total Balance	
$\frac{D}{F} = \frac{(x_F - x_B)}{(x_D - x_B)}$	0.6
Rectifying Section	
$\frac{L_{n+1}}{D} = \frac{(x_D - y_{n+2})}{(y_{n+2} - x_{n+1})}$	0.29
$V_{n+1} = D + L_n$	77.8 mol/hr
$\frac{L_n}{V_{n+1}} = \frac{1}{1 + L_{n+1}/D}$	0.23
Stripping Section	
$\frac{V_{n+3}}{B} = \frac{(x_B - x_{n+2})}{(x_{n+2} - y_{n+3})}$	1.1
$L_{n+3} = B + V_{n+4}$	83.9 mol/hr
$\frac{L_{n+3}}{V_{n+4}} = 1 + \frac{B}{V_{n+3}}$	1.9

stages. The stripping section operating line is on the left side of the feed point and characterizes fractionation by the stripping effect of the rising vapor toward the light component. By doing so, it frees up the heavy component from the light ones. The slope of each line is assumed equivalent to the liquid/vapor (L/V) molar ratio. This entails that the liquid (L) and vapor (V) molar flows are assumed to be constant in each section. In the ethanol/water example, the composition x_D and x_B were set arbitrarily at 0.61 and 0.083, respectively. Fixing those compositions and the number of equilibrium stages dictates the L/V ratio required for achieving that separation. In the McCabe–Thiele methodology, the equilibrium stages are established from a stair approach between the equilibrium curve and the operating line. As shown in Figure 2.21, the liquid and vapor leave the feed stage (i.e., $V\lambda_B = 68.5 * \lambda_B$) at composition x_{n+2} and y_{n+2} (0.25 and 0.56, respectively). The stairs defining stage $n+1$, $n+2$, and $n+3$ must match inlet/outlet composition with their boundaries defined by the equilibrium curve and operating lines.

While you can add other vessels to further purify the rich ethanol and water streams, this methodology would be complex and not economical. The above discussion was intended to demonstrate the principles driving multistage fractionation. In practice, multistage separation is conducted in towers where the vessels in the above

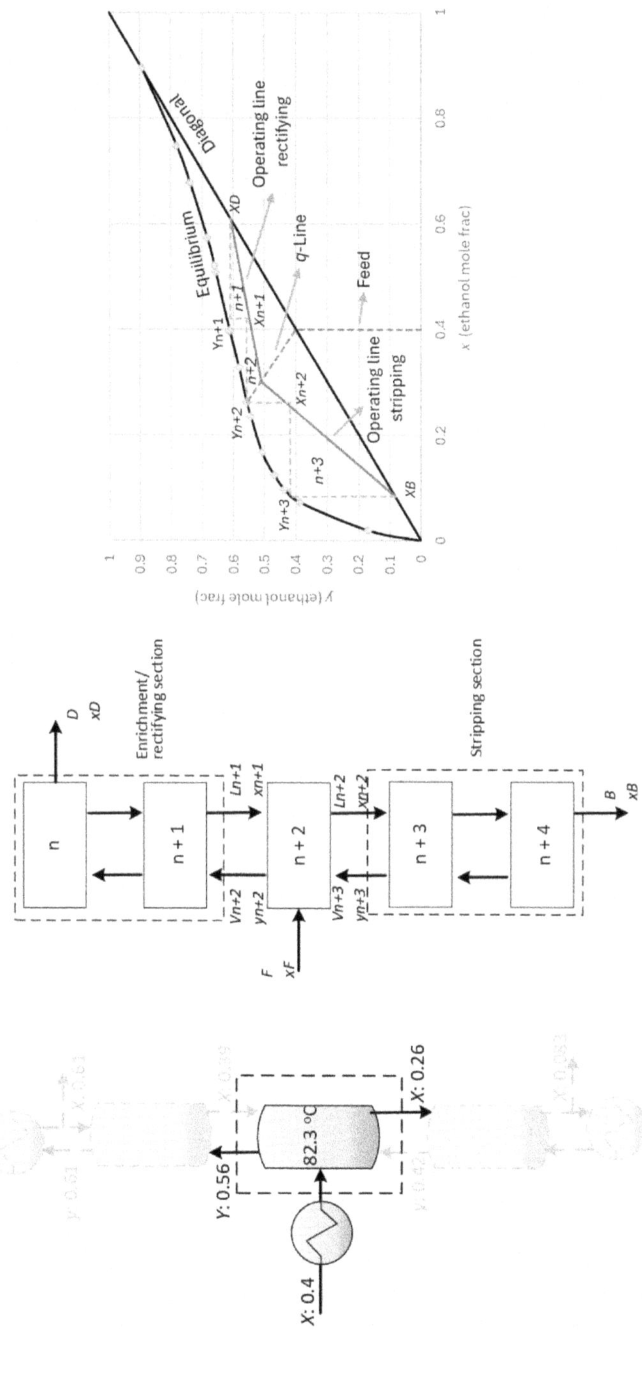

FIGURE 2.21 Ethanol/Water Multi-Stage Equilibrium McCabe–Thiele Evaluation.

TABLE 2.14
McCabe–Thiele *q*-Line

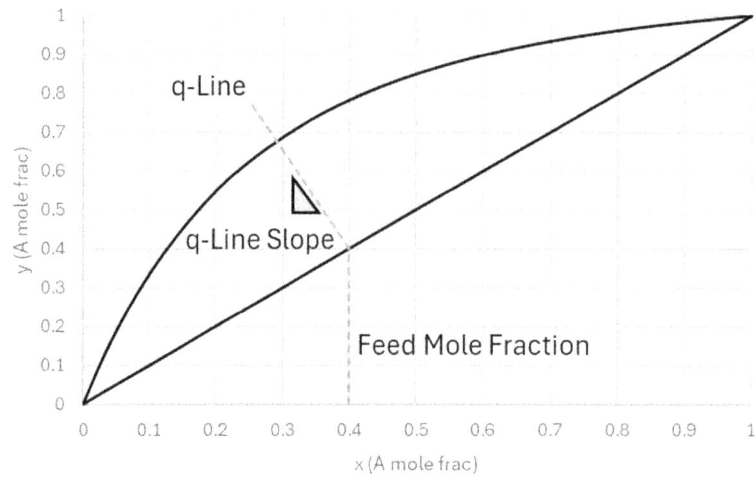

$$\text{q-Line Slope} = \frac{q}{q-1}$$

Cp_L or Cp_V : Heat capacity of liquid or vapor
T_{BP}: Boiling point
T_{DP}: Dew point
Λ: Latent heat

Feed Condition	***q* Calculation**
Saturated liquid	$q = 1$
Saturated vapor	$q = 0$
Vapor/liquid mixture	q = Feed Liquid Mole Fraction
Subcooled liquid	$q = 1 + \frac{Cp_L\left(T_{BP} - T_{Feed}\right)}{\lambda}$
Superheated vapor	$q = \frac{-Cp_V\left(T_{Feed} - T_{DP}\right)}{\lambda}$

discussion are replaced by mechanical devices intended to create different equilibrium stages. The functionality and characteristics of those internals are discussed in a later section.

Given common use of computer simulations, it may be argued that the above discussion provides little value for troubleshooting distillation towers. However, just looking at the equilibrium curve of a system may provide insights around problems that are sometimes related to operating conditions being thermodynamically infeasible. For instance, Figures 2.20 and 2.21 show that ethanol/water forms an azeotrope at 89% mol (95% vol). An azeotrope is defined as a vapor liquid equilibrium where the composition of the vapor is identical to the liquid. This is observed by the point where the equilibrium curve crosses or meets the diagonal on a *x–y* diagram or where the dew point curve crosses the boiling curve on a *T–x–y* diagram. Additional separation stages become ineffective once reaching the azeotropic composition as equilibrium imbalance driving mass transfer no longer exists between the downward liquid and rising vapor. On the ethanol/water system, the shape of the equilibrium curve is relatively fat and opens up below an ethanol composition of 20% mol. The curve changes in slope and curvature beyond that point and restricts the operating line of the rectifying section. As shown in Figure 2.22, operating the same *L*/*V* ratios used in the previous example would cause the rectifying section operating line to cross the equilibrium curve at an ethanol vapor composition of about 68% mol (Operating Line 1). In other words, attempting to run at those conditions would be associated with a much lower ethanol content than expected regardless of the number of separation stages provided. Relocating the *q*-line by feeding the 40% ethanol mixture at a liquid fraction of 95% (Operating Line 2 on the left side of Figure 2.22) would enable the separation to meet the azeotropic composition, assuming sufficient separation stages, but with a relatively high *L*/*V* ratio of about 0.90 or reflux ratio (*L*/*D*) of 8.8. Note that an *L*/*V* ratio of 1 (i.e., diagonal) is equivalent to total reflux operation (i.e., no distillate). As illustrated on the right side of Figure 2.22, shifting the *q*-line by preheating the feed mixture to a liquid fraction of 25% slightly reduces the *L*/*V* ratio to 0.87 or reflux ratio of 6.7. The above discussion demonstrates the importance of understanding the shape of the equilibrium curve and the impact it has on the feed conditions and location. An optimum design might look at creating other transition points through pump-around with side reboiler or condenser for taking advantage of the fat section of the curve and re-aligning the operating line with the equilibrium curve as it approaches the diagonal. Some ethanol plants would have a front-end extractive distillation unit for removing high-molecular-weight alcohols and feed the ethanol/water product tower at a lower composition such that the junction between the stripping and rectifying section operating lines occurs at or below 20% mol ethanol. This would provide more rangeability for the rectifying section operating line and allow a reduction in the *L*/*V* ratio. For example, feeding an ethanol mixture at 6% mol with a liquid fraction of 90% would reduce the *L*/*V* ratio to 0.78 or equivalent reflux ratio of 3.5.

The above discussion is counter-intuitive as one would expect a reduction in reflux ratio or condenser duty when feeding a liquid mixture. A higher light component composition in the feed should also reduce the reflux ratio and condenser duty. Mathias [53] demonstrated those concepts through analysis of the *q*-line on McCabe–Thiele diagram. Unlike the above results presented on the ethanol/water separation, he demonstrated that reflux ratios and reboiler duty are affected positively as the value of *q* increases (i.e., high liquid fraction to subcooled liquid). The case of ethanol/water as for other azeotropic mixtures can be classified as an exception that is driven by the shape of the equilibrium curve. The impact of the *q*-line reported by Mathias is summarized in Figure 2.23, where a binary mixture of component *A* and *B* having vapor/liquid equilibrium characteristic of light hydrocarbons is used to illustrate the concept; that is, where the rectifying and stripping section operating

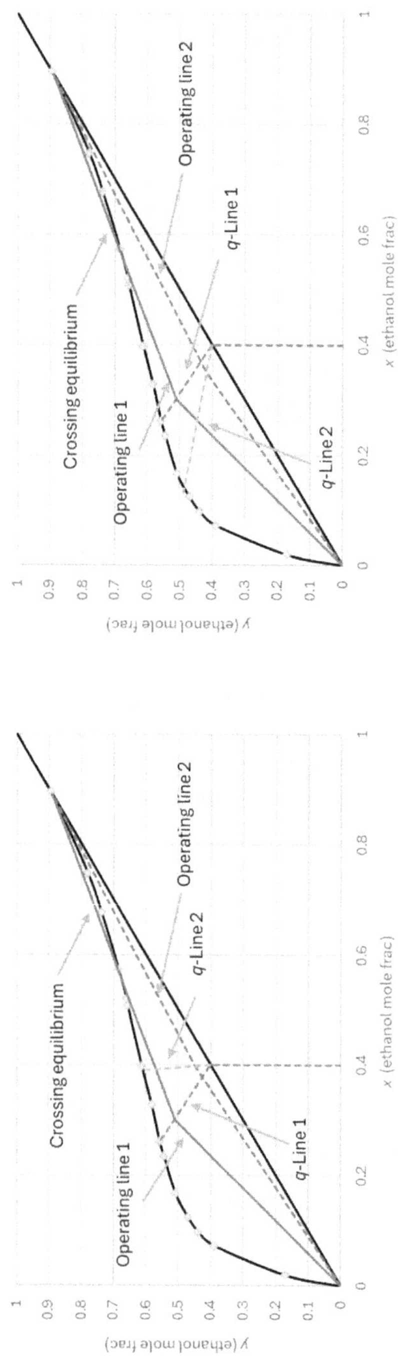

FIGURE 2.22 Feed Liquid Fraction Impact Evaluation by McCabe–Thiele Methodology.

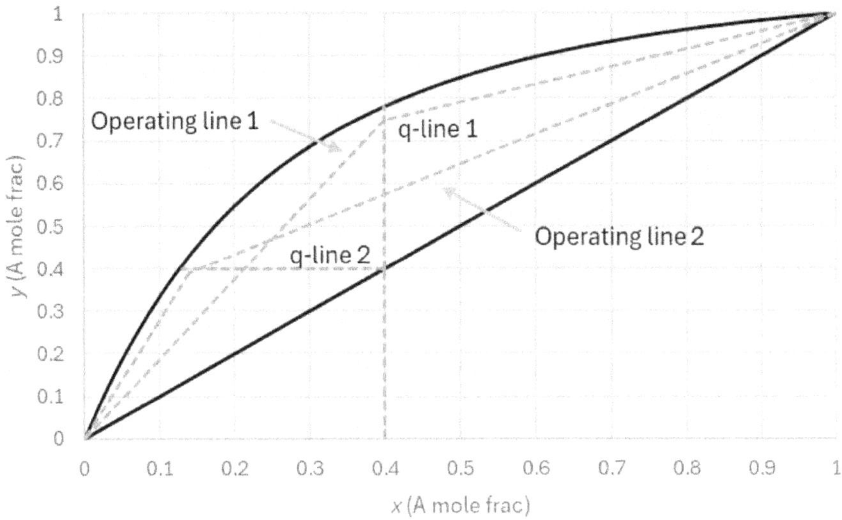

FIGURE 2.23 Impact of q-Line on Operating Lines at Minimum Reflux.

lines join each other near the equilibrium curve. Feeding a mixture containing 40% mol of component *A* as a saturated liquid results in a rectifying section operating line having a slope or *L*/*V* ratio of 0.42. The *L*/*V* ratio is then substituted in the overhead material balance for calculating the reflux ratio (*L*/*D*) to a value of 0.71. Assuming complete separation of *A* and *B*, meaning that $x_D = 1$ and $x_B = 0$, the reboiler duty can be calculated from the product of the vapor flow in the stripping section by the latent heat of component *B* (i.e., $V\lambda_B = 68.5 * \lambda_B$). Similarly, as for the reflux ratio, the value of V is calculated from the slope of the stripping section operating line or L/V ratio and the tower bottom material balance to a value of 68.5. Now, if the feed was to enter the tower as a saturated vapor (i.e., $q = 0$), following the same procedure as above, the reflux ratio would rise to 2.44 and the reboiler duty to $33.6 * \lambda_B$. The reboiler duty may look smaller, but the overall heat balance must account for the energy brought by the feed relative to a saturated liquid (i.e., 100*($y_{FA*}\lambda_A + y_{FB*}\lambda_B$ – where y_{FA} and y_{FB} are molar fraction of in the feed). The increase in reflux ratio is also associated with an increase in condenser heat duty.

The ethanol/water and light hydrocarbon systems discussed earlier were based on the feed composition and liquid fraction in alignment with the feed stage. In other words, the vapor and liquid composition were identical to the composition of the stage with or without the contribution of the feed. While a tower might be designed with feed in the optimum location, variability in plant operation from reactor yields to feed conditioning (pre-heating or cooling) may cause the feed to create a dilution effect and introduce some inefficiencies.

On light end hydrocarbons, the fatness of the equilibrium curve also suggests that the feed vapor fraction has more implication on the reflux and reboil ratios on high relative volatility systems than on the low ones. Regardless, some of those changes besides affecting tower hydraulics may cause exceedance in condenser or reboiler capacity. As such, maximizing tower capacity may require under certain conditions to operate with suboptimal thermodynamic conditions for satisfying some of the towers' physical constraints.

Components exhibiting strong interactions such as polar components like alcohols and organic acids have a tendency to form azeotrope. Figure 2.24 illustrates the equilibrium curves for the n-butanol/water and ethanol/benzene binary azeotropic system. As illustrated, butanol/water form a heterogeneous azeotrope. In such systems, the condensed azeotropic composition split into two immiscible liquids of composition different from the azeotropic composition. As illustrated for the *n*-butanol/water system shown in Figure 2.25, separation is achieved with two towers where the rich aqueous phase is refluxed to a tower where purified water is collected as bottom product with the overhead feeding a shared reflux drum at the azeotropic composition. Similarly, the organic phase refluxes a second tower where the purified organic component is collected as bottom product with the overhead azeotropic vapor condensed to the share reflux drum. The ethanol/benzene mixture forms a homogeneous azeotrope. The ethanol/water azeotrope was used to be broken using benzene, a practice that has been abandoned due to the carcinogenic nature of benzene. Less harmful hydrocarbon solvents are now used, while some ethanol plants are using molecular sieve as an alternative technology for dehydrating the ethanol/water azeotrope. Note that the azeotropic temperature of the ethanol/benzene system is lower than the boiling point of pure ethanol and benzene. Those systems are referred to as minimum boiling azeotropes. Maximum boiling azeotropes would have the azeotropic temperature above the boiling point of the pure components.

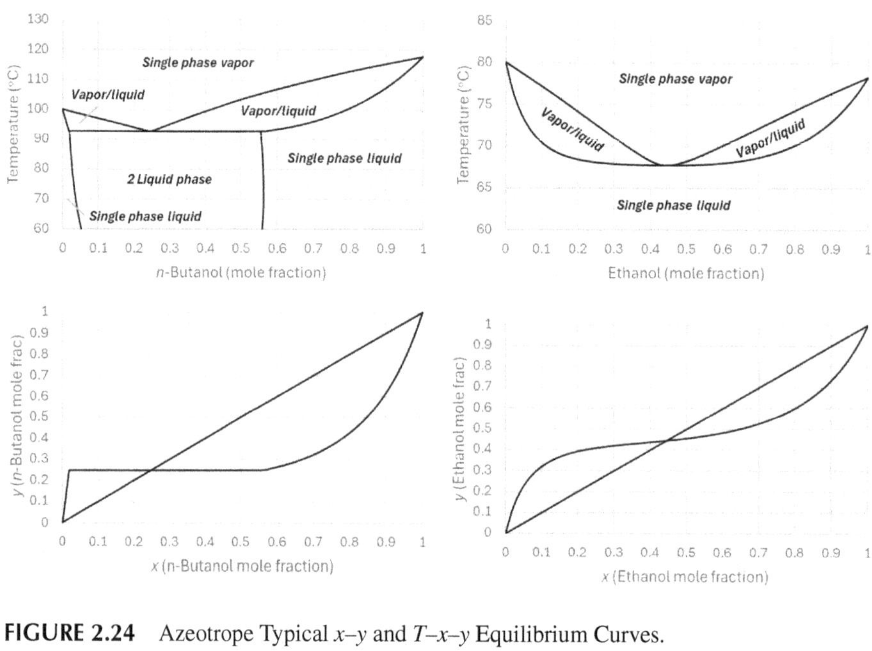

FIGURE 2.24 Azeotrope Typical *x–y* and *T–x–y* Equilibrium Curves.

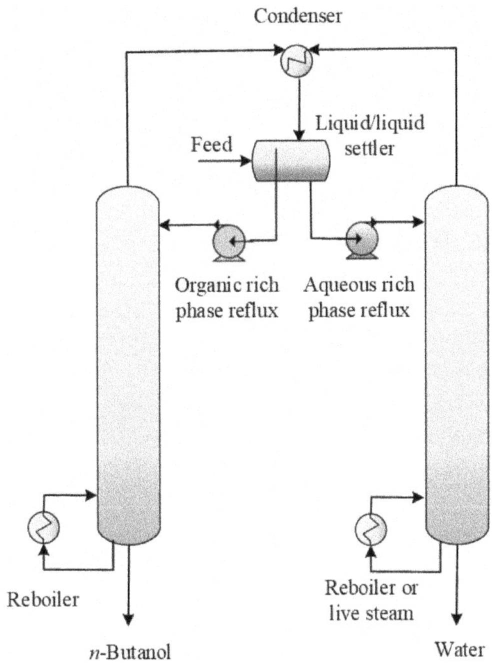

FIGURE 2.25 *n*-Butanol/Water Separation.

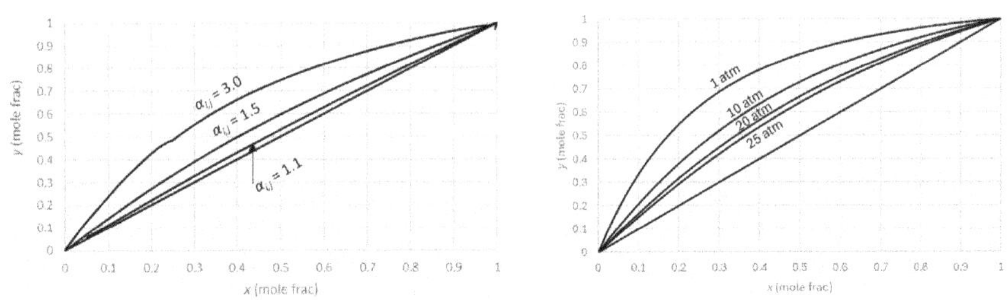

FIGURE 2.26 Impact of Relative Volatilities and Pressure on the Equilibrium Curve.

Light hydrocarbons' equilibrium curves have characteristics closer to ideality. As shown in Figure 2.26, the fatness of the curve is dependent on relative volatilities and pressure. The relative volatility $\alpha_{i,j}$ is defined by the ratio of the K-value of component i over the K-value of component j as per the following equations:

$$K_i = \frac{y_i}{x_i} \tag{2.97}$$

$$\alpha_{i,j} = \frac{K_i}{K_j} = \frac{y_i}{x_i}\frac{x_j}{y_j} \tag{2.98}$$

For binary mixture:

$$\alpha_{i,j} = \frac{K_i}{K_j} = \frac{y_i}{x_i}\frac{x_j}{y_j} = \frac{y_i(1-x_i)}{x_i(1-y_i)} \tag{2.99}$$

The relative volatility provides an indication on how close the components being fractionated are to each other, hence indicating the difficulty of the separation. As shown in Figure 2.26, low relative volatility equilibrium curves are close to the diagonal on the x–y diagram, hence suggesting a large number of equilibrium stages required and high reflux ratios. On the contrary, high relative volatilities make the equilibrium curve fatter, suggesting less required equilibrium stages and lower reflux ratios. Note that the relative volatility is influenced by the operating pressure, where lower pressure tends to expand the relative volatility, hence making separation easier.

While process engineers may not have or need to use equilibrium curves for evaluating a distillation tower, they should be aware of the curve characteristics and have a high-level picture in their mind, which will better guide their instincts in troubleshooting towers.

Efficiency of separation is an important concept in distillation. As for chemical reactions, meeting thermodynamic equilibrium can be a lengthy process and is not always feasible or practical. The efficiency of separation in distillation is a concept quantifying the deviation from the thermodynamic equilibrium. It has been quantified in a few different ways with the most practical one being the overall tower efficiency, that is, the ratio of the number of theoretical stages over the total number of actual separating devices:

$$Eff_{OV} = \frac{N_T}{N_A} \tag{2.100}$$

The topic of separation efficiency is relatively complex and beyond the scope of this discussion. Rigorous efficiency models are often constrained by the lack of measurements. Their application might be more geared toward the development of distillation hardware. Operation process engineers and process design engineers typically use overall efficiency, which are frequently known by experience. Nonetheless, understanding the concepts surrounding the development and basis of more rigorous efficiency models help in understanding the process variables and parameters that influence distillation efficiency.

If the actual stage vapor and liquid compositions were measured, they could be compared to the theoretical equilibrium composition and would provide the actual stage efficiency. This can be calculated for both vapor and liquid and is known as Murphree efficiency [54]:

$$Eff_{MV} = \frac{y_n - y_{n-1}}{y_n^* - y_{n-1}} \times 100\% \tag{2.101}$$

$$Eff_{ML} = \frac{x_n - x_{n-1}}{x_n^* - x_{n-1}} \times 100\% \tag{2.102}$$

where y^* denotes equilibrium with the actual liquid composition x_n, and x^* denotes equilibrium with the actual vapor composition y_n. y^* and x^* are also correlated to x_n and y_n by the following expressions:

$$y_n^* = mx_n + b \tag{2.103}$$

$$y_n = mx_n^* + b \tag{2.104}$$

Under constant liquid and vapor molar flows, a material balance around a separation unit gives

$$L\left(x_{n+1} - x_n\right) = V\left(y_n - y_{n-1}\right) \tag{2.105}$$

Assuming a linear equilibrium line

$$m = \frac{y_n^* - y_n}{x_n - x_n^*} \tag{2.106}$$

Combining Equation 2.105 with Equation 2.106 gives λ, the stripping factor:

$$\lambda = \frac{mV}{L} = \left(\frac{y_n^* - y_n}{x_n - x_n^*}\right)\left(\frac{x_{n+1} - x_n}{y_n - y_{n-1}}\right) \tag{2.107}$$

The above equations can be combined to express the vapor Murphree efficiency in terms of the liquid efficiency and stripping factor:

$$Eff_{MV} = \frac{Eff_{ML}}{Eff_{ML} + \lambda\left(1 - Eff_{ML}\right)} \tag{2.108}$$

Assuming linear operating and equilibrium curve, Lewis correlated tower section efficiency to the tray Murphree vapor efficiency:

$$Eff_{OS} = \frac{\ln\left[1 + Eff_{MV}\left(\lambda - 1\right)\right]}{\ln \lambda} \tag{2.109}$$

The above analysis is done at the stage level. One can make the analogy that a separation stage is assumed equivalent to a perfectly mixed enclosure. However, for cross flow trays, mass transfer occurs as the liquid travels from the inlet to the outlet of the tray in a plug flow fashion with some back mixing. As shown in Figure 2.27, that distance is defined as the flow path length. The vapor rises upward following a plug flow behavior. A vapor/liquid concentration gradient is then expected throughout the flow path length and the height of the vapor/liquid mixture. Breaking down the tray into infinitely small volume elements $dL \times dh$ can be compared as having small separation stages within a separation stage. The rate of mass transfer varies in all those smaller elements with the transfer rate driven by the departure from equilibrium. This would

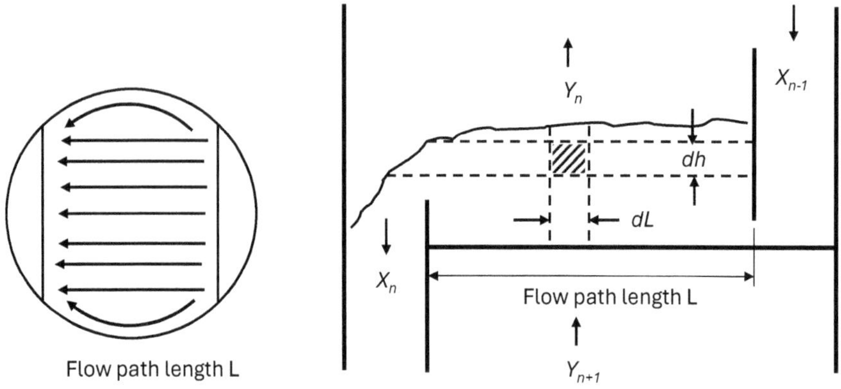

FIGURE 2.27 Tray Point Efficiency.

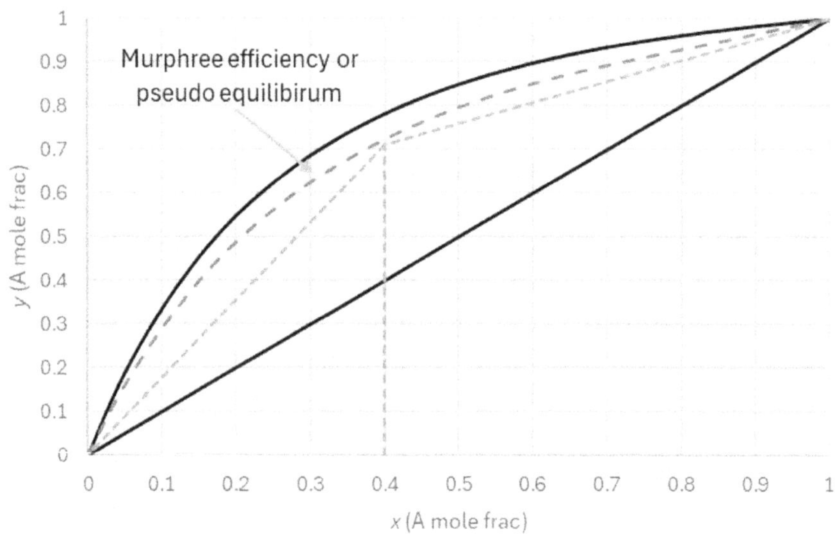

FIGURE 2.28 High-Level Murphree Efficiency Application to McCabe–Thiele.

suggest that the rate of mass transfer would be higher at the inlet of the tray and lower at the outlet and top of the froth. This concept is known as point efficiency [54] and demonstrates the significance of the flow path length on tray efficiency.

In practice, the Murphree efficiency can be applied at high level through a pseudo-equilibrium on a McCabe–Thiele diagram as shown on Figure 2.28. It can be assumed constant across the tower or split by section as it is not unusual to observe differences in efficiency between sections separated by transition point (i.e., feeds, draw, and pump-arounds). In computer simulation, the Murphree efficiency can be specified by tray or tray section.

As previously stated, calculating the separation efficiency requires knowledge of the stage by stage compositions, which are rarely available on an industrial operating tower. O'Connell (1946) [54, 55] developed an empirical correlation based on numerous test data done on bubble cap trays. This correlation is still widely used in the industry. As shown by Equation 2.110, it correlates the overall or section efficiency, depending on departure of physical properties between sections, to the liquid viscosity (μ_L in N s/m^2) and relative volatilities (α) of the key components. Liquid viscosity and relative volatility are evaluated at the average temperature between both ends of the section for which it is being evaluated:

$$Eff_{OV} = 9.06\left(\mu_L \alpha\right)^{-0.245} \tag{2.110}$$

The O'Connell correlation, although empirical, takes into consideration the properties of the fluid affecting mass transfer, primarily liquid viscosity, which dictates the resistance to mass transfer at the vapor/liquid interface. The relative volatility has been reported as influencing the significance of the liquid phase mass transfer resistance. Taylor and Duss [56, 57] have questioned the relevance of the relative volatility in the correlation. In their approach, they assumed that the O'Connell correlation described by Equation 2.110 must relate to the Murphree efficiency as described by Equation 2.109 or equivalent form. In their analysis, they concluded that using the relative volatility was consistent with total reflux conditions but resulted in deviations as operation departs from total reflux. Using the stripping factor λ was found to align better with the Murphree efficiency correlation. They modified the O'Connell correlation by replacing the relative volatility by the stripping factor as described by Equation 2.111:

$$Eff_{OV} = 0.503\mu_L^{-0.226}\sigma^{-0.08} \tag{2.111}$$

Where $\sigma = \lambda$ if $\lambda>1$ and $\sigma = 1/\lambda$ if $\lambda<1$. Except for the concept of point efficiency described in Figure 2.27, the above discussion and correlations on separation efficiency have no connection with the characteristics of the hardware use for maximizing mass transfer. Although the Murphree efficiency is implicitly related to the contribution of the hardware as it uses the actual liquid and vapor composition, it does not include any terms referring to the characteristics of the hardware. Separation efficiency is therefore only related to the properties of the fluids involved. Liquid viscosity and the stripping factor, which although not equivalent to the relative volatility, have similarities. At total reflux conditions (i.e., $V/L = 1$), the stripping factor is equal to parameter m, the slope of the equilibrium curve as described by Equation 2.106. The parameter m and relative volatility calculated from the equilibrium curves displayed in Figure 2.26 were plotted against each other for understanding any relationship between both parameters as they are evaluated from similar variables. As shown in Figure 2.29, there are significant departures between both parameters at low pressure. As pressure increases the range and variability of parameter *m* is significantly less in relation to the relative volatility.

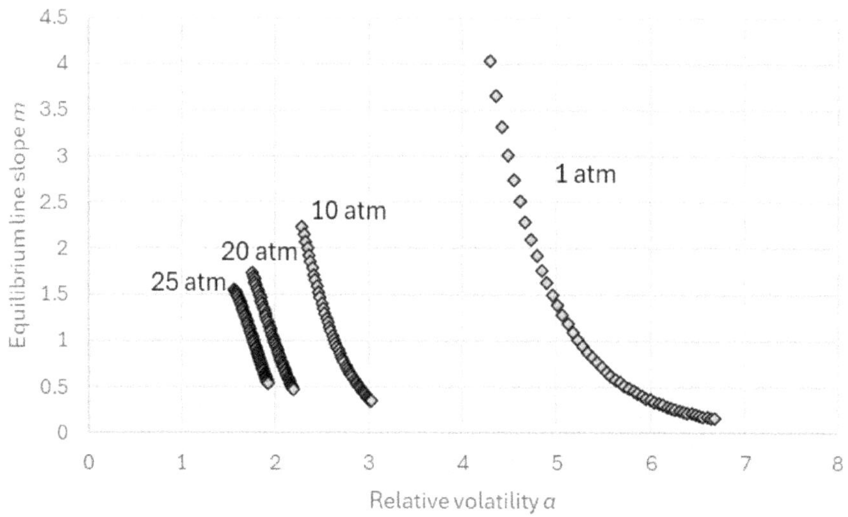

FIGURE 2.29 Parameter m as a Function of Relative Volatility.

Although not specifically addressed in the above correlations, it is understood that the hardware (i.e., trays or packing) must play a role in separation efficiency. One of the primary functions of the hardware is to maximize the vapor/liquid interfacial area. The concept of point efficiency as illustrated in Figure 2.27 suggests that the flow path length and froth or spray height would enhance tray efficiency. Kister [58] discussed the effects of tray design configuration on separation efficiency using test data generated by Fractionation Research Inc. (FRI) on sieve trays. In his analysis, he confirmed that flow path length is the primary element that impacts efficiency. For instance, for the same tower diameter, a two-pass tray would have a higher efficiency than a four-pass tray due to the difference in flow path length. However, unlike speculated above, a higher froth height does not seem to have any impact on tray efficiency providing that the tray operates below flooding and above weeping. Data suggest that tray efficiency is relatively flat within the stable operating range of the tray and starts dropping as hydraulic conditions approach flooding. Small hole diameter on a sieve deck is intuitively believed to enhance mass transfer due to the formation of smaller gas bubbles (i.e., higher vapor/liquid interfacial area). However, this effect is reported as being offset by the lack of turbulence generated with small hole diameter. In fact, Kister reported that hole diameter increases from 3 to 13 mm that provides slight improvement in efficiency. Further increases in hole diameters between 13 and 25 mm are, however, insensitive relative to efficiency. Holes larger than 25 mm exhibit a reduction in efficiency. Decreasing the fractional hole area is reported to provide a small improvement in efficiency. Summers [59] confirmed that cross flow trays, regardless of the tray deck design, would exhibit similar efficiencies providing they were operated within their design capabilities. He also reported that the primary tray geometry parameter influencing tray efficiency was the flow path length. Enhancement such as push valve would contribute to improving efficiency as they provide more definition on the flow path length as shown on the left side in Figure 2.27, hence minimizing peripheral stagnant areas.

The importance of vapor/liquid equilibrium data (VLE) is reported as critical, especially in low relative volatility systems. Kister [55] reported that a 3% error on the VLE of system having a relative volatility of 1.1 results in a 40–50% error on separation efficiency. The magnitude of the error drops significantly as relative volatility increases and becomes small at relative volatility above 2.0. It is therefore critical that efficiency assumed in engineering calculations for low-volatility systems be based on the same VLE for which the efficiency was established.

Unlike trays, packing can be characterized by vapor and liquid plug flow hydraulic profile. A separation unit is defined as an equivalent height of theoretical plate (HETP). High efficiency is characterized by small HETP and low efficiency by larger HETP. The number of theoretical stages used for trays is referred as a number of transfer unit (NTU) with packing.

Tray hydraulics. VLE property and separation efficiency define the vapor/liquid hydraulic circulation required for achieving a desired separation. In a tower, vapor rises vertically by pressure differential and liquid flows downward by gravitational forces. On a given tower diameter, there is a superficial vapor velocity where the drag forces equal or exceed gravitational forces exerted on liquid droplets. At that vapor velocity, the liquid can no longer flow downward regardless of the type of internals or tray spacing. Souders and Brown [60] established a correlation for estimating the vapor velocity causing liquid entrainment as described by Equation 2.112, where the value of K is determined experimentally:

$$u_V = K\left[\frac{(\rho_l - \rho_v)}{\rho_v}\right]^{1/2} \tag{2.112}$$

The Souders–Brown correlation has been widely used for scaling up knockout drum and selection of demister devices. The constant K in that application is being used as a vapor load factor, which is discussed in more detail later. Stupin and Kister [61] extended the Souders and Brown correlations, leveraging on experimental data obtained from commercial scale towers and published by FRI. They established two correlations which were found to fit experimental data. The smallest value of the two correlations is to be selected as the system limit:

$$C_{S1} = \left(\frac{1.4}{1+1.4\left(\Delta\rho/\rho_v\right)^{1/2}}\right)\left(\frac{4g}{C_d}\right)^{1/4}\left(\frac{\sigma\Delta\rho}{\rho_v^2}\right)^{1/4} - 1.4L_s \tag{2.113a}$$

$$C_{S2} = \left(\frac{1.12}{1+1.4\left(\Delta\rho/\rho_v\right)^{1/2}}\right)\left(\frac{4g}{C_d}\right)^{1/4}\left(\frac{\sigma\Delta\rho}{\rho_v^2}\right)^{1/4} \tag{2.113b}$$

$$C_s = \textit{Smallest between } C_{s1} \textit{ and } C_{s2} \tag{2.113c}$$

where $\Delta\rho$ is the difference between the liquid and vapor density $(\rho_l - \rho_v)$, and C_s is the tower limit or system limit C_{factor} as expressed by:

$$C_s = u_s \sqrt{\frac{\rho_v}{\Delta\rho}} \tag{2.114}$$

The C_{factor} represents the vapor load, with u_s being the vapor velocity, expressed in terms of the tower cross section, although some practitioners based u_s in terms of the active area of the tray. Neither approach is better than the other; what is critical is consistency. In troubleshooting and for convenience, it is preferable to express u_s in terms of the tower cross section. This avoids re-adjusting calculations whenever there is a change in active area, which would be typical of towers having different tray geometry. Tray vendors may report the vapor loading in different forms. For instance, VLOAD is a common way of reporting the vapor load. The VLOAD is very similar to the C_{factor} as it is a volumetric vapor load or equivalent to the C_{factor} multiply by the tower cross section (or active area depending on the basis):

$$VLOAD = CFS \sqrt{\frac{\rho_v}{\Delta\rho}} = C_{factor} * A \tag{2.115}$$

Less common is the F-factor which is similar to the C_{factor} but omit the $\Delta\rho$ term:

$$F_s = u_s \sqrt{\rho_v} \tag{2.116}$$

Understanding the concept of system limit, although important, is not likely going to help the troubleshooter in tray or packing sections as flooding will likely occur before reaching it. However, entrainment caused by vapor velocity exceeding the system limit can occur in area constraint by hardware, such as chimney trays and distributors. Those areas are frequently overlooked during the design phase relative to potential entrainment and can become a pinch point limiting capacity below the expected tray or packing flood point. As an example, tower simulations suggested creep capacity by transitioning from two feed points to a single feed location on a tower. The tower performance was significantly affected after the feed transition, with operation reverting to the original feed configuration. Hydraulic analysis around the feed distributors suggested that the feed transition caused vapor velocity above the system limit when considering the open area around the distributor hardware. The concept of system limit is also useful for design engineers as this provides a quick way of establishing the limitations on a tower capacity increase and delimits the boundary where a larger or parallel tower becomes necessary. Understanding the step-up in capital investments may restrict the expansion targets without extensive engineering at the project idea stage.

For both troubleshooters and process design engineers, it is important to understand what limits tray capacity. The typical hydraulic limitations of trays are illustrated in Figure 2.30 in a diagram frequently referred to as a tray operating window

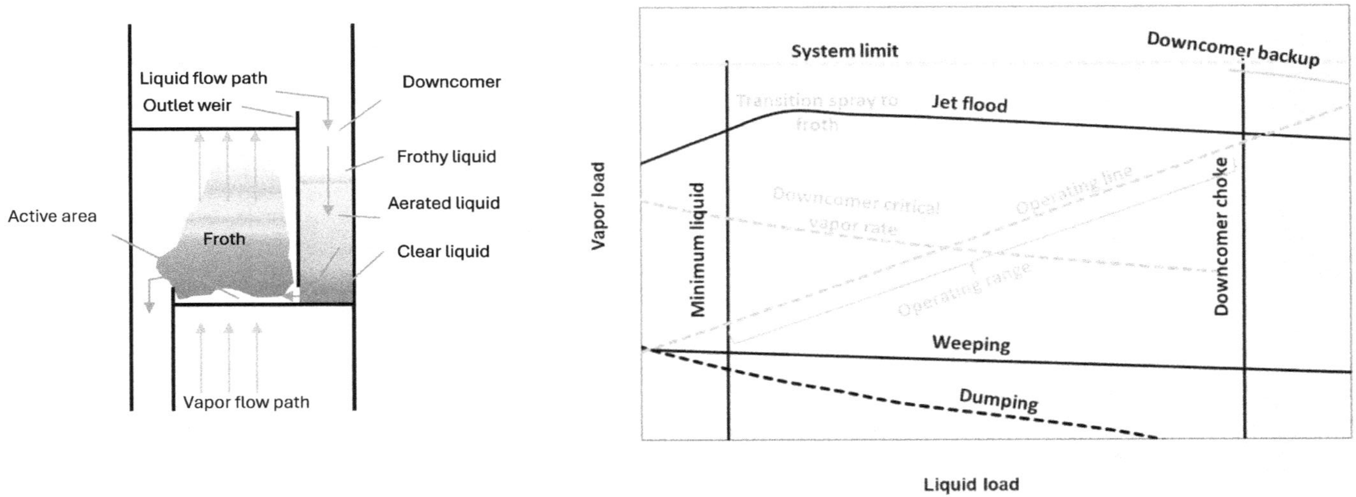

FIGURE 2.30 Tray Operating Window

[62, 63] or stability diagram [64]. As shown, the tray operating limitations are expressed in terms of the vapor load as a function of the liquid load. It is common practice to express the vapor load as VLOAD or C_{factor}, as previously described. The liquid load is typically expressed in terms of volumetric liquid rates.

The left side of Figure 2.30 illustrates a typical one-pass tray where liquid flows downward through downcomers and then enters the tray. As it travels across the tray, a vapor/liquid mixture is created from the momentum created by the vapor flowing through the open area of the tray deck. Depending on flow regime, this mixture is characterized as a froth, where the continuous phase is liquid and the disperse phase vapor droplets, or a spray with the vapor being the continuous phase and liquid droplets the disperse phase. The spray regime typically occurs at low liquid rates and although some towers operate in this mode, the froth regime is preferable as it is associated with less liquid entrainment. Jet flooding in the spray regime is defined by the vapor load at which liquid droplets are massively entrained to the tray above. In the froth regime, the froth height is a function of the vapor load. Below flooding, the volume above the froth is a clear vapor space. Jet flooding occurs as the vapor load increases, and the froth height reaches the tray above. Jet flooding is therefore a vapor-induced flood mechanism. The curve depicting jet flood limit in Figure 2.30 shows the transition from spray to froth as the liquid load increases by the change in the jet flood line curvature.

Downcomer backup is another flooding mechanism and is characterized by accumulation of aerated liquids in the downcomer that exceeds the tray spacing. Liquid travels down the tower through downcomers and must have sufficient static head to overcome the tray dry pressure drop, the froth height on the tray, the pressure drop across the downcomer apron, and downcomer friction losses. Downcomer backup flooding is a function of both liquid and vapor loads. In many cases, as illustrated in Figure 2.30, it is not the controlling flooding mechanism. However, at short tray spacing, or when the trays are plugged, downcomer backup can become the controlling flooding mechanism, occurring before jet flooding.

Downcomer choke, as illustrated in Figure 2.30, is a flooding mechanism that is mostly dependent on the liquid rate. It is characterized by the aerated liquid velocity at which friction losses in the downcomer and downcomer entrance exceed the capability of the downcomer in moving the liquid to the tray below. Summers introduced a maximum downcomer velocity curve that can be utilized for downcomer choke predictions [65]. In downcomer configurations with sumps, the restriction of the cross section between the tray deck and the sump may restrict the froth from accessing the downcomer. This behavior can also be referred to as downcomer choke.

While the focus is frequently on the high operating end of trays, turndown can also be an issue under some circumstances. Weeping for instance is characterized by insufficient vapor load for holding the liquid on the tray. Under those conditions, liquid falls to the tray below through the tray deck perforation and, as such, short-circuits the normal flow path on the tray. This behavior should be associated with a loss of efficiency. The extent of weeping increases as the vapor load is further reduced up to a point where most of the liquid falls through the tray deck perforation. This condition is normally associated with a severe loss of fractionation and is known as dumping. Summers introduced the concept of tray stability which is easier to calculate than weeping and can predict when tray efficiency may fall off at low vapor loads [66].

The minimum liquid rate can be dictated by three elements. The first two are dictated by weir loadings. At low weir loadings, the liquid head over the weir may get so small that the entire weir length might no longer be entirely wet, which results in maldistribution. Summers [63] stated that some practitioners suggested that weir loading below 0.5 gpm/in would cause the tray to perform poorly. A second element involves a transition from froth to spray at low liquid load. Summers and Sloley describe a low liquid load transition equation from the froth regime to spray regime and how to avoid the spray regime [67]. A third element dictating the minimum liquid loads involves the downcomer design relative to how it liquid seals itself. Figure 2.31 compares a conventional downcomer to downcomers typically found with high-capacity trays. Conventional downcomers are said to have a positive seal when the downcomer clearance is smaller than the weir heights. However, although this is valid for the froth regime, while there is a frothy pool on the tray, it may not be sufficient in the spray regime, where that pool is absent. A smaller clearance might then be required. In high-capacity trays, such as those shown in Figure 2.31, the downcomers have no mechanical barrier preventing the downcomer from draining its entire liquid inventory. As such, they require a minimum liquid load to ensure the presence of sufficient liquid head for sealing the downcomer. Unsealed downcomer results in vapor blowing through the downcomers, which short-circuits the vapor flow path and severely affect fractionation. The design intent of downcomers design used in high-capacity trays is to reclaim the active area occupied by the downcomers. This provides a gain in capacity by increasing the available active area. Conventional downcomers can also have a dynamic seal when the downcomer clearance exceeds the weir heights. This can happen with high liquid load applications where the downcomer clearance has to be relaxed for the benefit of the active area. Some designers would overcome this constraint by specifying downcomer sumps.

The downcomers are unsealed when starting up a distillation tower until sufficient liquid flow is established. There are different strategies for starting up a distillation tower, which depends on flowsheet configuration, streams available, and their conditions (i.e., vapor or liquid). Towers can be started from a vapor stream, which will

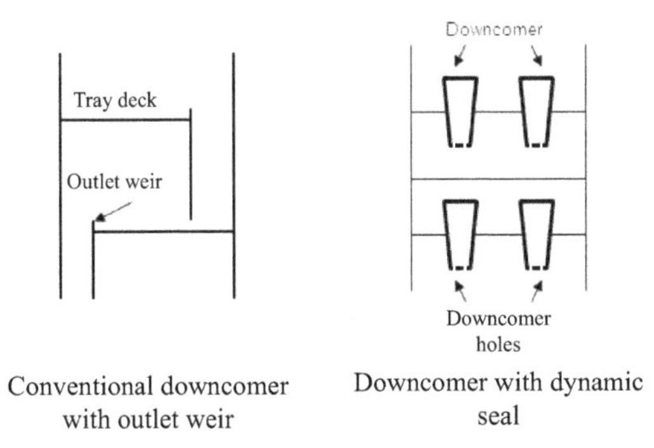

FIGURE 2.31 Conventional and Dynamic Seal Downcomers.

create vapor traffic from the feed point to the overhead condenser. Liquid circulation will then be established once sufficient liquid inventory is collected in the reflux drum, such that the reflux system can be commissioned. Until that point, there is no traffic below the feed point until liquid from the reflux system starts raining down the tower and accumulates in the tower sump. Heat to the reboiler can then be commissioned once sufficient liquid inventory builds up in the tower sump. At this point, if done correctly, circulation has been established throughout the tower. The tower can also be started from a liquid feed, which will create the opposite sequence as described earlier for a vapor feed. Liquid would first rain down the tower and then accumulate in the tower sump. Heat to the reboiler would then be started and vapor traffic would be established through the tower. The vapor would then condense in the overhead condenser and accumulate in the reflux drum allowing commissioning of the reflux system. In each startup mode, there will be a section of trays with unsealed downcomers, which would initially be exposed to vapor only, until the overhead condenser and reflux system get commissioned and provide sufficient liquid load. Similar to the system limit, if the vapor load is too aggressive, the vapor velocity inside the downcomer might exceed the velocity at which the liquid can descend freely into the downcomer. This limit is shown in Figure 2.30 as the downcomer critical vapor rate. However, unlike any other tray limitations, the downcomer critical vapor rate no longer applies once liquid seals the downcomer. A case study reflecting this limitation was discussed in Chapter 1 in the loss of downcomer seal in the depropanizer tower.

In the McCabe–Thiele discussion, it was established that the liquid/vapor mole rate ratio was constant between transition points. Assuming relatively constant molar volume, an operating line can be drawn on the tray operating window. The junction of the operating line with the tray limitations defines the operating range of the tray section. Rigorous simulation would show that operating points typically align well with the operating line but may, depending on variability of fluid properties, exhibit some scatter. Good practice would be to establish the operating line based on the controlling tray. It is also important to point out that the operating window with the operating line provides a quick way to estimate the tray section maximum capacity and understand the flooding mechanism involved.

The use of tray operating windows is extremely useful for troubleshooting as they provide a visual perspective as opposed to performing numerous tray rating calculations with tabular reports of data. They quickly provide a visual of the operating point location relative to tray limitations. It also tells the troubleshooter which limits are likely to cause the problem such that appropriate adjustments are made on the tower. Those diagrams, however, must be put together. Probably the most effective method for establishing the key operating limits for building a tray operating window would be to consult with the tray vendor as they could build those limits based on their design correlations. This is particularly convenient when dealing with proprietary technology as companies would not be willing to share their correlations but may agree to provide tray operating diagram based on vapor/liquid loads. Another methodology when dealing with conventional trays would be to build the operating limits with tray rating software. Note that tray rating software rarely provides the low operating end such as weeping and minimum liquid load. However, this can be overcome by assuming a minimum liquid load based on weir loading of 0.5 gpm/in and assuming 2:1 turndown on sieve trays, which can be extended to 4:1 on valve trays. Those limits can be adjusted based on experience as well. A third methodology would be to use published correlations for establishing the tray limitations. There are numerous correlations available in literature, which exceeds the scope of this book. It is important to state that tray operating window provides reference points relative to expected tray stability. However, those limitations may be associated with margins relative to the actual tray limitations and like any correlations may contain some degree of inaccuracy. The actual performance of the tower may provide a better indication as to where the exact limits are. As an example, the downcomer backup limit was exceeded on a tray section without efficiency lost nor pressure drop rise symptomatic of flooding. By plotting this new operating point on the tray window, the process engineer had provided a historical data point that could be used as reference in future operating mode.

The above discussion is based on overall tray hydraulic and neglects the impact of distribution of multi-pass trays. While designers try to minimize the number of tray passes, design guidelines in literature [68, 69] suggest limiting the liquid loads to 10–15 usgpm/in of outlet weir. Thus, at high liquid load the number of passes must be increased for meeting the above guidelines. Figure 2.32 illustrates the configuration of one-, two-, and four-pass trays. The number of passes is not limited to four, with six passes not being uncommon. Multiple downcomer designs are also used in high-capacity trays. Less common are three-pass trays as they are difficult to balance due to a lack of symmetry. Pass balancing is not a problem with single-pass tray. The center downcomer in two-pass tray redistributes liquid on the tray below and as such minimizes potential imbalance. However, the downcomers in multi-pass trays create a physical barrier that isolates some passes from each other. Vapor/liquid imbalance between the passes results in different *L*/*V* ratio, which affects the overall tray efficiency. As previously illustrated using the McCabe–Thiele diagram, the *L*/*V* ratio defines the slope of the operating line. For a fixed number of separation stages, differences in *L*/*V* ratio would result in varied level of fractionation. Literature [68] suggests that, although some passes might operate at more efficient *L*/*V* ratio, the benefit is far offset by the passes operating at suboptimal *L*/*V* ratios. As rate increases, the overall tray inefficiency is compounded by the drop in efficiency of passes approaching flooding earlier than others. Bolles [70] defined the concept

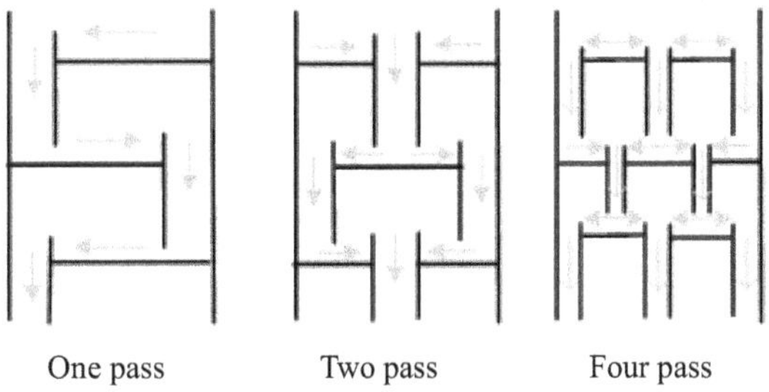

FIGURE 2.32 Tray Passes.

of distribution ratios, which is the ratio of the pass having the maximum L/V to the pass having the smallest L/V ratio. Bolles suggested that the distribution ratio be less than 1.2 for effective operation. Summers [69] indicated that common practice in the industry is to use a distribution ratio less than 1.1. Pass balancing is done at the design level through proper configuration of weir length and perforation on the tray deck. Along with the above, equal active area or equal flow path length is used as design strategy for balancing the passes.

Pass balancing goes beyond the tray design as feed distributors if not properly designed can induce maldistribution which will propagate through the tray section. Tray levelness may also influence distribution. Although of less consequences in single- and double-pass trays, a single tray off-level may induce maldistribution through an entire tray section. Startup transition in some cases can also induce maldistribution. For instance, asymmetrical operation of dual reboilers can create localized velocities, which may promote a preferential vapor flow path on multi-pass tray, especially on dry trays (i.e., no back pressure from the froth static head) during the startup transition or on tray designed with a high open area (i.e., low dry pressure drop).

Tray hydraulic performance is frequently correlated to pressure drop by process engineers. Pressure drop is a process variable that can be easily measured. Some towers have pressure drop instruments; others calculate the pressure drops from transmitter located in the bottom and top of the tower. Some towers measure pressure drops in different sections of the tower. It is important to understand the parameters that influence the tower pressure drop. Process engineers typically monitor pressure drop in the time domain and may erroneously diagnose flooding when observing a rise in pressure drop. Those rises are often the result of an increase in internal hydraulic circulation caused by factors such as higher feed rates and/or shift in feed composition. Understanding the parameters that influence tray pressure drop helps in recognizing how operational changes influence pressure drops. It also helps to distinguish between flooding and a normal or expected rise in pressure drop.

As described by Henry Kister's Distillation Design textbook [55], the total tray pressure drop is the summation of the dry pressure drop (h_d and the pressure drop through the aerated liquid $\left(h_l\right)$. The latter is the combination of the head of aerated liquid $\left(h_c\right)$ and residual $\left(h_R\right)$ pressure drop:

$$h_t = h_d + h_l = h_d + h_c + h_R \tag{2.117}$$

The dry pressure drop is caused by the loss of pressure as the vapor flows across the perforated area of the tray deck. It is calculated using a variation of the orifice equation:

$$h_d = K \frac{\rho_G}{\rho_L} u_h^2 \tag{2.118}$$

where h_d is the dry pressure drop in inches of liquids and K is the dry pressure drop coefficient. For a given tray geometry, K is a constant that is a function of the open area and the ratio of tray thickness to hole diameter. Summers and Cai provide a curve for K for sieve trays [71]. For a given separation, it can be assumed that transport properties are not varying significantly. As such, Equation 2.118 demonstrates that the dry pressure drop is function of the square of the hole vapor velocity (u_h).

The residual pressure drop (h_R) is related to liquid surface tension impeding bubble formation at the open area of the tray deck. The residual pressure drop is a function of hole diameter and properties and tend to be more predominant with small holes. It is characterized by the Bennett et al. correlation, which, as described by Equation 2.119, is not dependent on rates:

$$h_R = \frac{0.2165}{\rho_L}\left[\frac{\sigma^2(\rho_L - \rho_V)}{d_H}\right]^{1/3} \tag{2.119}$$

The head of aerated liquid is the froth height (h_f) multiplied by A:(ϕ_f)

$$h_c = \phi_f h_f \tag{2.120}$$

In the froth regime, the head of aerated liquid can be evaluated by the Colwell correlation [72]:

$$h_c = \phi_f\left[0.527\left(\frac{Q_L(1-f_w)}{C_d\phi_f}\right)^{2/3} + h_w\right] \tag{2.121}$$

The coefficient C_d is a function of the weir height (h_w) and the froth height over the weir (h_{fow}) and is given by:

$$C_d = 0.61 + 0.08\frac{h_{fow}}{h_w} \qquad \frac{h_{fow}}{h_w} < 8.135 \tag{2.122}$$

$$C_d = 1.06\left(1 + \frac{h_w}{h_{fow}}\right)^{1.5} \qquad \frac{h_{fow}}{h_w} > 8.135 \tag{2.123}$$

$$h_{fow} = h_f - h_w \tag{2.124}$$

The froth density is a function of the Froude (Fr) number and tray geometry

$$\phi_f = \frac{1}{\eta + 1} \tag{2.125}$$

$$\eta = 12.6\ Fr^{0.4}\left(\frac{A_B}{A_h}\right)^{0.25} \tag{2.126}$$

$$Fr = 0.37\frac{\rho_v u_B^2}{h_c(\rho_l - \rho_v)} \tag{2.127}$$

The above equations can be resolved numerically or by trial and error. Depending on time constraint, those can be tedious and time consuming to resolve when troubleshooting a tower. However, the intent here was to demonstrate the parameter that influences a tray pressure drop. As shown, the tray pressure drop is a function of transport property, tray geometry, vapor and liquid volumetric rates. Transport properties typically do not exhibit sufficient variability to influence changes in tray pressure drop. Tray geometry parameters are constant but can be altered by fouling. Making abstraction to fouling, only flow rates influence the tray pressure drop variability.

Associating flooding to pressure drop should first be assessed by plotting pressure drop as a function of feed rates and vapor rates. The tower overhead vapor rates can be easily estimated by summing up the reflux and distillate rate. For the tower bottom, the pressure drop can be plotted as a function of the reboiler heat stream flow rate, which can be measured or obtained from process simulation. As illustrated in Figure 2.33, comparison with historical data would reveal if the tower is experiencing premature flooding. The left side of Figure 2.33 shows a different behavior in the tower pressure drop when plotted as a function of feed rate, after an upset that caused a temporary feed interruption. On re-start, the rise of pressure drop suddenly exhibited a steeper climb as rates were increased when compared to the pre-upset conditions. This could be symptomatic that something got damaged during the upset. As discussed later, such suspicion should be validated with a gamma scan prior to making a decision of shutting down the tower for inspection. In this case [73], the tower was equipped with four-pass trays and the gamma scan showed off-center downcomer flooded on one side while the other side looked normal. Maldistribution was suspected to have been initiated on re-start as the tower operated two reboilers, one base loaded and one trimmed for meeting the tower bottom specifications. On this particular re-start, the reboiler operation sequence was significantly different from previous operation. The significant change in reboiler duty ratios in the early stage of startup is believed to be responsible for initiating maldistribution as the trays were still dry, hence providing less back-pressure in the absence of liquid froth. The plot on the right side of Figure 2.33 illustrates a significant step change when comparing pre and post shutdown. Again, this could be symptomatic of mechanical

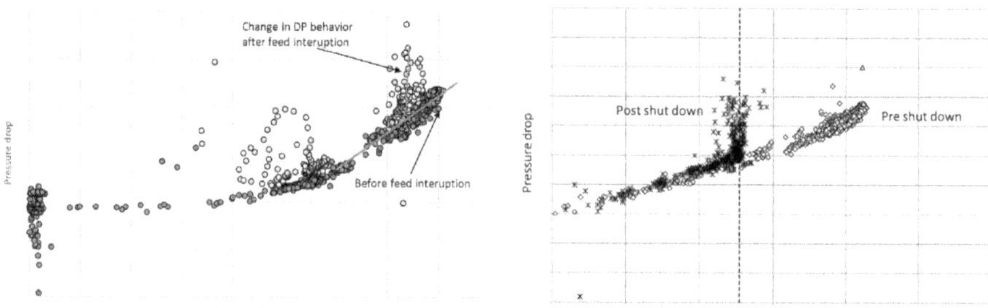

FIGURE 2.33 Historical Pressure Drop Comparison [75].

Reproduced with Permission from Distillation in Ethylene Plants – An Operators' Overview, André Bernard, 2022 AIChE Spring Meeting

damage, pluggage caused by shut down debris, or the presence of hydrates in cryogenic services. As for the previous example, gamma scans should be conducted for validating some of those suspicions and for deciding if a tower outage is required.

Pushing the tower to rates never previously experienced may make the analysis relative to historical data blurry. The data may suggest normal extrapolation of the pressure drop versus rate curve or exhibit a significant change in curvature, which would be symptomatic of flooding. By definition, jet flooding occurs when the froth height equals or exceeds the tray spacing. Based on the above correlations, approach to flooding should not cause a significant shift in the pressure drop plot curvature. However, as previously discussed, tray efficiency is expected to drop when approaching flooding. Therefore, mild flooding will be associated with a drop in separation efficiency before being noticed on a pressure drop/rate plot. Rigorous pressure drop calculations can always be performed, although the above pressure drop plots and efficiency checks should be sufficient for requesting validation using gamma scans. It is important when conducting gamma scans on a tower to maintain stable operation throughout the entire period required for conducting the scan. The tower should also be operated below flooding conditions, ideally just below rates for which flooding is suspected. A flooded tower scan would not provide any indications relative to the origin of flooding. Tray operating window, while not entirely accurate, may provide valuable clues relative to the flooding mechanism. This would also provide directions for area of the tower to be scanned (i.e., active area vs. downcomers).

It may be desirable to validate tray performance by comparing pressure drops reported on datasheets to actual tower pressure drop. Given that rates might differ from the datasheet, some practitioners may use the above correlations or rating software for conducting this evaluation. It is important to point out that tray pressure drop correlation does not account for the vapor head; that is, height of clear vapor between the froth and the tray above. While this might be relatively negligible on tray basis at low pressures, it can be significant on tall towers and at high pressure. The tower pressure drop is the summation of each tray pressure drop, including the vapor head.

Packing hydraulics. Unlike trays, packings have no downcomers and, as such, they utilize the full tower cross section for fractionation. The capacity of packing is therefore limited by flooding at the high end and minimum wetting rate or distributor turndown at the low end. Flooding can be calculated from vendor rating software. It can also be approximated by the generalized pressure drop correlation (GPDC), which provides good approximation for random packing. The original GPDC was developed by Sherwood and Eckert and has been modified over the years by a few practitioners with the latest changes attributed to Eckert and followed by Strigle [74]. It is important to understand the version of the GPDC chart used, as associated guidelines relative to flooding may vary. The GPDC is a graphical correlation relating the capacity parameter (CP), the flow parameter F_{lv}, and pressure drop per feet of packing:

$$CP = C_s F_p^{0.5} \nu^{0.05} \tag{2.128}$$

where ν is the liquid kinematic viscosity (centistoke). F_p is the packing factor, which is specific to packing size and geometry. The values of F_p can be found in literature or obtained from the vendor. C_s is the C_{factor} as described in Equation 2.114. The flow parameter F_{lv} on the abscissa of the GPDC chart entirely differs from the F-factor described by Equation 2.116 and is described by Equation 2.129:

$$F_{lv} = \left(\frac{L}{V}\right)\sqrt{\frac{\rho_v}{\rho_l}} \tag{2.129}$$

The Strigle GPDC correlation applies strictly to random packing. Kister and Gill [75, 76] developed a modified GPDC chart for structured packing as illustrated in Figure 2.34.

The low end of packing operation is limited by the wetting rate and liquid distributor turndown. In most cases, the liquid distributor is the controlling limit, where maldistribution and channeling are expected when operating below turndown. The turndown limit depends on the distributor design. As a rule of thumb for through distributor with drip tubes, two inches static head over the drip tubes should be a minimum. The minimum wetting rate recommended for structured packing and grid [61] is 0.2–0.3 gpm/ft^2. The minimum recommended wetting rate for random packing [55] is 0.5–2 gpm/ft^2.

Maldistribution in packed tower due to operation below turndown during startup transition is likely. However, unless aggressive fouling occurs during the transition, distribution is expected to re-establish itself as rate rise above turndown.

Reboilers provide heat required for sustaining vapor circulation in a distillation tower. There are different reboiler configurations as illustrated on Figure 2.35 with thermosiphon reboilers being the most commonly used in the CPI. Less common are forced thermosiphon, kettle reboilers, and internal reboilers. Reboilers' configuration

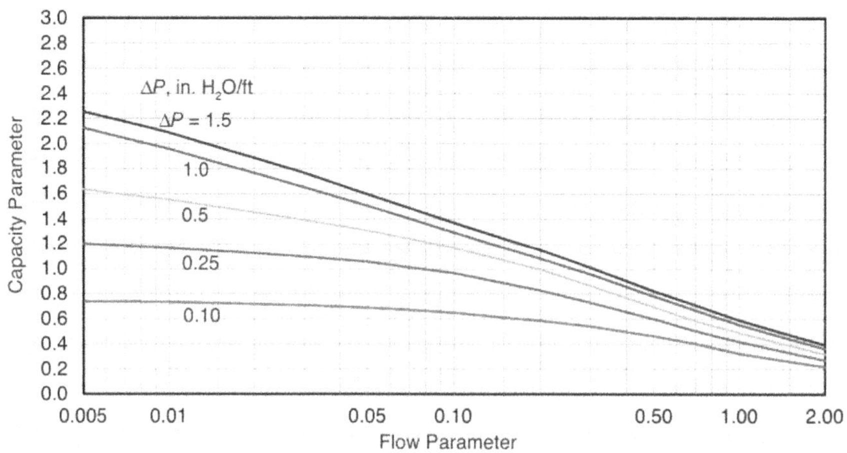

FIGURE 2.34 GPDC for Structured Packing.

Reproduced with Permission from figure 2.2, Realistically Predict Capacity and Pressure Drop for Packed Columns, Henry Z. Kister, Jeffrey Scherffius, Khashayar Afshar, Emil Abkar, *Chemical Engineering Progress*, July 2007

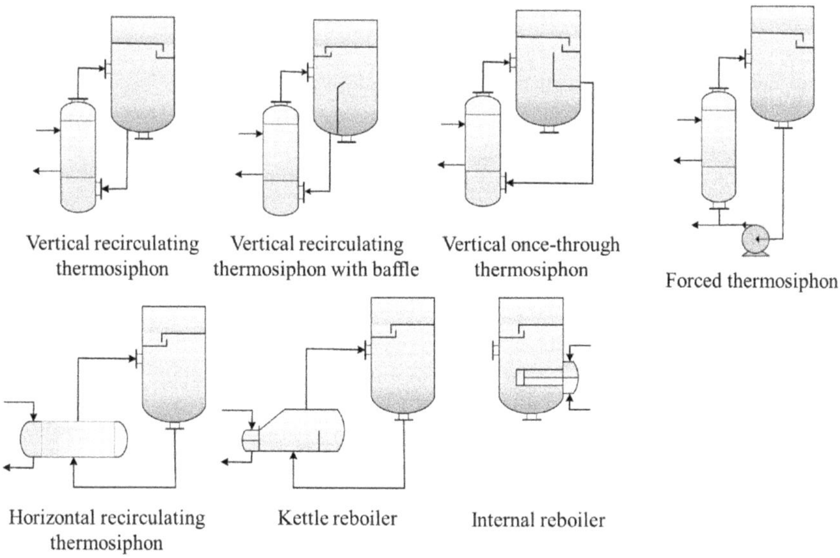

FIGURE 2.35 Different Reboiler Configurations Typically Encountered in the CPI.

selection should be based on service and applications but is often influenced by the type of industry or organization, which would tend to favor a specific configuration. This bias toward a specific configuration can be explained by experience and established design practices. As a reference, Sloley [77] indicated that the petroleum industry tends to favor horizontal thermosiphon; the chemical industry seems to prefer use vertical thermosiphon, with petrochemicals using a mix of both horizontal and vertical thermosiphon. Those trends are mostly explained by the nature of the fluids being processed, where petroleum refineries handle dirty fluids prone to fouling, the chemical industry processing mostly clean fluids and petrochemicals plants processing a mix of both. Kister [64] provides a description of the different reboiler configurations and their typical applications. In general, the fouling propensity of the heating and process fluids, plot space available, and required heat transfer area are factors to consider when selecting a reboiler configuration. Overall, vertical thermosiphon configuration is the most common type used. Their application can however be limited by excessive tube length, where high tower skirt would increase capital cost especially on large towers. Fouling heating media, high heat duty requirements, or high required heat transfer area would tend to favor horizontal thermosiphon configuration. On the downside, horizontal reboiler requires more plot space and piping than their vertical counterparts.

Figure 2.35 provides a high-level description of some less common reboiler configurations. Forced circulation reboilers are to be considered with viscous liquids or highly fouling process streams. They are also used in fired heater application, where a minimum continuous flow is required for avoiding high TMT. Kettle reboilers, unlike thermosiphons, vaporize most or all of the process stream. Neglecting potential entrainment, the reboiler return is a vapor only stream. High vaporization rates make them subject to fouling. There is also a potential accumulation of heavy components, which may have boiling points above the heat stream temperature. The accumulation of such material may stall the reboiler on a temperature pinch, if the reboiler is not regularly purged to prevent accumulation of those heavy components. High vaporization rates and flow regimes inside the kettle give them lower heat transfer coefficient when compared to thermosiphon reboilers. They are typically relatively large units which consume substantial plot space. Internal reboilers may offer the advantage of being compact. However, in large towers, it may require relatively large manways for installing or removing them. Some might be installed in the tower while building the tower. That is, if part or the full tower is built on the plant site, it must be understood that reboiler replacement will not be an option if significant failure occurs over the years. On small towers, a removable tube bundle can be inserted in the tower sump. The same can be applied to larger towers, providing that the heat transfer area is relatively small. Fouling propensity is another consideration, where internal reboilers would not be recommended in fouling service, unless the tower is operated in batch mode where frequent cleaning is possible. From an operation perspective, internal reboilers cause sump liquid level difficulties due to the turbulence, which creates a frothy liquid environment inducing noise on level instruments and possible liquid carryover into the trays or packings.

In thermosiphon reboiler, the vapor fraction of the process fluid increases as it travels through the length of the tubes. This result in flow regime transition as illustrated in Figure 2.36 [78]. It is understood that the liquid at the surface of the tower sump is at its boiling point. The static head exerted on the liquid entering the reboiler however causes subcooling. As such, the liquid is first preheated to its boiling point. As it reaches boiling, small bubbles are created and dispersed in the rising liquid, which is identified as the bubbly flow regime. As more bubbles are created, they coalesce together with the flow regime transitioning to slug and churn. Further vaporization results in annular flow where a liquid film still covers the tube wall and rises from the shear effect of the vapor traveling upward. Vapor becomes the continuous phase between the liquid film covering the tube inner surface, with gas velocity entraining liquid droplets. Further evaporation leads to the mist flow regime where the tube wall gets dry and the small liquid droplets carried by the vapor velocity out of the tube. For best performance, the preheating zone should be limited to a minimum as it consumes heat transfer area. It also involves sensible heat transfer which has lower heat transfer performance than boiling. The loss of liquid film at the tube wall in the mist flow regime is associated with significant deterioration in heat transfer performance. This flow regime should therefore be avoided. The flow regime is dependent on the extent of vaporization and circulation. High static head will drive high circulation and reduce the extent of vaporization. In some systems, excessive static head causes boiling point suppression at the inlet of the reboiler and may result in capacity limitations. Contrarily, low circulation rate increases the extent of vaporization and may affect performance if a significant fraction of the exchanger tubes operates in the mist flow regime. Some designers have specified orifices on the inlet piping of the reboiler to offset boiling point suppression caused by high static head. However, this requires to be extremely precise on the design. Designing the right circulation can be challenging as it may be influenced by different operating modes,

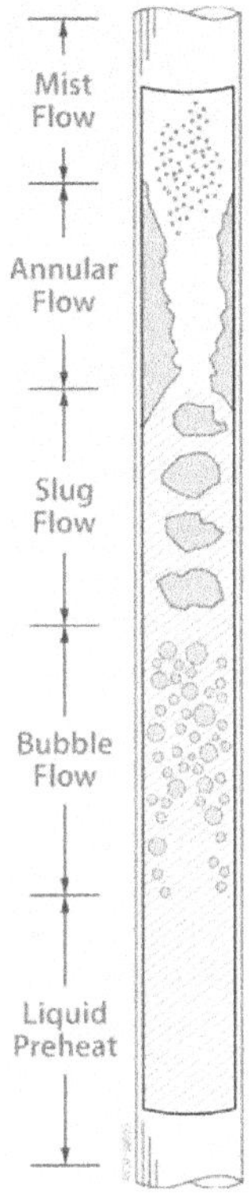

FIGURE 2.36 Reboiler Tubes Flow Regime [81].

Reproduced with Permission from figure 2.2, E. Van Doorn, Henry Z. Kister, Stable Thermosiphon Reboiler Operation: With or Without a Valve in the Inlet Piping, Hydrocarbon Processing, February 2019

such as startup, reboiler end/beginning of run in fouling services, and variability in tower feed composition due to feedstocks or reactor yields variability. Van Doorn and Kister [79, 80] indicated that thermosiphon reboiler should be designed for 10–40% vaporization. The lower end can be extended to 5% vaporization with fluid having significant density differential between liquid and vapor (e.g. water). The controlling design case would typically aim for 20–25% vaporization. Given all the variability, in the absence of safety constraints, some designers prefer using an inlet throttle valve instead of an orifice plate as it provides more flexibility. This provides the ability to adjust the pressure drop at the inlet of the reboiler such that the circulation rates align with recommended level of vaporization. Using process simulation, the level of vaporization can be correlated to the reboiler outlet temperature.

Thermodynamics dictates that heat travels from hot to cold. Simple principle but sometimes missed in troubleshooting heat exchangers. While this may have little implication for towers operating near atmospheric pressure, it has more relevance in high-pressure applications, especially during startup where the initial mixture may get heavier at low pressure due to flashing of light components. Figure 2.37 [81] illustrates the boiling curve of a mixture expected to feed a reboiler at the initial tower pressure (Pinitial), and at steady-state operating pressure (Psteady State). Assuming that the feed stream initially enters the tower and fills the tower sump near atmospheric pressure at a temperature of about –37 °C; about 96% of the mixture would flash before reaching the sump. Expecting to use the reboiler heat for pressurizing the tower would result in a temperature pinch with the liquid supplying the reboiler. The liquid fraction remaining from the feed mixture (i.e., 4%) would have an initial boiling point over 72 °C at the operating pressure, which significantly exceeds the heat stream temperature of 8 °C. This startup strategy had been tried unsuccessfully in some plants and resulted in numerous reboiler stalls and unnecessary draining of the tower and reboiler sump for purging heavy liquid out. It may be argued that the boiling point of the mixture in the reboiler sump will transition as feed comes in and vapor gets generated from the reboiler. However, that transition was evaluated by Bernard [81] who demonstrated by material balance that the reboiler feed mixture will get heavier as heat is introduced to the reboiler. As vapor gets generated by the reboiler, the liquid returned to the sump in a thermosiphon reboiler is heavier than at the inlet of the reboiler. The liquid returned mixes with the feed, which causes the mixture supplying the reboiler to get heavier. At the temperature pinch, the reboiler stalls and stops generating vapor, hence preventing pressurization of the tower. The above discussion is relative to startup but can also occur during a plant upset where the tower may lose pressure. Under those circumstances, the mixture supplying the reboiler will likely get heavier due to flashing of light components and result in a

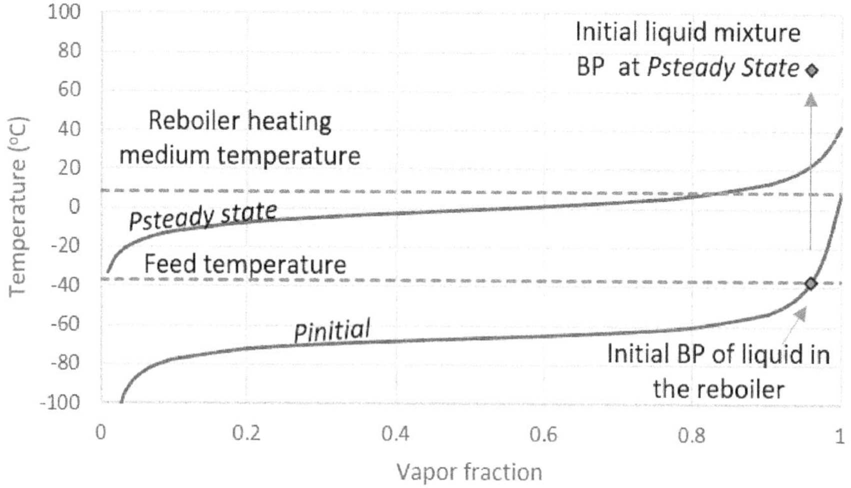

FIGURE 2.37 Impact of Pressure on Multicomponent Mixture Boiling Point [84].

Reproduced with Permission from figure 2.3, Distillation Startup Considerations in Ethylene Plants, André Bernard, 2021 AIChE Spring Meeting

reboiler stall. Good practice following a trip which causes a tower slump and pressure loss is to drain the sump prior to re-starting the unit.

As previously discussed for thermosiphon reboilers, the density differential between the reboiler inlet and outlet drives the circulation. However, on vertical thermosiphon, a minimum critical vapor velocity is required to offset gravitational forces on the liquid film and droplets. Below that critical velocity, a fraction of the liquid is dropping down, which can be compared to an internal reflux. This condition is known as flow reversal and is likely encountered during startup transition or operation at low rates. The Wallis flow reversal criterion is a dimensionless vapor velocity used as an indicator of the potential for flow reversal and is described by Equation 2.130:

$$u_g^* = \frac{u_g \rho_g^{0.5}}{\left[g_n d_i \left(\rho_l - \rho_v\right)\right]^{0.5}} > 0.5 \tag{2.130}$$

where:

$$u_g = \frac{x_g \dot{m}}{\rho_g} \tag{2.131}$$

u_g^* is the Wallis flow reversal criterion which is based on the superficial tube velocity u_g. The Wallis flow reversal criterion must be above 0.5 for avoiding flow reversal. Bernard [82] evaluated the impact of an inlet throttle valve on the reboiler outlet liquid boiling curve and the potential for flow reversal on a demethanizer pre-stripper reboiler heated by condensing propylene refrigerant (C3R) vapor in an ethylene plant. The left side of Figure 2.38 shows 90% vaporization with a reboiler inlet butterfly valve open at a 20° position. The Wallis flow reversal criterion was estimated at 0.38, suggesting flow reversal conditions. The reboiler outlet liquid boiling curve illustrated was estimated to have higher density than the liquid at the reboiler inlet by 25%. As such, this heavier liquid will migrate to the bottom of the reboiler and accumulate. This will eventually cause a temperature pinch with the heat stream and stall the reboiler. The thermosiphon was re-evaluated with the butterfly valve put in the fully open position as illustrated on the right side of Figure 2.38. With this configuration, the Wallis flow reversal criterion remains below 0.5 but rises to 0.46. Most importantly, the vaporization dropped to about 15% with the reboiler outlet liquid boiling curve slightly heavier but relatively similar to the reboiler inlet. Therefore, despite flow reversal conditions, a temperature pinch and reboiler stall are less likely. Although this example may suggest irrelevance of the inlet throttle valve, the initial startup rate would result in vaporization below 5% and may cause difficulty in starting the reboiler. At those early startup conditions, pinching the inlet throttle valve to a 40° position would increase vaporization to 15%. Knowing that the reboiler still operates in flow reversal mode, the vaporization should be kept to a minimum such that the boiling curve of the returning liquid will not cause a temperature pinch. As rate increases, the valve should then be put in a fully open position.

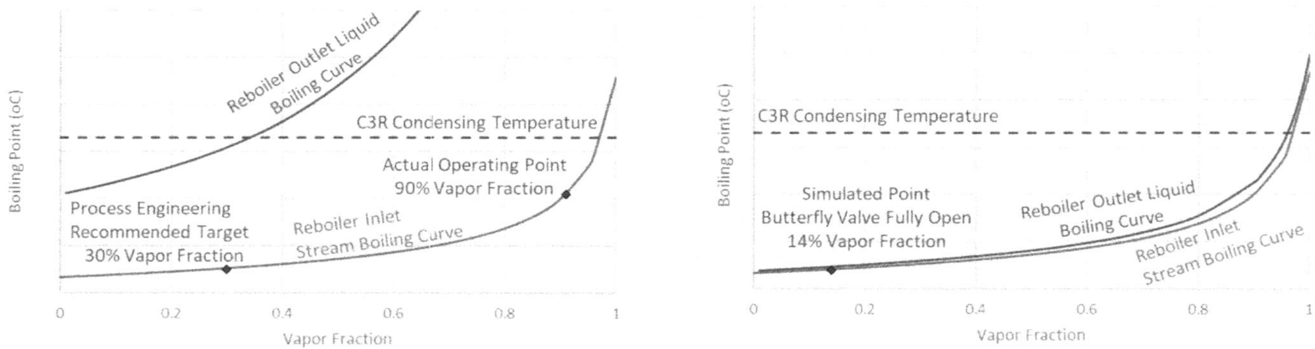

Inlet Butterfly Valve: 20° Position
Wallis Criteria = 0.38

Inlet Butterfly Valve: Fully Open
Wallis Criteria = 0.46

FIGURE 2.38 Reboiler Liquid Inlet and Outlet Boiling Curves versus Inlet Butterfly Valve Position [84].

Reproduced with Permission from figures 2.7 and 8, Distillation Startup Following a Plant Revamp with Focus on Reboiler Operation, André Bernard, 2023 AIChE Spring Meeting

Recovering reboiler performance from flow reversal and accumulation of heavies in the reboiler can be accomplished by draining some of the liquids from the reboiler. This can be tricky on a single reboiler and may cause a tower slump. On a two-reboiler operation, it is best to pinch hard on the inlet valve of the reboiler being drained. Doing so maximizes drainage of the heavy liquid accumulated in the bottom of the reboiler. During this operation, the other reboiler is providing most of the heat for the tower operation and minimizes the operational impact. The process can then be repeated on the other reboiler. Flow reversal can also be prevented or mitigated by using a lift gas when operating in flow reversal. The lift gas will compensate for the gas velocity deficiency and ensure operation above flow reversal. It does not need to be continuous and can be applied to push the heavy liquid back in the tower when noticing drops in performance.

Temperature pinches and reboiler stalls caused by flow reversal are frequently confused with film boiling. Although the symptoms are similar, the mechanism is different. Film boiling is characterized by the formation of a vapor film on the tube wall. This film creates an additional heat transfer resistance, as heat from the tube wall must travel to the gas–liquid interface for creating gas bubbles. In nucleate boiling, the gas bubbles are generated at the tube wall and dissipate in the liquid phase leaving the tube wall wet. Film boiling typically occurs at high heat flux and high-temperature differential between the hot and cold stream. Published rules of thumb quoted by Kister [64] suggest that to avoid film boiling the temperature differential across the tube wall should not exceed 32–38 °C (90–100 °F) or alternatively a heat flux not exceeding 136,140–340,355 kJ/hr m^2 (12,000–30,000 BTU/hr ft^2). The above rules of thumb are, however, reported to be simplified as they do not take into consideration the physical properties and exchanger geometry. Hagan and Kruglov [83] demonstrated how the temperature differential (i.e., tube wall temperature – tube bulk process temperature) affects the heat flux, using rigorous heat exchanger rating software. Figure 2.39 generated by Hagan and Kruglov [83] as an example shows the heat flux as a function of the temperature differential and shows how it relates to the transition from nucleate boiling to film boiling. Of particular interest is the transition zone where, unlike expectation, increasing the temperature differential is associated with a reduction in heat flux. Once in film boiling, rise in temperature differential has initially no influence on the heat flux. Further increase in temperature differential eventually causes the heat flux to rise. Although not ideal from a heat transfer perspective, some exchangers are designed to operate in the film boiling regions.

While the symptoms of film boiling may be similar to those observed when experiencing flow reversal, the cure is different. As illustrated in Figure 2.39, the temperature differential is the primary cause for transitioning outside nucleate boiling. Once in the transition zone, reduction in heat flux, similar to level experienced in nucleate boiling, is ineffective at high-temperature differential. The tube wall temperature must be reduced below the critical differential temperature where the exchanger transitions from nucleate boiling. A higher-temperature differential than design or normal steady state is a situation that is prone to occur on a reboiler from which heat is supplied from interstage refrigeration compressor vapor streams in the early stage of startup. Under those conditions, it is not unusual for the compressor to operate in false load mode, that is, with anti-surge control valve open, which ensures a safety

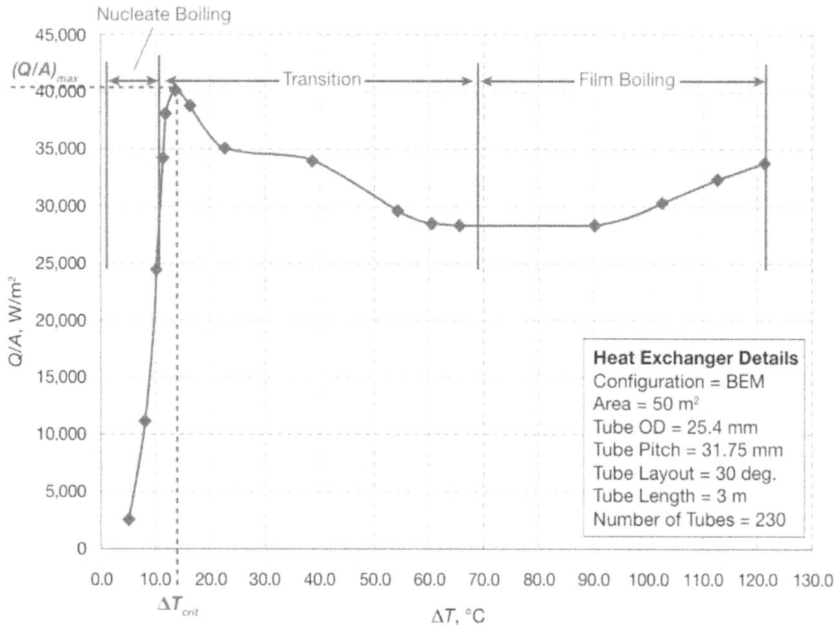

FIGURE 2.39 Heat Flux/Temperature Differential Impact on Nucleate versus Film Boiling in Vertical Thermosiphon Reboiler [85].

Reproduced with Permission from figure 2.1, Understanding Heat Flux Limitations on Reboilers Design, Hagan M. Kruglov V., *Chemical Engineering Progress*, November 2010

margin from surge. Although the heat of compression is minimized using liquid quench or coolers within the minimum flow circuit, the degree of superheat carried through the compressor can be higher than at steady state. Therefore, this may cause higher-temperature differential in reboilers that may result in film boiling and cause shortfall in capacity. This problem can be alleviated by adding desuperheating station on the heat streams supplying the reboiler such that the temperature differential remains below the critical value.

The performance of reboiler involves the ability of the exchanger in transferring heat from the hot to the cold side. On a macro level, this is defined by the relationship between the heat duty, the temperature difference, the heat transfer area, and the heat transfer coefficient as described by Equation 2.132:

$$Q = UALMTD \tag{2.132}$$

where Q is the heat duty, A is the heat transfer area, U is the overall heat transfer coefficient, and LMTD is the log mean temperature difference. The LMTD for countercurrent flow heat exchangers is given by:

$$LMTD = \frac{\left(T_{hot\ in} - T_{cold\ out}\right) - \left(T_{hot\ out} - T_{cold\ in}\right)}{\ln\left[\left(T_{hot\ in} - T_{cold\ out}\right) / \left(T_{hot\ out} - T_{cold\ in}\right)\right]} \tag{2.133}$$

Depending on exchanger geometry, correction factors may apply to the LMTD value. U is the overall heat transfer coefficient and depends on flow rates, flow regimes, exchanger geometry, material of construction, and fluid thermal properties. While not always the case, but if properly instrumented, temperatures should be measured on the inlet and outlet streams of the reboiler along with the heat stream flow rate. Those measurements allow calculation of the overall heat transfer coefficient over time and provide a high-level indication on the exchanger performance. The U value is flow-dependent and as such would vary with load. However, for a unit operating at relatively constant rate, downward trend of the U value is an indication of performance degradation typically caused by fouling.

The heat duty is dictated by fractionation requirement but is easily measured from the stream hot side by extending Equations 2.132 to 2.134 when condensing is involved:

$$Q = UA\,LMTD = m\left[\left(Cp\,\Delta T\right)_{Superheat} + \lambda + \left(Cp\,\Delta T\right)_{Subcool}\right] \tag{2.134}$$

Or Equation 2.135 when only sensible heat is involved:

$$Q = UA\ LMTD = m\,Cp\,\Delta T \tag{2.135}$$

There are two operating strategies in reboilers where heat is provided by condensation of the hot stream. As illustrated in Figure 2.40, the heat transfer area can be manipulated by back flooding the reboiler with condensate when the control valve is downstream the exchanger. In this case, the pressure is equivalent to the supply pressure of the hot stream with the associated condensing temperature. The heat contribution from the desuperheating and subcooling terms in Equation 2.134 is small compared to the latent heat λ. With the condensing temperature and process temperature being relatively constant, the effective heat transfer area used for condensation is the main term balancing the left and right side of Equation 2.134. When the heat stream control valve is upstream of the exchanger and condensate traps prevent vapor blow through, the heat transfer area is constant. Assuming negligible change

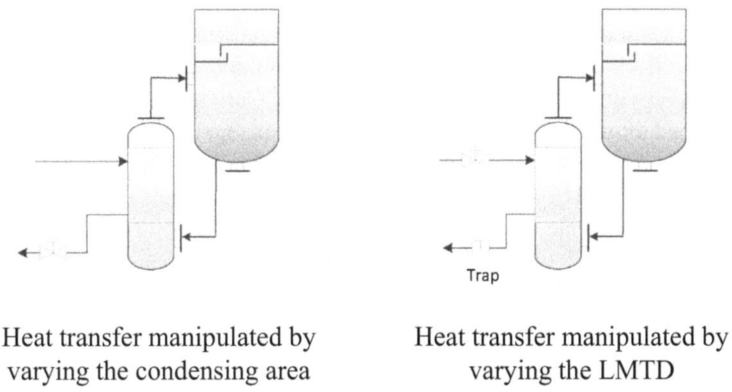

FIGURE 2.40 Reboiler Heat Stream Control.

in heat transfer coefficient means that balancing Equation 2.134 is done by varying the LMTD, that is, variability in the hot stream condensing temperature driven by change in exchanger chest pressure.

In single-phase heat streams, the heat duty is strongly dependent on the flow m as the heat capacity Cp is relatively constant and the ΔT may exhibit little variation if operating close to a temperature pinch, that is, with the heat transfer area A constant and little variation in LMTD; Equation 2.135 is primarily balanced by variation in the overall heat transfer coefficient.

Condensers are used for condensing the overhead vapor into a liquid stream, which is primarily used for refluxing the tower and, in most cases, providing a liquid distillate. Partial condensers are used when a vapor distillate is desired. As for reboilers, there are many different configurations of condensers. However, their operation is significantly less complicated than thermosiphon reboilers. They can be oriented horizontally or vertically, with the cold stream on the tube or shell side. Some can be integrated inside the tower and provide additional separation stages when oriented vertically. Similar to reboilers, the cooling stream heat duty can be based on sensible or latent heat. The cooling streams depend on the required condensing temperature, which is also linked to the tower operating pressure. Cooling water is widely used when appropriate and would typically flow in the tube side of the condensers due to the inherent fouling nature of this stream. In cryogenic applications, liquid refrigerant provides cooling through evaporation in kettle-type exchangers with the process flowing through the tube side of the exchanger.

Condensation at the tube wall is described by two mechanisms. Film condensation, being the predominant mechanism found in heat exchangers, is described by a liquid film covering the metal surface. The film is known to get thicker as it travels through the length of the exchanger tubes, which creates a resistance to heat transfer. Dropwise condensation is a second mechanism, where liquid droplets are formed at the tube surface and disengage, leaving the tube wall relatively dry. Dropwise condensation is reported to have heat transfer coefficient that could be ten times higher than film condensation [27]. Although dropwise condensation would be desired, it is practically impossible to achieve as condensing vapor easily wets the tube surface. The use of enhanced heat transfer surfaces has been successful in improving heat transfer over smooth surfaces by reducing the thickness of the liquid film. Those technologies enabled tighter temperature approach between the hot and cold stream. Core in kettle exchangers is another technology that provides close temperature approach between the hot and cold stream and uses significantly less plot space than conventional shell and tube exchangers. Core in kettle configuration contains a brazed aluminum plate fin core exchanger inside the shell of the exchanger. Those exchangers still operate in film condensation but have a high heat transfer area per exchanger shell. ALPEMA [84] states that those exchangers can have a total heat transfer area per volume of 1,000–1,500 m^2/m^3 in comparison to conventional shell and tube having 40–70 m^2/m^3. The use of core in kettle exchangers is however limited to clean service and found good application in cryogenic services.

Kister [64] described a third mechanism where vapor condenses as small droplets (0.1–40 μm) that are suspended in the vapor phase. This mechanism can be compared to fine droplets of water in suspension in moist air, when the temperature drops below the dew point. The vapor/liquid mixture in such a case is called a fog, where it is difficult to separate the liquid from the vapor phase. Therefore, those liquid particles would tend to be purged with non-condensable resulting in loss of valued chemical and/or release to the environment. Depending on the characteristics of the liquid, it could also cause other problems where it may accumulate. Fog formation is promoted when heat transfer is significantly higher than mass transfer. In general, the resistance to heat transfer is fundamentally less than for mass transfer. However, the imbalance can be amplified by operating with large temperature differential between the cold and hot streams.

It is not unusual in small chemical plants to locate the condenser above the tower such that reflux and liquid distillate flow can be driven by gravity. This eliminates the need for a reflux drum and associated pumps. This design strategy is impractical in refineries and large petrochemical plants due to the size and weight of the equipment. Condenser, reflux drums, and pumps are therefore required and located at or near grade. This adds to the complexity and reliability of operation.

Another aspect of cooling water that is frequently overlooked is the presence of dissolved oxygen. In most cases, cooling water is supplied from a close-loop system where hot cooling water exiting heat exchangers is cooled down in atmospheric cooling towers before returning to the network of heat exchangers that are part of the cooling water network. Cooling towers work on the principle of humidification, which states that water circulating counterflow or cross flow to a stream of air cools down due to water evaporation as it saturates the air stream with water. By doing so, the water stream gets saturated with oxygen. The solubility of oxygen decreases as the temperature of the water rises in condensers, which can cause accumulation of non-condensable. The reduction in cooling water pressure in elevated exchanger may further compound this problem as pressure also influences oxygen solubility. This phenomenon can limit the capacity of condensers if the non-condensable accumulates in the exchanger. High point facilitating non-condensable accumulation and low cooling water velocities are factors that may promote accumulation of non-condensable.

Process control of distillation towers fulfills four objectives: meeting product specifications, minimizing losses, optimizing energy usage, and maintaining tower stability against external disturbances. Typical distillation controls are illustrated in Figure 2.41, where the overall balance is described by Equation 2.136:

$$F = D + B \tag{2.136}$$

The molar component balance for component i is expressed by Equation 2.137:

$$Fz_i = Dx_{Di} + Bx_{Bi} \tag{2.137}$$

The tower feed flow rate and composition are the outcomes of upstream units and are not control variables. They are the primary disturbance that the control system must overcome for keeping the product streams within specification targets. As stipulated by Equation 2.136, for a given feed rate, only the distillate or the bottom stream flow rate can be manipulated as one depends on the other. Furthermore, for

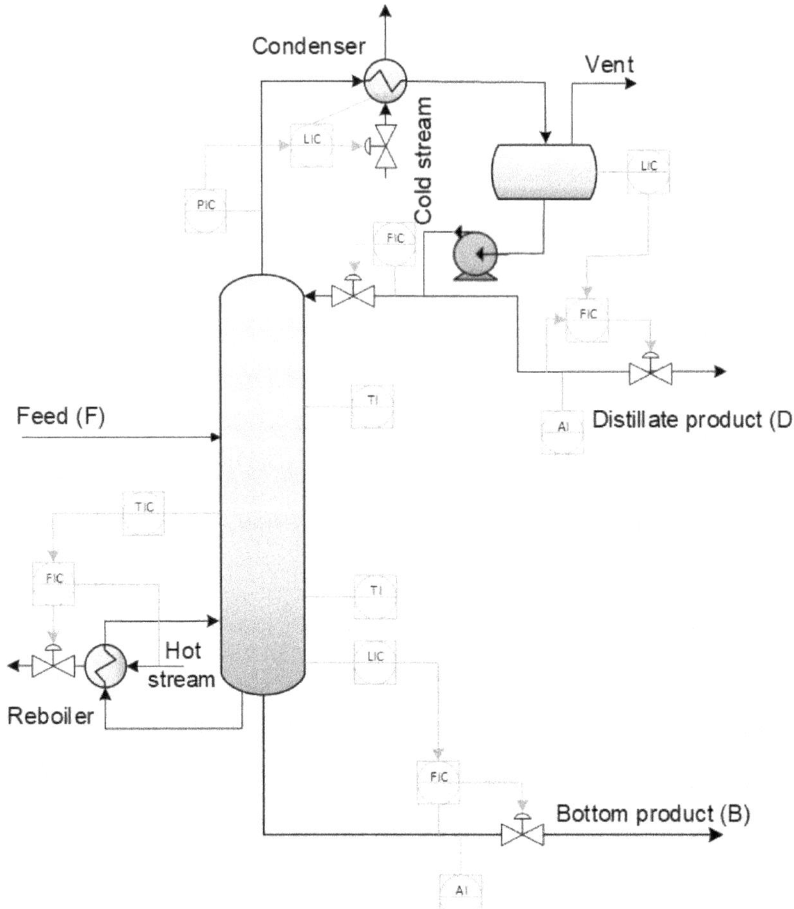

FIGURE 2.41 Distillation Tower Typical Control [75].

Reproduced with Permission from figure 2.15, Distillation in Ethylene Plants – An Operators Overview, André Bernard, 2022 AIChE Spring Meeting

a targeted distillate molar fraction x_{Di} Equations 2.136 and 2.137 can be combined together for providing a relationship between the feed flow rate and compositions to the required distillate rate:

$$D = F\frac{(z_i - x_{Bi})}{(x_{Di} - x_{Bi})} \tag{2.138}$$

This satisfies the material balance but does not address fractionation requirements. As described in the previous section, the desired level of fractionation for a fixed number of equilibrium stages is achieved by circulating the appropriate vapor/liquid molar flow rates in the tower. This involves evaporating a fraction of the bottom liquid to provide the required vapor rate in the tower and refluxing a fraction of the condensed overhead vapor for supplying the required liquid rate. This operation is closely related to the energy balance of the tower. As illustrated in Figure 2.41, the heat balance is regulated by the hot stream flow to the reboiler. The control variable is typically the residual molar composition of the light key component. While analyzers can be used, this composition is frequently inferred to a tray temperature, which remote-cascades to the reboiler hot stream flow controller. The control strategy illustrated in Figure 2.41 shows the reflux on flow control. It is known that the reflux ratio is a function of the number of separation stages and feed composition. The number of separation stages is indirectly related to separation efficiency and can be considered constant when operating the tower within the hydraulic capability of the internals. For a given feed composition, the reflux ratio remains constant and is not affected by feed rates. This leaves the feed composition as variable influencing the required reflux ratio. One common mistake is to control the reflux flow in a similar fashion as the reboiler by remote SP of the reflux flow controller from a tray temperature. This practice exceeds the degree of freedom of the tower energy balance. An increase in reflux rate will likely be compensated by an increase in heat from the bottom reboiler. The two controllers (reboiler and reflux) will likely fight each other causing tower instability. To avoid those instabilities, it is not unusual to have the reflux on flow control at SP manipulated by the process operator as required. This practice is not necessarily energy-efficient. Expanding on Equation 2.138 and the above discussion, the reflux mole rate can be obtained by multiplying Equation 2.138 by the reflux ratio (L/D). The reflux ratio is therefore a function of the feed composition $f(Z_i)$. Figure 2.42 illustrates the relationship between the reflux ratio and the feed composition for an ethylene fractionator. It also illustrates the expected feed composition based on conventional steam cracker feedstocks. As such, the reflux mole rate can be expressed in terms of the reflux ratio/feed composition function and distillate as described by Equation 2.139:

$$Reflux = f(z_i) * D = f(F) * F \frac{(z_i - x_{Bi})}{(x_{Di} - x_{Bi})} \tag{2.139}$$

Note that the ethylene fractionator reflux ratio curve is relatively linear between feed composition equivalent to propane and ethane feeds. With DCS capabilities, the reflux flow can then be regulated based on Equation 2.139. However, it is critical to understand that changes in feed flow rate and/or composition are associated with lag. The feed condition, vapor, or liquid affects the tower dynamics differently. Composition or rate changes from a vapor feed will first be sensed by the overhead system, while a liquid feed will first disturb the bottom and reboiler. As such, adjusting the reflux flow rate from changes in a vapor feed would cause a smoother transition. The distillate level to flow cascade will self-regulate the material balance and the reflux adjustment will eventually be compensated by the bottom reboiler. Conversely, a liquid feed will first be experienced by the tower bottom and reboiler. A longer lag would be expected before it affects the tower overhead. Reflux adjustments

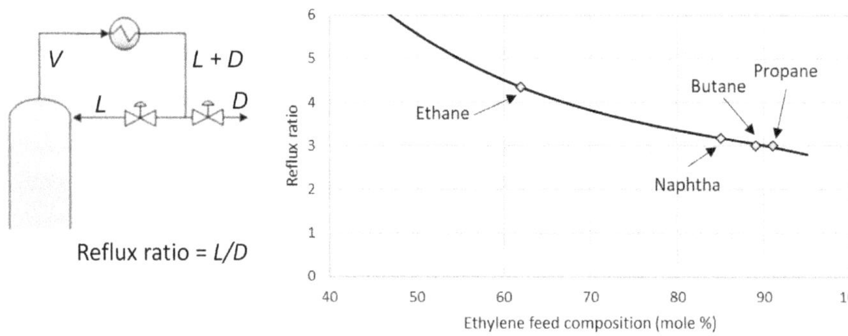

FIGURE 2.42 Ethylene Feed Composition Impact on Ethylene Fractionator Reflux Ratio [75].

Reproduced with Permission from figure 2.4, Distillation in Ethylene Plants – An Operator's Overview, André Bernard, 2022 AIChE Spring Meeting

based on feed changes should therefore be delayed until the transient condition is sensed in the overhead system. Premature action in such cases will cause tower instability and may result in losing level or flooding the reflux drum.

As stated earlier, the primary source of disturbance comes from the feed flow rates, conditions (temperature and pressure), and composition. Those conditions affect control of the tower in different ways. Large composition differences significantly affect the required reflux and reboil ratio. This typically results in a shift of the temperature profile. Bernard [65] illustrated this effect using process simulation by comparing the predicted reboiler duty and control tray temperature for ethylene fractionator feeds of 60% and 80% mole ethylene. Those compositions reflect differences in ethylene cracker feedstocks from ethane to mix feed of naphtha/propane/butane. Although such large shifts in feed composition are rare, the evaluation highlights the potential impact on tray temperature control and its dynamics. Figure 2.43a illustrates the reboiler duty as a function of tower bottom ethylene mole fraction and control tray temperature for a given feed rates and distillate purity. As illustrated, the bottom ethylene mole fraction varies linearly with the reboiler duty for both feed composition with the slope increasing at higher ethylene content. The temperature plot shows a range of sensitivity between composition, temperature, and reboiler heat duty. This range is reduced by about 1 °C when transitioning from 60% to 80% ethylene feed. The process gain as defined by the ratio of change of the control variable (tray temperature) over the manipulated variable (reboiler heat duty) rises by a factor of 4 when increasing the tower ethylene feed composition from 60% to 80% mole. To maintain operational stability, such a change in dynamic behavior may require different tuning parameters for the controller. Most of all, the shift in feed composition would require an SP change on the temperature controller for meeting the tower bottom specification. As illustrated in Figure 2.43b, the change in required SP is also explained by the impact the feed composition has on the temperature profile for a given residual ethylene mole fraction in the tower bottom. One interesting

(a)

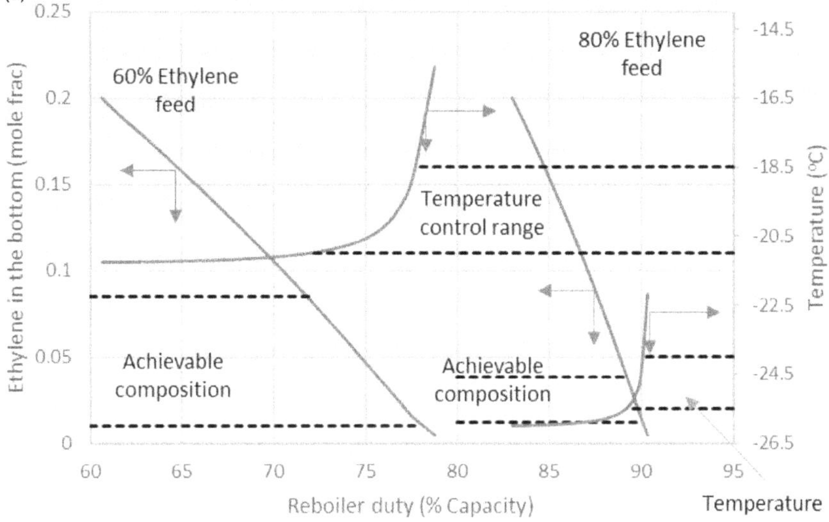

(b)

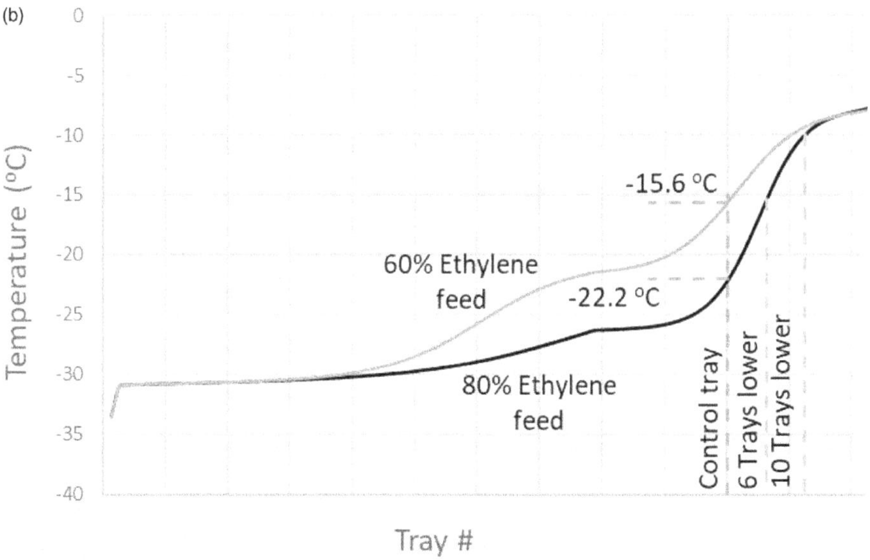

FIGURE 2.43 Ethylene Feed Composition Impact on (a) Temperature Control and (b) Temperature Profile [75].

Reproduced with Permission from figure 2.17, Distillation in Ethylene Plants – An Operator's Overview, André Bernard, 2022 AIChE Spring Meeting

observation from Figure 2.43b is that while the control tray temperature seems to provide good sensitivity relative to its location in the tower, the temperatures are outside the control range established in Figure 2.43a. This discrepancy is explained by a targeted residual ethylene mole fraction of 0.005 while Figure 2.43a suggests a residual ethylene controllable mole fraction range of 0.012–0.023 for an 80% ethylene feed and 0.01–0.085 for a 60% ethylene feed. Below the above residual ethylene control range, there is a loss of temperature sensitivity that makes this control strategy ineffective. Figure 2.44 illustrates the relationship between different control tray temperature's location and the residual ethylene mole fraction for an 80% ethylene feed. As shown, the targeted residual ethylene mole fraction is in the flat portion of the curve. Moving the control tray location slightly improves the controllable range. However, the locations of thermowells are fixed by design. Additional thermowells would require an outage and often require post-weld heat treatment on process vessels depending on their fabrication and operating conditions. The use of analyzers and advanced process control (APC) becomes more appropriate when targeting very low residual concentration of the light key components. It is important to point out that every separation has its own characteristics. Figure 2.45 illustrates the reboiler duty as a function of tower bottom residual methane molar concentration and control tray temperature for a demethanizer pre-stripper. As shown, the behavior is quite different from what has been described for the ethylene fractionator. The reboiler duty no longer follows a linear relationship as it did with the above ethylene fractionator example. Instead, it follows an exponential relationship with the residual methane concentration. The temperature behavior is also opposite to the ethylene fractionator example, with the control tray temperature losing sensitivity to heat duty beyond 93% of the reboiler capacity. Figure 2.45 suggests temperature control for residual

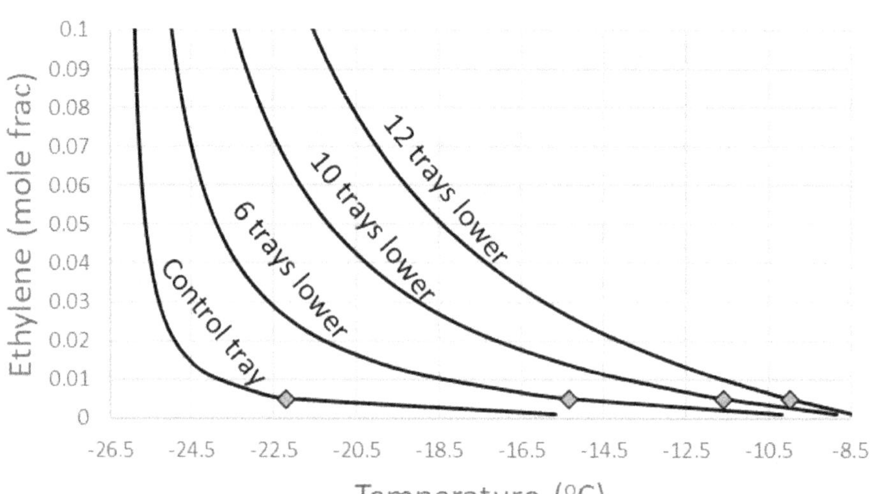

FIGURE 2.44 Ethylene Fractionator Control Tray Temperature Location and Set-Point versus Tower Bottom Residual Ethylene.

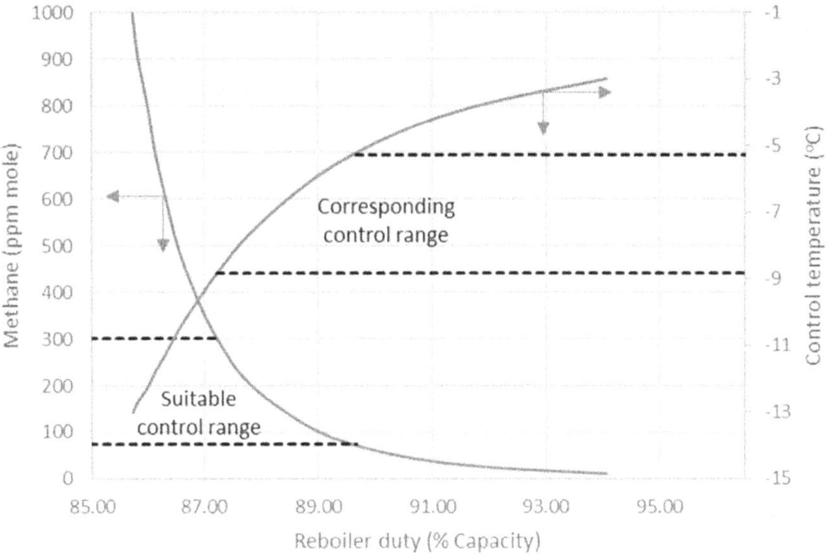

FIGURE 2.45 Demethanizer Pre-Stripper Reboiler Duty versus Residual Methane and Control Temperature [75].

Reproduced with Permission from figure 2.16, Distillation in Ethylene Plants – An Operators Overview, Andrée Bernard, 2022 AIChE Spring Meeting

methane concentration between 90 and 300 ppm mole. As for the ethylene fractionator example, targeting a cleaner bottom stream may require to ratio the reboiler heat with the tower feed if the feed composition is constant, or with the use of analyzers and APC adjusting the ratio based on feed rates and composition.

Figure 2.46 illustrates the temperature profile of a tower where the control tray temperature is located in a section of the tower that does not have any sensitivity. The temperature profile is flat regardless of the operating mode of the tower. In addition, a slight increase in temperature in the bottom last few trays indicates the presence of a heavy tail. As such, an increase in temperature SP would push this heavy tail higher up in the tower and create unnecessary internal hydraulic circulation. Operation process engineers and process operators rarely get to see or understand the full temperature profile of the tower as they are limited to the temperature indication provided in DCS. The total tower temperature profile can only be determined by process simulation, which is discussed in the next section.

For fixed conditions and composition, feed flow rate variations affect a tower operation from a hydraulic perspective. Although it may affect lags, required reflux ratios, control tray temperature should remain similar. Automatic controls may however become ineffective for feed flow rates resulting in internal hydraulic circulation outside the internals' (tray or packing) stable operating range. For instance, low rates experience during startup transition may result in hydraulic circulation below the turndown of the internals. Those conditions should be recognized by process

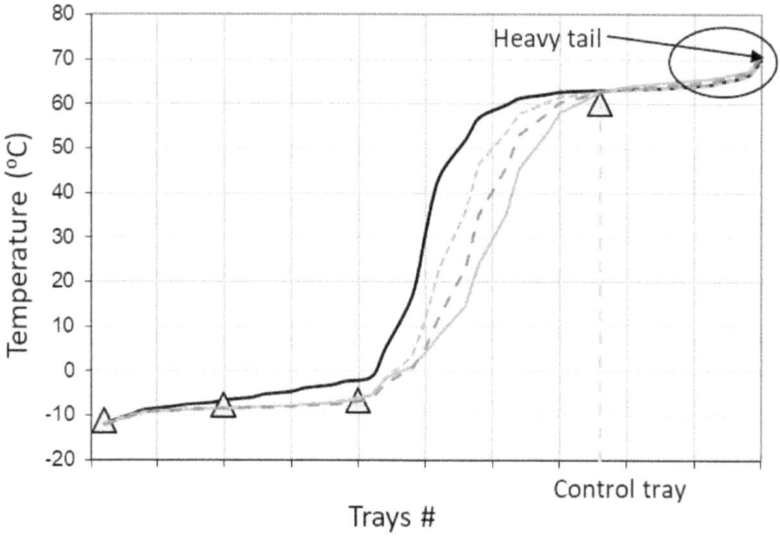

FIGURE 2.46 Insensitive Control Tray Temperature.

engineers who should provide operating conditions allowing stable operation. Below turndown, the tower should be false loaded hydraulically to bring conditions above minimum stable limits. Hydraulic false load is defined as operation at higher reflux ratios for satisfying the minimum operating conditions of the internals rather than meeting product stream specifications. This can be accomplished by decoupling the reboiler heat stream flow controller from the tray control temperature. The reboiler heat stream flow rate should then be controlled on flow control with SP calculated for exceeding the internals turndown limits. The reflux flow rate SP must also be calculated for aligning with those conditions. It is often required to raise the reboiler heat stream and reflux flow rate by step increment, as large change may imbalance the tower. This is typically observed by starvation or flooding of the liquid inventory in the tower sump and/or reflux drum. High feed flow rate, exceeding the capability of the internals, would, on the other hand, be associated with flooding. Flooding is typically associated with a loss of efficiency and higher pressure drop. This can be observed by difficulties in meeting product stream specifications. As the tower approaches flooding, the reduction in efficiency may cause the control system to aggravate the conditions as it may struggle in meeting SPs. For example, the control tray temperature may release a larger output to the reboiler heat stream flow controller. With the reflux on flow control, the additional heat load to the tower may cause an off-specification excursion of the distillate product, to which the process operator would respond by increasing the reflux flow rate. This would aggravate the extent of flooding with further loss in efficiency.

The feed vapor fraction may have implications on the tower hydraulic and controls. In many cases, two-phase feeds are the result of flashing across control valves between systems of different pressures. In such an arrangement, the vapor fraction of the feed is not expected to exhibit any significant changes. However, in other systems where feed preheaters are used, often as part of heat integration strategies, variability in feed vapor fraction can be significant. Large variation in feed vapor fraction can shift hydraulic loadings between sections separated by the feed point. This is illustrated by the impact that the feed vapor fraction has on the *q*-line on the McCabe–Thiele diagram. Figure 2.22 illustrates this behavior with reference to the feed liquid fraction. Change in *q*-line changes the slope of the operating line (L/V ratio), hence the internal circulation. A higher feed liquid fraction must be compensated by a higher reboil ratio. Contrarily, a lower liquid fraction will cause a higher vapor rate in the rectifying section, which needs to be compensated by a higher reflux ratio. From a control perspective, this variability will be associated with changes in control temperature SP and reflux rate or reflux ratio (if ratioed to the feed). From an operation perspective, varying the feed vapor fraction can be intentionally used for balancing loads between the section separated by the feed. For instance, if the rectifying section operates near flooding and spare capacity is available in the stripping section, reducing the feed vapor fraction will offload the rectifying section, hence providing a boost in capacity.

The sensitivity of the control and manipulated variable is rarely evaluated at the design level. The process design engineer would run tower simulation with stream purity set for meeting operation and sales requirement. However, as discussed earlier, neglecting this analysis may cause operational and control difficulties. Poor thermowells' location may render temperature control impossible. Insufficient instrumentations such as absence of reflux flow metering make it impossible to effectively control and assess tower performance. Although most process control engineers are chemical engineers by formation, they rarely spend time for fully understanding the unit they are attempting to control. The typical practice is to create a step change on the manipulated variable and observe the response of the control variable. Those are referred to as bump or step testing. Collaboration between the process and process control engineers should be encouraged with the use of process simulation for clearly understanding the impact of operating conditions on the tower temperature and composition profiles.

Achieving the above objectives requires stable operating pressure and closure of the material and energy balances. The tower pressure is critical due to its influence on many parameters involved in distillation, such as relative volatilities between the components, densities, condensing and boiling temperatures. Unstable pressure would cause variability in energy balance and hydraulic loadings resulting in unstable tower operation and could also result in premature flooding. Moreover, the operating pressure has a significant impact on tower capacity. While some practitioners claimed capacity increase by operating at higher pressure, others claim the opposite. The relationship between pressure, relative volatilities, and densities is primarily responsible for this gap. While higher operating pressure increases densities, which would intuitively reduce volumetric traffic for the same mass balance, the gain can be offset by the change in relative volatilities. As previously discussed, increasing pressure is typically associated with a reduction in relative volatilities, hence higher required reflux ratio. Liu and Jobson [85] suggested that increasing pressure reduces the required tower diameter or provides capacity upside in low pressure systems ($P <$ 3 bar). On high-pressure system ($P >$ 7 bar) they stated that it would require a larger tower diameter or capacity reduction on an existing tower. Their analysis suggested little impact at intermediate pressure (i.e., $3 < P < 7$ bar). The ability to reduce operating pressure is limited by the cooling stream supplying the overhead condenser. With cooling water, operation may take advantage of colder weather to operate the tower at lower pressure when deemed beneficial. There is less opportunity for optimization with refrigeration systems as refrigerant temperatures are dictated by the refrigeration system. Energy optimization may be considered at low rates but are typically limited by the refrigeration compressor's low operating limits, which involve false loading through anti-surge valves. Sloley [86] describes different pressure control strategies, which are based on adjusting the material or energy balance. For instance, in towers withdrawing a vapor distillate product, pressure can be controlled by the draw rate. Adjusting the energy balance is typically done by varying the cold stream flow rate and/or varying the cold side heat transfer area. The heat transfer area can also be manipulated from the process side. This would be typical with cooling water exchangers as regulating the cooling water flow may promote fouling resulting from lower tube velocities.

In process control, the location of the transmitter providing the process variable to the primary controller depends on the value product being recovered. For distillate products, it is preferable to have the pressure transmitter located on top of the tower or overhead line. However, for convenience and ease of access for maintenance, the pressure transmitter can also be located at the reflux drum. Locating the pressure transmitter in the tower bottom is not desirable with a distillate product, as the pressure drop will cause pressure fluctuation where the product is being withdrawn. Location in the tower bottom is recommended when the value product is collected in the bottom.

Condenser's operations are closely involved in tower pressure control scheme based on energy. Unlike reboilers, condenser operations are relatively trouble free. Operational and pressure control problems associated with condensers are frequently related to accumulation and difficulties in displacing non-condensable. Among the different pressure control strategies, Kister [87] indicated that the hot vapor bypass schemes account for about one-third of problems reported. As for accumulation of non-condensable, those problems are typically associated with poor or inadequate design. Bernard [88] discussed a case where over surface of core in kettle condensers on a parallel ethylene fractionator resulted in pressure control difficulties. The parallel ethylene fractionator was sharing a product drum with the primary ethylene fractionator, and by doing so imposed a minimum operating pressure for keeping the overall system in balance. The excessive heat transfer area by itself would not have been a problem if the tower had been decoupled from the primary tower. Stable pressure controls would have been achieved by operating at a lower pressure than originally targeted.

Simulation is a valuable if not essential tool for troubleshooting distillation tower performance. Towers have limited number of instrumentations. Other than a few temperature and pressure measurements at specific location, most of the measured process variables are external to the tower. There are no physical measurements providing internal hydraulic circulation, detailed temperature, composition profiles, and vapor/liquid properties. Graphical methods such as McCabe-Thiele and

Ponchont-Savarit can be used for estimating some of those process variables but are not as accurate, limited in terms of information provided, and are more tedious to use. Those methods are limited to internal vapor/liquid molar rates estimates and stage temperature (from T–x–y diagram). Being methods developed for binary systems, they are limited to prediction of light and heavy key components composition. They also require knowledge of the vapor/liquid equilibrium data. Process simulations, on the other hand, when properly configured can provide a relatively accurate picture of the tower operation ranging from stage-by-stage temperature, pressure, and flow rates, to compositions, thermal and transport properties. Simulation outputs can be used for calculating process loads, which can be plotted on tray operating windows or used as inputs in tower internals (i.e., trays or packing) rating software. Running simulations at two different feed rates provides the data required for plotting an operating line on tray operating windows, which helps in estimating the expected operating range of trays. Doing so also helps in identifying the most likely flood mechanism. Operating lines for different operating modes can also be established; for example, the impact of feed composition or feed vapor fraction on the operating line and tower expected capacity can be evaluated. A simulation provides a non-destructive test ground for evaluating operation strategies and potential root causes affecting performance.

To be effective, simulations must provide accurate predictions of the many process variables influencing the performance of a distillation tower. Inexperienced users are frequently overwhelmed by the many inputs required before they can hit the run icon. To speed up the work, they typically accept all the defaults present in the simulator package. Simulation is a tool to help and should not take away the process engineering knowledge and thinking process. Before jumping in setting up a simulation, a simplified process schematic of the tower with associated ancillaries and all process tags available from the data historian should be put together. Any stream laboratories analysis should be gathered. Reviewing tower drawings will help in locating the thermowells which will become handy for tuning the simulation temperature profile. The tower schematic should include the tray count, feeds, and product draw locations. At this point, there should be sufficient data for performing a rough material balance. The material balance can be documented on a piece of paper or digitally on a Microsoft Excel® worksheet. There are advantages in using spreadsheets such as Excel, as the material balance can be easily changed for different conditions and, if properly filed, easily retrievable. Furthermore, many data historians have connectivity with Microsoft Excel through Object Linking and Embedding (OLE). The above data gathering exercise provides most of the initial data required for configuring a simulation. The streams analyses should provide an initial guess on the components that should be included in the simulation.

The next step involves selecting the vapor/liquid equilibrium method. Equation of state (EOS) and activity coefficient methods are the most commonly used with organic mixtures [89] [90]. EOS such as Soave–Redlish–Kwong (SRK) and Peng–Robinson (PR) are the most popular with hydrocarbons. Polar components, such as organic acids and alcohols, are known to exhibit strong interaction due to hydrogen bonding and as such exhibit significant deviations from ideal mixture in the liquid phase. Azeotropic equilibriums are commonly the result of such interactions.

Activity coefficient methods such as Wilson, NRTL, and UNIQUAC are more suited to deal with the non-ideality resulting from those interactions in the liquid phase.

Without going into a deep dive of fundamental thermodynamics, understanding the origins and principles of vapor liquid equilibrium calculation methods helps in troubleshooting simulations that are incapable of providing realistic match with plant data. Thermodynamically, a mixture is in vapor/liquid equilibrium when the vapor fugacity (f_i^V) equals the liquid fugacity (f_i^L). This equality is described by Equation 2.140 in terms of fugacity coefficient and Equation 2.141 in terms of activity coefficient. At equilibrium, the system pressure is equivalent to the vapor pressure of the mixture for molar composition equivalent to x_i and y_i. The concepts of fugacity and activity are often confusing and misunderstood by many process engineers because they cannot be physically measured. It has been introduced by thermodynamicists as a correction for non-ideality:

$$f_i^V = y_i \varnothing_i^V P = x_i \varnothing_i^L P = f_i^L \tag{2.140}$$

$$f_i^V = \gamma_i^V y_i f_i^{oV} = \gamma_i^L x_i f_i^{oL} = f_i^L \tag{2.141}$$

For gases following the ideal gas law, the fugacity coefficients $\varnothing_i^V$ is equal to 1. In such cases, the vapor fugacity becomes equal to the partial pressure (x_i) as defined by Raoult's law, shown by Equation 2.142. For ideal liquid solution, the activity coefficient γ_i^L is equal to 1 $\left(\gamma_i^L = \frac{P_i}{x_i P_i^*} = 1\right)$. This then satisfies the right-hand side of Raoult's law, which correlates the partial pressure to the liquid molar fraction (x_i) and vapor pressure $\left(P_i^*(T)\right)$ of component *i*:

$$P_i = y_i\ P = x_i\ P_i^*(T) \tag{2.142}$$

Raoult's law is generally valid when $x_i \rightarrow 1$, with the Antoine equation typically used for evaluating the component vapor pressure. Therefore, when applicable, it provides a vapor/liquid equilibrium relationship. As gases and liquids depart from ideality, the vapor and liquid, respectively, fugacity coefficients are no longer equivalent to unity and must be evaluated. Classical thermodynamics define the vapor fugacity coefficient as a function of pressure, temperature, and volume as described by Equation 2.143:

$$\varnothing_i = exp\left\{\frac{1}{RT}\int_V^\infty \left[\left(\frac{\partial P}{\partial n_i}\right)_{T,V,n_j} - \frac{RT}{V}\right] dV - \ln Z\right\} \tag{2.143}$$

The pressure term in Equation 2.143 can be substituted by a *P–V–T* relationship such as SRK, PR, or others. As an example, using SRK, defined by Equation 2.144:

$$P = \frac{RT}{v-b} - \frac{a \propto}{v(v+b)} \tag{2.144}$$

where parameter a and b are:

$$a = 0.4278\frac{R^2T_c^{2.5}}{P_c} \text{ and } b = 0.0867\frac{RT_c}{P_c} \tag{2.145}$$

and

$$\propto = \left[1 + m\left(1 - T_r^{0.5}\right)\right]^2 \tag{2.146}$$

$$m = 0.480 + 1.574\omega - 0.176\omega^2 \tag{2.147}$$

where R is the universal gas constant, T_c and P_c are the critical temperature and pressure, T_r is the reduced temperature $\left(T/T_c\right)$, and ω the acentric factor. The expression of the vapor fugacity coefficient can be reduced to Equation 2.148 after combining Equations 2.143 and 2.144:

$$\phi_i = \exp\left((Z-1)\frac{B_i}{B} - \ln(Z-B) - \frac{A}{B}\left(\frac{2A_i^{0.5}}{A^{0.5}} - \frac{B_i}{B}\right)\ln\left(\frac{Z+B}{Z}\right)\right) \tag{2.148}$$

The constant A_i, B_i, A, and B are obtained from the following expressions:

$$A_i = 0.42747 \propto_i \frac{P_{ri}}{T_{ri}^2} \tag{2.149}$$

$$B_i = 0.08664\frac{P_{ri}}{T_{ri}} \tag{2.150}$$

$$A = \sum_{i=1}^{C}\sum_{j=1}^{C} y_i y_j A_{ij} \tag{2.151}$$

$$A_{ij} = \left(A_i A_j\right)^{0.5} \tag{2.152}$$

$$B = \sum_{i=1}^{C} y_i B_i \tag{2.153}$$

A_{ij} can be expressed in terms of the binary interaction parameter k_{ij} as shown below:

$$A_{ij} = \left(1 - k_{ij}\right)\left(A_i A_j\right)^{0.5} \tag{2.154}$$

The intent of the above demonstration is not for the reader to perform those calculations but to understand the data being used by process simulators for resolving them. EOS are used for predicting the vapor/liquid equilibrium composition, the enthalpies, and molar volumes. As described by the above, solving those equations relies on pure components' acentric factor, critical pressure, and temperature. While the parameter A_{ij} can be calculated from critical temperature and pressure described by Equation 2.152, the use of binary interaction parameters calculated from experimental data is preferred especially for light gases and non-ideal systems.

It may be argued that since pure component acentric factor, critical temperature, and pressure are constant, there should not be any concerns. Unfortunately, the number of those constants varies depending on sources. Literature [91, 92] reports a case where simulation of a distillation tower for the recovery of 1,3-butadiene predicted erroneous results. 1,2-Butadiene, less volatile than 1,3-butadiene, is expected to be recovered in the bottom of the tower with 1,3-butadiene recovered as distillate product. The simulation erroneously predicted recovery of 1,2-butadiene in the distillate product. The cause of this problem was related to incorrect acentric factor, critical pressure, and temperature used by the simulation program for 1,2-butadiene. Differences in pure components constant is not unusual. Table 2.15 compares the acentric factor ω and critical constants for ethane and ethylene extracted from different references. The above anomalies bring an element of caution when using binary interaction parameters extracted from literature or different simulation software. As per the above discussion, binary interaction parameters are calculated from measured vapor/liquid equilibrium composition at temperature T and pressure P using Equations 2.144–2.154. As such, they are calculated based on pure components, ω, P_c, and T_c values. Binary interaction parameters k_{ij} use in simulations must therefore

TABLE 2.15
Ethylene and Ethane Critical Temperature, Pressure, and Acentric Factor versus Data Source

	Ethylene		
Reference	T_c **(K)**	P_c **(psia)**	**Ω**
NIST data (Aspen)	282.34	731.1439	0.086407
Properties of Gases and Liquids, 5th Edition, Poling B, Prausnitz J, O'Connell J.	282.34	725.904	0.087
Equilibrium-Stage Separation Operations in Chemical Engineering, Henley E, Seader J.	282.72	742.2	0.0872
VLE data at high pressure, Ohe	282.7	733.53	0.084
	Ethane		
	T_c (K)	P_c (psia)	Ω
NIST data (Aspen)	305.37	708.3972	0.10016
Properties of Gases and Liquids, 5th Edition, Poling B, Prausnitz J, O'Connell J.	305.32	701.568	0.099
Equilibrium-Stage Separation Operations in Chemical Engineering, Henley E, Seader J.	305.2	709.8	0.1064
VLE Data at High Pressure, Ohe	305.5	708.54	0.097

be consistent with the pure component constants (i.e., ω, P_c, T_c) from which they originate.

Literature [93] states that EOS provide poor predictions on liquid densities. Many simulators use an API correlation for pseudocomponents and the Rackett model for real components as default method. Literature [93] states that liquid densities can be calculated by two methods, both based on the principle of corresponding states. The first methodology involved prediction of the liquid density at saturation pressure followed by a correction for the actual pressure. The Rackett and COSTALD methods are reported as belonging to this category. The first methodology is also reported to be limited to reduced temperatures of 0.98. Above T_r of 0.98, liquid density prediction can be done through EOS such as Lee–Kesler. Literature [93] reports that COSTALD and Lee–Kesler methods are appropriate for high-pressure hydrocarbons applications.

Similar demonstration can be done for equilibrium of polar components which deviate from ideality in the liquid phase due to hydrogen bonding. As stated earlier, such mixtures are typically modeled with activity coefficients methods such as Wilson, NRTL, or UNIQUAC. As for EOS, like SRK and PR, activity coefficient-based methods rely on experimental data for evaluating associated binary interactions parameters. Unlike SRK and PR, activity coefficient-based methods require calculating pure components' vapor pressure. As stated earlier, the Antoine equation is frequently used for that purpose. As shown by Equation 2.155, the Antoine equation also relies on coefficients (*A*, *B*, and *C*) that are part of the simulator databank. As for critical properties and acentric factors, those coefficients can also exhibit variability depending on their source:

$$log_{10} P^* = A - \frac{B}{T + C - 273.15} \tag{2.155}$$

The takeaway from the above discussion is that simulation of vapor/liquid equilibrium relies on a consistent set of data. Differences, while in many cases expected to be small, would likely exist between simulators using different pure component databanks. This is why, as previously discussed, fractionation efficiency is specific to the VLE data for which it has been calculated. Calibrating the model fractionation efficiency is discussed below.

Once the vapor/liquid equilibrium method is selected, it is a good idea to generate a binary equilibrium curve. This is straightforward for binary system as most process simulators can generate a *T–x–y* curve. For multicomponent system, a pseudo binary system can be created by running a distillation simulation. An *x–y* and *T–x–y* plot can be created by lumping light components together from each separation stage. This may help, especially with azeotropic equilibrium, in providing realistic estimate during the first few runs.

Selecting the number of theoretical stages without knowing the efficiency is a challenge and requires a trial-and-error process. At this point, the objective is to get a converge run such that it be used for further tuning. To do so, using a global efficiency between 40% and 70% can be done as a first guess. Note that the material balance conducted before configuring the simulation becomes handy as it provides some initial conditions that may facilitate converging the simulation run. Once a converge run is obtained, the O'Connel correlation described by Equation 2.110 can be used for providing a better efficiency estimate. At this point, for simplification, the same efficiency should be applied to the entire tower. In doing so, the property used for the O'Connel correlation should be averaged. Note that when working with global efficiency (i.e., theoretical stages/tray count), the feed points need to be re-adjusted when changing the stage count. Some may prefer using the Murphree efficiency as they can input the actual tray count and just vary the efficiency. Note that this practice reduced the area between the pseudo equilibrium curve and diagonal on the McCabe–Thiele diagram and as such restricts the range of convergence. This could be more problematic on systems having low relative volatility such as ethane/ethylene and propane/propylene. Braswell et al. [94] evaluated how the Murphree efficiency can cause deviation from fundamental thermodynamics. For instance, mathematically, using the Murphree vapor efficiency described by Equation 2.101, the actual vapor composition would be less than predicted by the equilibrium. This means that on a *T–x–y* diagram, the vapor composition would fall in the two-phase area. The predicted vapor temperature leaving the tray would then be lower than the dew point. In reality, it is possible that the vapor temperature leaving the tray might not match the dew point, but it cannot be lower. The pseudo equilibrium illustrated in Figure 2.28 is consistent with how chemical engineering students are introduced to the concept of Murphree efficiency. However, Braswell commented on an efficiency pinch when mathematically computing the Murphree efficiency. When the Murphree vapor efficiency is less than 1, an efficiency pinch will occur as illustrated in Figure 2.47. Based on the above discussion, best practice would be to use the global efficiency concept when conducting process simulations of distillation tower.

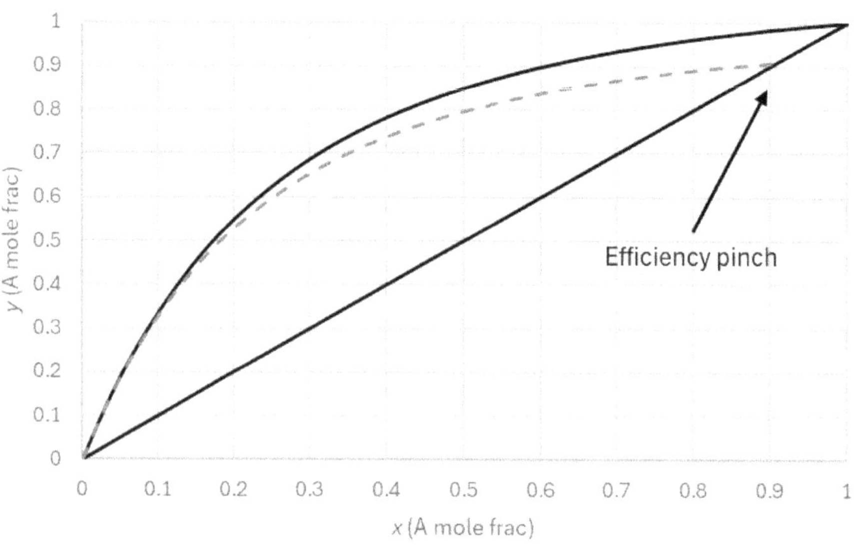

FIGURE 2.47 Murphree Efficiency Pinch.

With a converged solution, the next step is to tune the number of theoretical stages (or efficiency) to match the tower performance. In a first step, the stage count is manipulated for matching the reflux rate and reboiler duty. In some cases, a good match can be obtained with the same global efficiency throughout the tower. Obtaining a good match on one with a significant deviation on the other is an indication that the tower may exhibit different efficiency between top and bottom. The temperature profile is another indication that should be used for balancing the simulation. It is not unusual that differences in efficiency occurred around transition points, such as feed, product draws, and/or pump-around circuit. Significant difference in tray geometry between tray sections, especially around flow path length, could likely cause differences in efficiencies. Once a decent match is obtained, the simulation model should be used for plotting the reflux ratio as a function of the distillate specification. As illustrated in Figure 2.48, the plot on the left shows good agreement with the simulation, while the plot on the right suggests that the distillate composition is insensitive to the reflux. In fact, it trends in the opposite direction, where higher reflux ratio seems to increase the residual concentration of heavy components. This is symptomatic of a tower not working properly potentially due to flooding. Gamma scans would therefore be highly recommended prior to proceeding any further with troubleshooting.

Thermal and transport property predictions are another aspect that needs to be considered. Default property methods in commercial simulators are not always appropriate for the mixture being evaluated. Given that one of the main uses of process simulators is for providing tray hydraulic loadings, error in density predictions would mislead troubleshooting relative to operating point on tray operating windows. Good accuracy on liquid surface tensions is also critical when performing rigorous flood calculations. Viscosities are frequently required in rating software for assessing downcomers. Commercial simulators typically contained several property methods for each property being evaluated. Consulting experts may help in selecting the appropriate property methods. Trial and error of different property methods and comparison to reliable data is another methodology. Reliable data can be in the form of using the original plant heat and material balance from a trusted licensor and using the data as benchmark. Another methodology would be to search literature and ensure the selected property method matches the data. As a last resort, those properties can be measured in laboratory, which could be challenging, at times depending on the operating conditions, for example, for high-pressure systems operating under extreme temperatures.

Simulation outputs provide a lot of data that can be cumbersome to analyze. Many commercial simulators have OLE compatibility, which gives the ability to import simulation results in a spreadsheet environment. Good practice would be to generate a stage-by-stage table, which contains all the variables required for tray or packing ratings. These would normally include for each theoretical stage the correlated tray number, temperature and pressure, molecular weight, mass and/or molar flow rate, density, and viscosity for both vapor and liquid. Surface tension should also be reported for the liquid phase. Additional columns can be configured for calculating the vapor and liquid loads in the form for which tray windows have been developed (e.g., liquid volumetric rate vs. C_{factor}). By doing so, process loads can be plotted on the tray window for the case study being evaluated. Additional worksheets can

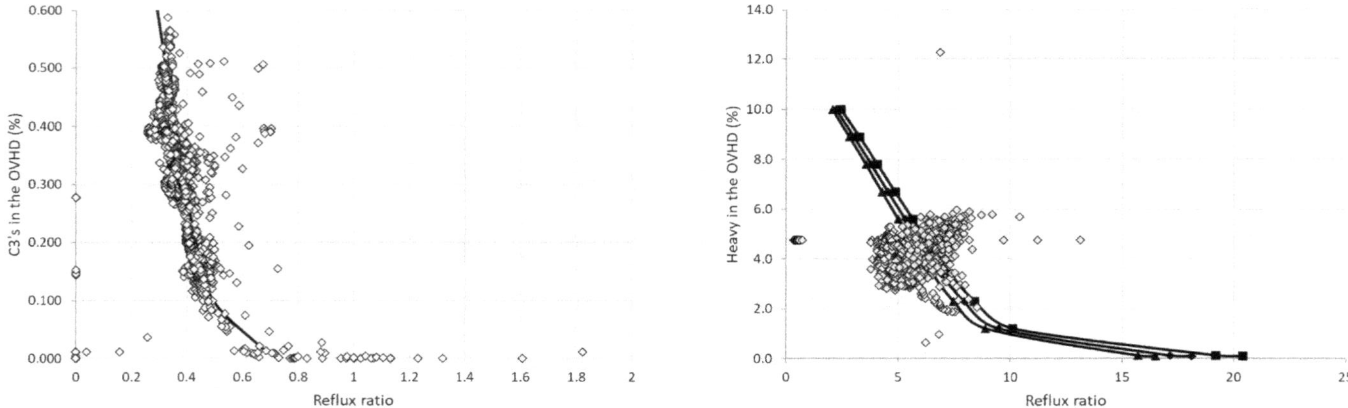

FIGURE 2.48 Distillate Composition as a Function of the Reflux Ratio [75].

Reproduced with Permission from figure 2.9, Distillation in Ethylene Plants – An Operator's Overview, André Bernard, 2022 AIChE Spring Meeting

be used for importing stage-by-stage molar composition. From this data, a pseudo equilibrium curve can be plotted by summing up the light component mole fraction in the liquid phase (x axis) and vapor phase (y axis). A McCabe–Thiele plot can be sketched out with the operating lines defined by joining the points (x_n, y_{n+1}) and the steps representing each stage by the succession of points $(x_n, y_n), (x_n, y_{n+1})$ as illustrated in Figure 2.49 for a column with three feeds. As previously discussed, the McCabe–Thiele diagram provides a visual representation of the tower operation which helps in identifying potential pinch points and operating conditions reaching thermodynamic limitations. The diagram illustrated in Figure 2.49 suggests, for instance, that the feed points are in a non-optimal location. The change in slope of the operating line shows the actual location of the feeds as calculated by the simulation compared to those calculated by the q-line. It is clear that the ideal thermodynamic location for the feeds, as illustrated by the q-lines, does not match the actual location. Not feeding in the optimum location causes re-mixing or dilution at the feed tray, which is compensated by higher reflux and/or reboil ratio, hence higher hydraulic loadings.

A worksheet should also be included for a tower material balance, including stream composition and properties. It is not unusual in troubleshooting to gather trays or packing rating sheets. However, the tower material balance basis for those rating sheets is not always clearly stated, making it difficult to relate to the operation mode.

As described earlier, the development of tower simulation models can be time consuming and tedious. Proactive facilities would already have developed those models for evaluating operation or for design purposes. If not already developed, with the plant experiencing significant throughput setbacks due to unusual tower

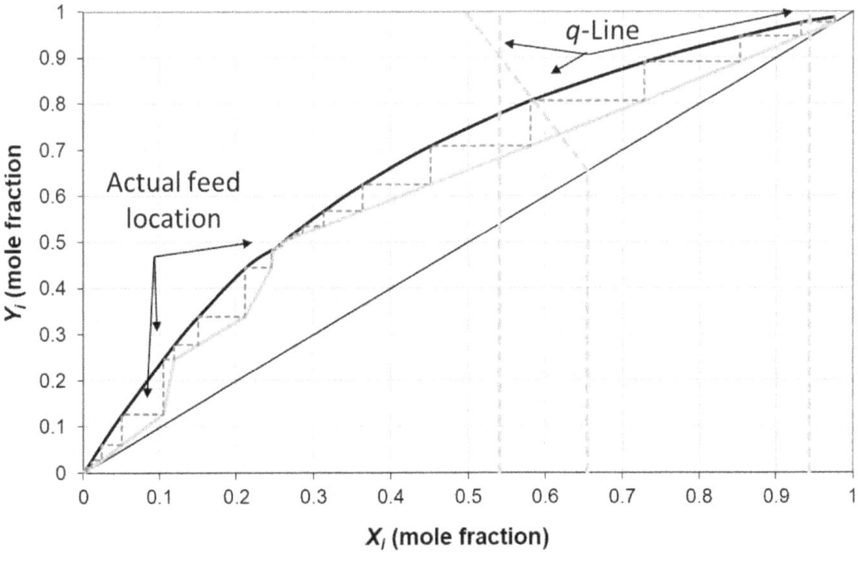

FIGURE 2.49 McCabe–Thiele Diagram Build from Process Simulation.

performance, model development would still be time well spent. This would minimize or eliminate guessing root causes, which can often consume more time and resources not counting unnecessary downtime if the tower is brought down for inspection. Once developed, those models should be stored in a location where they can be easily retrievable and re-used when experiencing other problems. Process engineers should also be encouraged to use those models for assessing tower performance and optimizing process conditions.

In distillation troubleshooting, simulation models are not limited to the tower. Modeling can be extended to more rigorous reboiler thermosiphon in conjunction with the tower.

Gamma scan is a technique based on absorption of radiation when gamma rays emitted from a source travel through a medium, with unabsorbed radiation measured by a detector. The location of the source and detector defines a linear path followed by the gamma ray, often referred to as the scan line or chords. The amount of unabsorbed radiation can be calculated from the following expression:

$$I = I_0 \, e^{(-\mu \, \rho \, x)} \tag{2.156}$$

With I the intensity of radiation reaching the detector (count/s), I_o the radiation intensity at the source (count/s), ρ the density of the medium (lb/ft^3), x the thickness of the medium, and μ the absorption coefficient (ft^2/lb). Gamma scans typically use radioactive sources between 500 and 2,500 keV (kiloelectronvolts). Above 200 keV, the energy of the gamma ray becomes independent of the chemical composition and the term μ can be neglected in Equation 2.156 [95–97]. The radiation absorption then becomes a direct function of the density and thickness of the material.

Gamma scans are extremely useful for troubleshooting distillation towers. One important aspect in performing gamma scans is to measure a clear vapor space in the tower, which becomes a reference point. The tower overhead space or above the reboiler return are typical areas where a clear vapor space is present. Good collaboration between the process engineer and technician performing the scan is also critical. The position of the source and detector will define the scan line and requires direction from process engineering relative to what is being evaluated (e.g., active area, downcomer). Common mistakes are to request gamma scans and let the technician set the equipment without much direction. Review of the tower and internal hardware (i.e., trays, distributor) drawings shall be done to ensure alignment with the information that is being sought and understanding of internal obstruction which will interfere with the scan. Review of access is also important; the locations of platform might not accommodate the desired scan line, which may require erecting scaffolding. External obstructions such as nozzles and piping must also be taken into consideration. Failing or disregarding the above may result in scan lines crossing downcomers and active areas, which would not provide any value in the troubleshooting initiative.

Gamma scans should not be the first step in a troubleshooting exercise. Understanding the deviations from normality is the first step. Whether the tower is not meeting product specifications or experiencing pre-mature flooding is a typical deviation requiring troubleshooting. H&MBs and hydraulic evaluation shall be conducted first as they would provide clues relative to the most likely root causes. Comparison with historical data will reveal if there is a deviation from normality or if the tower is being pushed outside previous experience. The case study presented in Chapter 1 is a good example of using a qualitative gamma scan for confirming suspicions of unsealed downcomers. In that example, while heat and material balances and hydraulic loadings seemed to fall within pre-outage experience, the tower bottom was irresponsive to reboil rate. The hydraulic loads, although within the stable region of the tray operating windows, were close to a low limit. The theory of unsealed downcomer was a speculation that could not be proven theoretically, and this problem had no previous history. For the unsealed downcomer speculation to be valid, a gamma scan would have to show unloaded trays in the bottom section of the tower. As such, scanning the active area of the tray was sufficient for validating the theory and proper action was taken on the tower for sealing the downcomers. In many cases, changes in tower behavior follow an abnormal transient condition, whether intentional such as re-start following a shutdown or non-intentional when recovering from a plant upsets. Startup malpractice as for the case study in Chapter 1, could result in poor tower performance

From a qualitative perspective, gamma scans can be used for validating the mechanical integrity of the internals such as all trays that are accounted for. Poor performance caused by flooding would be characterized by low recorded radiation intensity (i.e., high absorption) with the scan not returning to clear vapor between the space delimited by two trays. Gamma scan of a flooded tower does not provide useful information other than confirming that the tower is flooded. In such case, feed rate should be reduced for bringing the tower out of flooding conditions and gamma scans re-done. Figure 2.50 illustrates a typical gamma scan conducted on a tower equipped with single pass trays. A qualitative analysis suggests damage or potential missing tray manway on tray 13. While other trays can be located at the

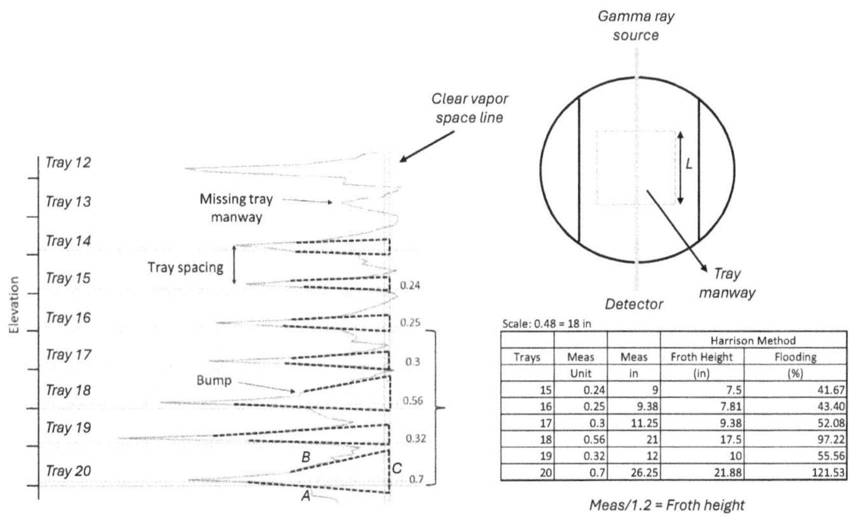

			Harrison Method	
Trays	Meas Unit	Meas in	Froth Height (in)	Flooding (%)
15	0.24	9	7.5	41.67
16	0.25	9.38	7.81	43.40
17	0.3	11.25	9.38	52.08
18	0.56	21	17.5	97.22
19	0.32	12	10	55.56
20	0.7	26.25	21.88	121.53

FIGURE 2.50 1 Pass Tray Active Area Gamma Scan Example.

radiation absorption peak, one may question why those peaks are not all equivalent. The maximum absorption is closer when comparing trays 14–17 than with trays 18 and 19. This highlights the importance of reviewing drawings and identifying potential source of interference such as nozzles or any other components that may cause interference. From a high-level qualitative observation, drawing conclusions might be speculative. At first site, for the non-experienced, one may conclude that this scan does not suggest signs of flooding. A quantitative analysis can be conducted using the Harrison method for froth height estimation. Figure 2.50 illustrates how the method can be applied using a single-pass tray as example. The method as described by Kister [96] involves identifying the clear vapor space line. The second step involves drawing tangent lines from each side of the peaks as illustrated by the dashed black lines A and B on tray 20 of Figure 2.50. The length of the vertical line connecting the tangents at the clear vapor line, shown as *C* on Figure 2.50, is then measured. The froth height can be estimated by dividing the length of the vertical line connecting the two tangents by 1.2. The process is repeated on trays 14–19 illustrated in Figure 2.50. When a bump occurs on the scan, as observed on tray 18, the tangent must be drawn from the bump. The Harrison method can be accomplished on a paper copy of the scan using a ruler or as illustrated in Figure 2.50, on an Excel spreadsheet on which the lines drawn are scaled using a reference. Tray spacing is a reliable reference, by measuring the vertical distance between the maximum radiation absorption resulting from the tray panel location. The percentage of flooding can then be calculated by dividing the estimated froth height by the tray spacing. The table on the right side of the scan illustrated in Figure 2.50 summarizes the calculations that can be included in the Excel spreadsheet. By doing this, it can be concluded that tray 20 operates beyond flooding and that tray 18 is also close to flood, while other trays operate with significant margin from flooding. This suggests that the tower operates at the early stage of flooding and may or may not show signs of performance degradation yet. However, further increase in feed rates would be associated with flooding propagation upward from tray 20 and 18, which should negatively impact fractionation efficiency.

Speculation on the missing tray manway can be validated using Equation 2.156. The ratio of the intensity of radiation I between a good tray and the suspected tray can be correlated to the difference in metal thickness that the gamma ray must travel across as described by Equation 2.157:

$$x_1 - x_2 = \frac{1}{\rho}\ln\left(I_1 / I_2\right) \tag{2.157}$$

Assuming a square tray manway of dimension L and a tower diameter D, x_1 is equivalent to D with the associated intensity of radiation I_1. The intensity of radiation of tray 13 would be I_2 with an associated metal thickness of D–L. Therefore, the speculation of a missing tray manway can be validated if $x_1 - x_2$ is equivalent or close to the tray manway dimension L.

The example illustrated in Figure 2.50 shows a single scan line and therefore provides information relative to the froth height across this specific line. A single scan line would therefore be insufficient for assessing the potential for maldistribution or channeling on the tray. This would require running a few parallel scan lines across the flow path of the tray. By doing so, the profile of the froth across the flow path length can be estimated and drawn on a tower tray diagram. Kister [96] introduced this practice and named those diagrams Kistergram.

As previously discussed, distribution on multi-pass trays is critical for good efficiency. Performing gamma scans of each active area of multi-pass trays is the only way that actual distribution can be assessed. Quantitative estimation of the froth height can then be used with rating programs or flood correlations for back calculating the *L*/*V* ratios of each pass. When possible, downcomers should also be scanned as they could be initiating flooding or maldistribution.

Gamma scans of packed towers are also useful in troubleshooting. For those inexperienced, absorption of the gamma rays by the mass of packing makes it difficult to draw conclusions. A dry scan before starting the tower can provide a baseline for the level of radiation absorption caused by the mass of packing. Still, without a baseline dry scan, a gamma scan can reveal qualitative information relative to missing packing, dislodge bed supports, damaged distributor, and areas of high density suggesting fouling or flooding. In fouling service, it is recommended to conduct annual or semi-annual gamma scan. A scan of the clean tower shortly after startup will provide a baseline for the froth density. Subsequent scan throughout the run length of the tower will provide qualitative information relative to the progress and location of fouling material accumulation. For tower exceeding 4 ft in diameter, it is recommended to run four scan lines that form a rectangular grid. Between 2 and 4 ft, triangular scan lines should suffice and a single scan line on tower of diameter smaller than 2 ft [95]. Kister [98] extended the application of pack tower gamma scans from qualitative to quantitative by extending Equation 2.156 with the term ρx subdivided in tower wall, packing, and liquid. Some calibration is required by calculating the tower wall component in the clear vapor space. This methodology does not rely on a dry scan of the tower.

2.1.3.2 Absorption

Absorption is very similar to fractionation as it is a vapor/liquid mass transfer operation. It also takes place in tower with internals similar to those used in distillation. In both applications, the internals must promote a high gas/liquid interfacial area. In an absorber, an environment where the liquid is the continuous phase and the vapor dispersed is more favorable than the opposite where the vapor is the continuous phase. While the intent is to maximize the vapor/liquid interfacial area, having the vapor as the dispersed phase maximizes contact of the solute with the solvent rather than having them confined in the bulk of a continuous gas phase. Unlike fractionation, mass transfer is unidirectional from vapor to liquid and rather than being based on vapor/liquid equilibrium; it is driven by solubility of gases in a solvent. Although reported as optimistic, O'Connel extended his efficiency correlation to absorption. Both O'Connel's correlations (i.e., fractionation and absorption) involve a liquid viscosity term, which is associated with the liquid-phase resistance to mass transfer.

Absorption is widely used for removal of acid gases such as CO_2 and H_2S for sweetening natural gas. It is also used in carbon capture for CO_2 removal in flue gas of fired equipment. Amines and caustic solutions are typically used in those applications. The viscosity of the solvent limits the mass transfer rate. As such, absorption calculations or simulation based on equilibrium tends to be optimistic where most of the gas transfer to the liquid phase happens within the first theoretical separation stage. Many practitioners therefore favor a rate-based approach in simulation rather than equilibrium stages. The rate-based approach relies on the mass transfer coefficient.

2.1.3.3 Liquid/Solids and Liquid/Liquid

Liquid/solid separations are based on density difference between the solids and the fluid as well as particle size. The magnitude of the density differential and particle size distribution would typically dictate the appropriate type of equipment for conducting the separation. This principle applies to settling, flotation, cyclones, and centrifuges. Solids can also be separated by filtration, which depends less on density differential but more on particles size. The filter media provide a mechanical barrier for capturing the solid particles.

Settlers, also called decanters, use gravitational forces for separating solids from a liquid stream. Density differential and particle sizes dictate the size of a settler. They are widely used in water and wastewater treatment, but also in hydrocarbon service where removal of gross solid particles is required. While they may not fully be removing solid particles, they would offload downstream filters when used together and extend their run length. The design of liquid/solid settler is based on the terminal velocity of the solid particles, which is obtained from a force balance around a particle free falling in a fluid. As illustrated in Figure 2.51, gravity exerts a downward force on the particles, which at equilibrium is counter balanced by frictional drag forces applied by the fluid on the particles:

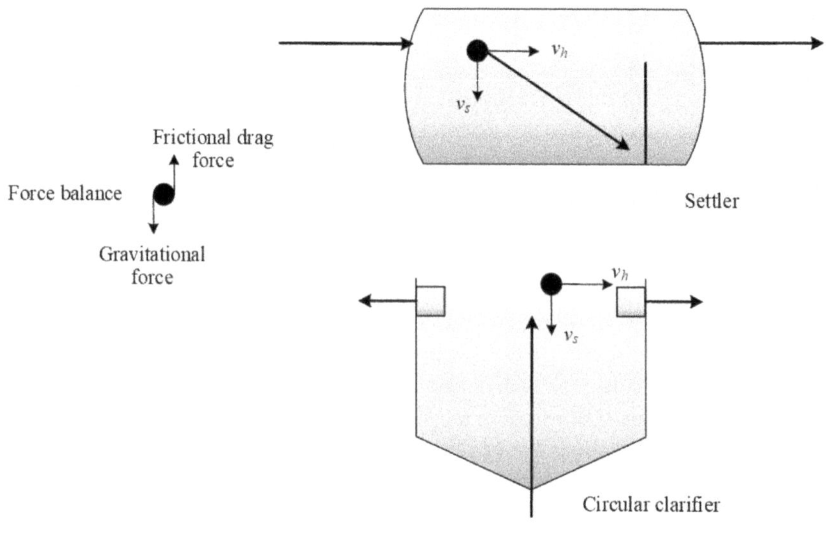

FIGURE 2.51 Settlers/Decanters.

$$Gravitational\ force = (\rho_s - \rho_{fluid}) gV \quad (2.158)$$

$$Drag\ force = \frac{C_d\ A\ \rho_{fluid}\ v^2}{2} \quad (2.159)$$

where C_d is a drag coefficient, V the volume of the particle, A the horizontal projected surface of the particles, and v the velocity. At equilibrium, the gravitational and the drag forces are equivalent with the associated velocity corresponding to the terminal velocity. Assuming spherical particles of diameter d and combining Equations 2.158 and 2.159 provide a relationship for the terminal velocity v_s:

$$v_s = \left[\frac{4}{3} \frac{g(\rho_s - \rho_{fluid}) d}{C_d\ \rho_{fluid}} \right]^{1/2} \quad (2.160)$$

The drag coefficient depends on the flow regime surrounding the particles but can be approximated by Equation 2.161 for spherical particles and Reynolds numbers less than 1×10^4:

$$C_d = \frac{24}{Re} + \frac{3}{\sqrt{Re}} + 0.34 \quad (2.161)$$

For Reynolds number less than 0.3, the drag coefficient can be approximated to a value of 24, which after substituting in Equation 2.161 gives Equation 2.162, known as Stokes' law:

$$v_s = \frac{g(\rho_s - \rho) d^2}{18\mu} \quad (2.162)$$

For a given fluid, the Stokes' law shows that the terminal velocity is influenced by the density differential between the particles and the fluid along with the particle's diameter. Figure 2.52 illustrates the impact of particle diameter and fluid flow rates through a settler of fixed dimensions. Assuming constant particles and fluid densities, the terminal velocity depends only on particle diameter. The fraction of particles settling down is determined by comparing the terminal velocity v_s to the fluid horizontal velocity v_h across the settler. Particles with terminal velocity superior to the fluid horizontal velocity will settle, while those of lower terminal velocity will likely be entrained. Therefore, at low rate, a settler will remove more particles than when operated at higher rates.

Designing or troubleshooting a settler performance requires understanding the particle's diameter distribution, the density of the solids forming the particles, and the total mass fraction of the particle contained in the liquid stream. Those characteristics are not always fully understood and frequently overlooked. Without data, design would be based on assumptions. Settling poor performance can be directly related to the validity of those design assumptions. In which case, proper

Part size	Weight distribution
(mm)	%
2.36	21.8
1.4	10
0.6	11.5
0.3	12.6
0.15	9
0.075	10.6
0.038	14.5
0.001	10

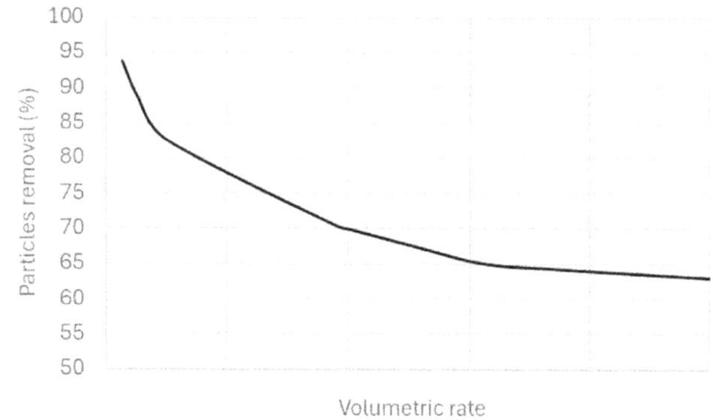

FIGURE 2.52 Impact of Particle Size and Flow Rate on Settling.

troubleshooting would involve sampling of the streams associated with the settler operation. Sampling by itself is a challenge in such system as the sampling methodology is critical for getting a representative sample. Collecting samples from a nozzle oriented perpendicular from the flow will preferentially extract liquid. A proper sampling nozzle would be facing the stream with the velocity inside the sampling nozzle equivalent to the stream velocity, which is referred as isokinetic sampling [99–101]. The fraction of solids collected would be less than in the process stream if the sample collection is done at a nozzle velocity higher than the process stream. Contrarily, if sampled at lower velocity, the fraction of solids would be higher. The same principles apply to gas/solid systems. It is fair to assume that many sample systems do not meet the requirements for isokinetic sampling. This obviously adds an element of complexity for troubleshooting liquid/solid settlers. Nonetheless, alternatives such as settling experiments (jar testing) in laboratory can be used to test sensitivity of the settler design relative to particle size distribution and fraction. Backward correlation to the actual system performance can then be approximated.

The required volume of settlers can take substantial plot space, which may often constrain their size. Removal capabilities can sometimes be increased with internals such as a series of angle baffles, which provide mechanical obstacle for the particles, thus limiting their ability to follow the process fluid flow path but also to agglomerate. Chemicals can also be used to promote particle agglomeration, hence contributing to increasing their diameters and improving settling capabilities.

Some particles, such as coke, are porous in nature and can contain adsorbed gases within their pores. This contributes to reducing the bulk density of the particles. As described by the Stokes' law, reduction in density differential, in turn, reduces the terminal velocities, thus affecting the settler performance. Depending on the magnitude of gas adsorption, the particles will eventually float when the density differential becomes negative.

The Stokes' law also applies to liquid–liquid separation for immiscible liquids. In such application, the diameter d of the dispersed phase bubbles is used in Equation 2.162 with the fluid associated density for ρ_s. As for solids, the extent of separation depends not only on the density differential and bubble diameter but also on a critical mass of the dispersed phase fluid beyond its solubility; that is, for separation to occur, the two fluids must be present at composition creating two distinct liquid phases. Sufficient volume of the immiscible fluids must be present for establishing a stable liquid–liquid interface.

Problems frequently associated with liquid–liquid separation are caused by operation below the immiscibility concentration, or small density differential between the two fluids. Those problems are inherent to the fluid properties and can be resolved by addition of compatible streams that will enhance separation. Gasoline–water separation in ethylene plants using ethane as feedstocks is a good example of low gasoline/water ratio and small density differential. Bernard [102] demonstrated by simulation that recycling 10% of a debutanizer bottom stream rich in benzene would increase the density differential by about 14%. The rate of the recycle stream also increases the gasoline/water ratio by a factor of about 3.4. While recycles contribute to increasing loads on process units, they could enhance separation that would otherwise result in poor performance. However, caution must be exercised with recycle streams as they also contribute to increasing residence time and concentrating impurities that may participate in chemical reactions responsible for accumulation of fouling material in process vessels. Emulsion is another problem frequently encountered in liquid–liquid separation. The presence of coke particles, sulfur components, and high pH often contributes to emulsions in gasoline–water separation. When possible, this problem should be worked at the source by preventing streams promoting emulsions from entering the settler. pH control is also critical in such systems. The use of chemicals as emulsion breaker is a common practice. Dosage must be done correctly as literature [103] has reported cases where overdosage has contributed to the emulsion rather than breaking it.

As for solids, the droplet diameter of the dispersed phase influences settling. For small droplet, this problem is frequently solved with a coalescing element, which provides a physical barrier promoting the droplets to meet and grow in size.

2.1.4 Heat Exchangers

As previously stated, a chemical process can be subdivided into reaction and separation. Both activities however involve the transfer of heat, which is done through heat integration between process streams or provided from utilities. The most common utilities being steam, cooling water, and in some cases may include refrigeration systems and thermal fluids. Heat exchangers involved in boiling and condensing service have been discussed in detail in the distillation section. The reader is encouraged to refer to that section, while the following discussion will be limited to high-level heat transfer concept and understanding how to apply the performance and troubleshooting of heat exchangers. The focus will be on shell and tube heat exchangers, given their wide use in the CPI.

While the first principle of thermodynamics is about energy conservation, the second principle states that heat travels from hot to cold. Heat can be transferred by three mechanisms: conduction, convection, and radiation. Radiant heat and convection have already been discussed in the fired equipment section. The heat transfer mechanism in heat exchangers is driven by convection and conduction. A process stream can gain or lose heat without phase changes, where heat is correlated to the fluid heat capacity and temperature difference as previously described by Equations 2.132, 2.133, and 2.135.

The overall heat transfer coefficient in a shell and tube exchanger includes convection on the shell and tube side as well as conduction across the tube thickness. Figure 2.53 illustrates the parallel between heat transfer across two fluids separated by a solid barrier (exchanger tube thickness) and its electrical equivalent. In this system, the resistance to heat transfer is equivalent to three resistances in series. Convection resistance occurred in the fluids on each side of the barrier while conduction resistance occurred across the thickness of the tube material. Convection results from fluid movement on both sides of the barrier, its resistance being defined by the inverse of the convection coefficient, which depends on fluid velocity, thermal, and transport properties. The conduction resistance is defined by the inverse of the material thermal conductivity.

The overall heat transfer coefficient is related to the individual convection and conduction coefficients by the following expression:

$$\frac{1}{U} = \frac{1}{h_{shell}} + \frac{D_o - D_i}{2k} + \frac{1}{h_{tube}\left(\frac{D_i}{D_o}\right)} \tag{2.163}$$

where D_o is the outside tube diameter, D_i is the inside tube diameter, h_{shell} is the heat transfer coefficient on the shell side, and h_{tube} is the heat transfer coefficient on the tube side. The constant k is the thermal conductivity of the material separating the fluid between the shell and tube side.

Fouling is the most common mechanism affecting performance of heat exchangers. Fouling is characterized by a solid layer accumulating on the tube wall of exchangers. This can occur on the tube and/or the shell side. That layer adds an

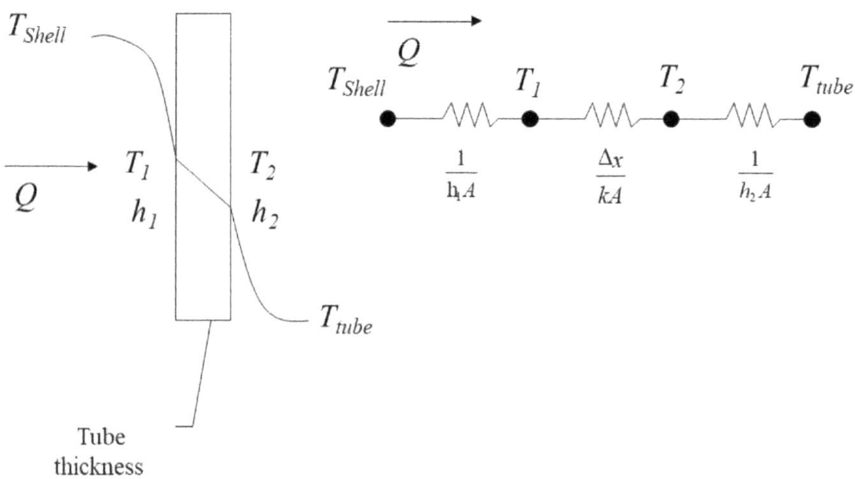

FIGURE 2.53 Overall Heat Transfer Coefficient.

additional resistance to heat transfer. This would be characterized in Figure 2.53 by adding another conductive resistance. Fouling material accumulating on the tube wall is typically of polymeric nature in chemical plant and as such has low thermal conductivity. The accumulation of solids on the tube wall, besides impacting the thermal conductivity, increases pressure drop. This could be inconsequential from a heat transfer perspective, if flow rates are not affected. However, for relatively fixed supply pressure systems like cooling water, an increase in pressure drop is associated with a reduction in flow rates. For fully turbulent flow in tubes, Dittus and Boelter [27] have empirically correlated the heat transfer coefficient to the Reynolds and Prandtl number as described by Equation 2.164:

$$h_{tube} = 0.023 \frac{k}{D_i} Re^{0.8} pr^{0.4} \tag{2.164}$$

The Reynolds number brings a flow dependency on the heat transfer term. The reduction in tube velocity also favored settling of any solids contained in the stream. Furthermore, the drop in velocity reduces the shear forces exerted by the fluid on the tube wall, which contributes to keeping the surface clean. In summary, fouling impacts heat transfer by adding thermal resistance to conduction, reduces the fluid heat transfer coefficient due to flow reduction with those conditions, further contributing to fouling by settling and reduction in fluid shear forces on the tube wall.

Temperature differential between the hot and cold streams also plays a role in fouling. High TMT will promote fouling reaction in the metal/fluid boundary layer.

Corrosion is another mechanism largely responsible for malfunctioning of heat exchangers. Fouling can also promote corrosion with under deposit corrosion. Corrosion can be mitigated by chemical programs, typically done around pH control or material selection. Caution must be exercised when changing tube bundle metallurgy as some more resilient metallurgy is often associated with lower thermal conductivity. Cross-contamination between streams is one major consequence of corrosion, which may impact product quality.

The lack of instrumentation is one of the major challenges for evaluating the performance of heat exchangers, as it requires inlet/outlet temperatures and flow rates on at least one side of the exchanger. When sufficient instrumentation is available, the overall heat transfer coefficient of critical heat exchangers should be monitored over time to understand performance degradation over time. In high fouling service, parallel exchangers should be provided such that one can be taken offline for cleaning while operating on the other one.

2.1.5 Hydraulics and Rotating Equipment

It has been stated earlier that chemical processes can be divided into two main sections: reaction and separation. However, the movement of material between operating units is also important and, in many instances, it can constrain capacity. Gases and liquids travel between units through pipes and fittings by pressure differential. Flow rates can be regulated by a control valve when sufficient positive pressure differential exists between units. When insufficient or in cases of negative pressure differential, compensation is done using pumps or compressors. Inconsistency in pressure profiles is frequently driven by process requirements. For example, yields in ethylene steam cracking furnaces are positively influenced by low back pressure, which is inconsistent with the recovery section operating pressure that is dictated by refrigeration temperature. A compressor is therefore required for reconciling the pressure discontinuity between the reaction and recovery sections. Pumps are used for reconciling the pressure profile for liquids and compressors for gases.

There is a tendency in the operating world to refer to centrifugal pumps or compressors as fixed capacity without giving much consideration to the system characteristic curve. People are often surprised when starting an extra pump and getting less than the expected additional design flow. The piping system along with the pump head curve has to be analyzed to clearly understand what one or several pumps can deliver in a given system. This step is typically thoroughly done during the design stage. However, it is unusual for operation to use such analysis when troubleshooting or monitoring unit performance. This kind of analysis can be used for estimating flow rates and evaluating rotating equipment performance.

2.1.5.1 System Curve

A system curve is the correlation between flow rate and pressure differential between two points. It is mathematically characterized by the Bernoulli's Equation [104], which originates from a macroscopic energy balance:

$$\underbrace{\Delta\frac{1}{2}\frac{\langle v^3\rangle}{\langle v\rangle}}_{\text{Kinetic Energy}} + \underbrace{\Delta\Phi}_{\text{Potential Energy}} + \underbrace{\int_{P1}^{P2}\frac{1}{\rho}dP}_{\text{Gibbs Free Energy}} + \underbrace{\hat{W}}_{\text{Mechanical Work}} + \underbrace{\hat{E}_v}_{\text{Friction Loss}} = 0 \tag{2.165}$$

- Friction losses represent the rate at which mechanical energy is irreversibly converted into heat. This includes losses across pipes and fittings.
- Mechanical work is usually assumed at zero when calculating the system curve of a pipe segment. It is considered when estimating pumping power by restricting the system boundary to the pump inlet and outlet.
- The Gibbs free energy is a thermodynamic property that characterizes the pressure differential. The Bernoulli's equation assumes an isothermal system or flow at constant temperature.

- The potential energy represents the elevation difference between the piping segment starting and destination point.
- Piping systems are typically designed to operate in a turbulent flow regime. Under those conditions, the velocity profile is flat, and the kinetic terms of the Bernoulli's equation can be reduced to ½$\langle v^2 \rangle$. The kinetic term can also be taken out of the equation by converting the piping system to a single equivalent diameter. By using a single equivalent diameter, the velocity at the starting point equals the velocity at the exit.

Using the above considerations, the Bernoulli's equation can be reduced to a more user-friendly and understandable form:

$$\Delta P = \rho g h_L - \rho g \Delta H \quad (2.166)$$

where h_L, the number of velocity head, equals $-g\hat{E}_v$ ($\hat{E}_v$ being negative) and is defined by:

$$h_L = Kv^2/2g \quad (2.167)$$

K in the above equation is the resistance coefficient, which represents the total frictional resistance of the piping system. Values of K for fittings such as elbow and tees are tabulated in many handbooks. Pipe K values are correlated to the Darcy friction factor f by:

$$K = fL/D \quad (2.168)$$

Note that caution must be exercised as some references provide the Fanning friction factor, which is represented by the same character f. The Darcy friction factor is by definition four times the Fanning friction factor [105]. In the fully turbulent regime, f is independent of flow rates and depends only on pipe diameter and pipe material roughness. Crane Technical Paper 410 [106] provides a diagram, which correlates the Darcy friction factor f to pipe diameter and roughness for fully turbulent flow. Churchill's equation [107] correlates the friction factor with the Reynolds number, the pipe diameter, and the relative roughness from the laminar to the fully turbulent flow regime. The Shacham's equation [107], less complex, correlates the friction factor to the same parameters as the Churchill equation but only covers the transitional to fully turbulent flow regime.

Combining Equations 2.166 and 2.167 provides a relationship between the pressure differential and the fluid velocity:

$$\Delta P = \frac{\rho K v^2}{2} - \rho g \Delta H \quad (2.169)$$

Or in terms of flow rate:

$$\Delta P = \frac{8\rho K Q^2}{\pi d^4} - \rho g \Delta H \quad (2.170a)$$

With proper care to engineering units, Equation 2.170a becomes:

$$\Delta P = \frac{2.7957\times10^{-8}\rho K Q^2}{g_c d^4} - \frac{g}{144 g_c}\rho\Delta H \tag{2.170b}$$

where:

ΔP – System pressure drop [=] psi
ΔH – Elevation difference [=] ft
ρ – Fluid density [=] lb/ft^3
Q – Flow rate [=] usgpm
g_c – English units conversion factor between lb_f and $lb_m = 32.174\ ft\ lb_m/\left(sec^2 lb_f\right)$
d – pipe diameter [=] ft

For pipe and fitting components in series, the K value of Equations 2.170a and 2.170b is the summation of the individual K values. To be consistent with the simplification that enables cancellation of the kinetic energy term, all K values have to refer to a single diameter. Components of different diameter can be expressed to the reference diameter by the following expression:

$$K_a = K_b\left(\frac{d_a}{d_b}\right)^4 \tag{2.171}$$

where d_a is the reference diameter and d_b the diameter of the pipe segment being converted to the equivalent diameter d_a.

Evaluation of the overall K value on a parallel system is similar to evaluating the equivalent resistance of an electrical circuit as shown on Figure 2.54. The friction loss term of Equation 2.170a is similar to the electrical relationship that correlates the potential difference (E) to the electrical current (I) and system resistance (R):

$$E = RI \tag{2.172}$$

$$\Delta P = \frac{8\rho K Q^2}{\pi d^4} \tag{2.173}$$

where ΔP, Q, and K are respectively the hydraulic equivalents of E, R, and I. A slight difference from the traditional Kirchoff law has to be introduced in calculating the overall K value because of the exponent involved in the flow rate in Equation 2.173. The relationship for evaluating the overall K value is demonstrated below by combining a material balance with Equation 2.170a

$$\Delta P = \frac{8\rho K_{Total} Q_{Total}^2}{\pi d^4} = \frac{8\rho K_1 Q_1^2}{\pi d^4} = \frac{8\rho K_2 Q_2^2}{\pi d^4} \tag{2.174}$$

$$Q_{Total} = Q_1 + Q_2 \tag{2.175}$$

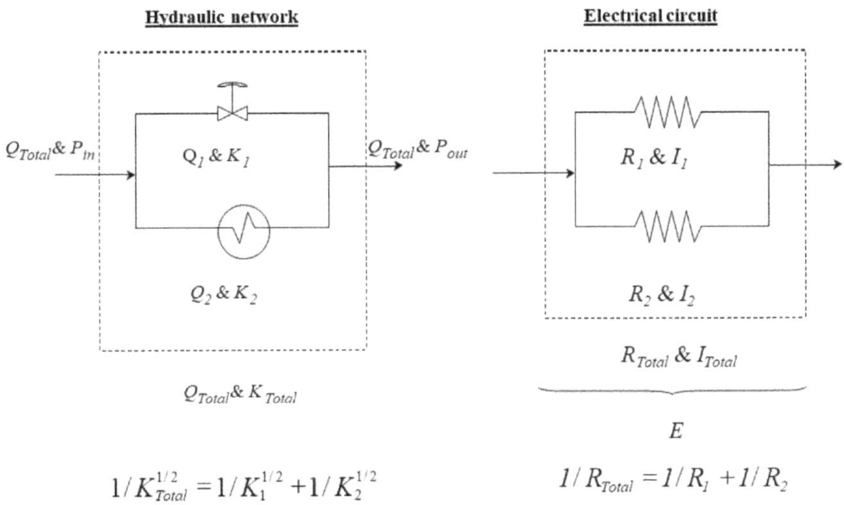

FIGURE 2.54 Electrical Analogy.

By combining Equations 2.174 and 2.175:

$$1/K_{Total}^{1/2} = 1/K_1^{1/2} + 1/K_2^{1/2} \tag{2.176}$$

The system curve is represented by the plot of the pressure differential (ΔP) as a function of flow rate (Q). As described by Equations 2.170, 2.170a, 2.173, and 2.174, this relationship is equivalent to a conventional binomial expression $f(x) = ax^2 + bx + c$. The head curve is half a parabola that could only cover the positive flow axes side as the flow rate and resistance coefficient K can only be positive. The coefficient b in the equation is null and c represents the static head. The parabola gets closer to the vertical axes as the K value increase (a coefficient). Example of system curves is illustrated in Figure 2.56 as discussed later with centrifugal pump head curves.

Control valve *K* value. The flow resistance caused by a control valve is normally characterized by its flow coefficient C_v. Valve vendors would typically provide a valve curve correlating the percentage opening to the flow coefficient C_v. By definition, the flow coefficient is the flow rate of water at 60 °F that will cause 1 psi pressure drop across the valve and is described by the following relationship:

$$C_v = Q\sqrt{\frac{G_f}{\Delta P}} \tag{2.177}$$

where:

G_f = Specific gravity of the fluid at flowing temperature
Q [=] usgpm
ΔP [=] psi

A relationship between C_v and the resistance coefficient K can be obtained by merging Equations 2.170a and 2.177. As the pressure loss across a valve is caused only by friction, the static head expression is not involved in Equation 2.170b

$$K_{Valve} = \frac{g_c d^4}{1.7455 \times 10^{-6} C_v^2} \tag{2.178}$$

The piping resistance being constant, regulation of a control valve provides variability in hydraulic resistance. A more appropriate form of Equation 2.170 for system that includes a control valve should be:

$$\Delta P = \frac{2.7957 \times 10^{-8} \rho Q^2}{g_c d^4} * \left(K_{Valve} + K_{System}\right) - \frac{g}{144 g_c} \rho \Delta H \tag{2.179}$$

Compressible flow. The above equation applies to incompressible fluids but can also provide reasonable accuracy with compressible fluids providing that the calculated pressure drop (ΔP) is less than 10% of the inlet pressure with the density based upon either the inlet or outlet conditions. Those equations also apply if the calculated ΔP is greater than 10% but less than 40% of the inlet pressure with the density based on the average between the inlet and outlet conditions [106]. For greater pressure drop, flow/pressure drop correlations are based on adiabatic or isothermal flow. API 521 [108] recommends the isothermal method for relief systems, with the exception of cryogenic conditions, where the adiabatic method is believed preferable. Crane [106] suggests a simple equation for adiabatic flow as described by Equation 2.180:

$$w = 1.111 \times 10^{-6} Y d^2 \sqrt{\frac{\Delta P \rho_1}{K}} \tag{2.180}$$

where w is the mass flow rate in kg and Y is the expansion factor, which depends on the heat capacity ratio, the K value and the pressure ratio are $\left(\frac{\Delta P}{P_1}\right)$ as described by the charts in Crane [106] appendices A-22. The appendices also provide the value of Y for the critical pressure ratio, which limits the value of Y at sonic velocity.

Equation 2.181 [109, 110] provides a correlation for the isothermal method:

$$\frac{fL}{D} = \frac{MW g_c P_1^2}{G^2 RT}\left[1 - \left(\frac{P_2}{P_1}\right)^2\right] - ln\left(\frac{P_1}{P_2}\right)^2 = \frac{1}{M_1^2}\left[1 - \left(\frac{P_2}{P_1}\right)^2\right] - ln\left(\frac{P_1}{P_2}\right)^2 \tag{2.181}$$

M_1 is the inlet Mach velocity as described by:

$$M_1 = \frac{v_1}{\sqrt{\frac{RT}{MW}}} \tag{2.182}$$

where v_1 is the inlet gas velocity and G in Equation 2.181 is the mass velocity. Equation 2.181 is implicit in terms of flows versus pressure and must be resolved by trial and error or by using numerical methods.

Flow pressure network. Simple hydraulic segments can be resolved using the above equations on paper or in spreadsheets. Most process simulators have pipe blocks that can also be used for simple pipe segment hydraulic. Process simulators provide the advantage of calculating transport properties and can also handle two-phase flow, which is useful when evaluating thermosiphon circuits. However, the complexity arises when dealing with networks involving several users. There is commercial software that is specifically designed for resolving flow pressure network. Most of them are generally good with single-phase single-component systems. However, caution is advised as simplification in mixing rules and flash calculation can result in significant errors when handling multicomponents and/or two-phase systems. It is therefore recommended to validate properties and flash calculation with a process simulator. There is a tendency to model large network systems like cooling water in a single model. This often results in convergence difficulties and/or challenges for doing sensitivity in localized part of the system. Bernard [111] described how a large model can be broken down into smaller subsection models. The methodology requires two iterations where each submodel is run with inlet and outlet pressure estimate first for calculating the overall K value of the submodels. The submodels are then simulated as single heat exchangers, configured with the calculated K value, in a simplified global model. Running the global model then provides the inlet/outlet pressures of each submodel, which can be run a second time for providing detail flow distribution.

2.1.5.2 Pumps

Pumps are used for reconciling pressure discontinuity between a low- and a high-pressure point within a process. The fluids moved by pumps are characterized by relatively high and constant densities [112]. In chemical plants, they are mostly used for moving liquid stream and slurry. Pumps can be divided into two main categories, positive displacement and rotodynamic. Positive displacement pumps move a definite volume of fluid trapped in a chamber. The chamber is filled from the inlet and released at a higher pressure at the discharge. There are two subclasses of positive-displacement pumps. Reciprocating pump where the fluid fills a stationary cylinder that gets pressurized and released by the reciprocating action of a piston plunger. Rotating pumps are the second subclass where the chamber moves from the inlet to the discharge. A gear pump is an example of rotating positive displacement pumps. Positive displacement pumps are typically used for moving viscous fluids. Centrifugal pumps are rotodynamic, as they are configured to convert mechanical energy from a motor or turbine to kinetic energy by rotating an impeller which transmits motion to a fluid. The majority (around 90%) of pumps installed and operated in the CPI are centrifugal pumps [113]. The following discussion will therefore be dedicated to centrifugal pumps.

The addition of a centrifugal pump adds an extra degree of freedom in a hydraulic segment. That degree of freedom is defined by a characteristic performance curve, which is a plot of the total dynamic head (TDH), expressed in feet of liquid against the flow rate (typically in usgpm). The total dynamic head is the difference between the pump discharge (P_d) and suction pressure (P_s) expressed in head of liquid. The actual operating point on a pumping system occurs at the intersection of the system and pump curve. To make the system curve consistent with the pump curve, it has to be evaluated at the suction and the discharge. Using the system illustrated in Figure 2.55 as an example, the system curve plots the difference between P_d and $P_s(\Delta P)$ against the flow rate. Equation 2.170a is used for evaluating both the suction and discharge piping. It is not unusual to have the pump relatively close to a suction drum, in such cases it is reasonable to neglect frictional losses and only consider the static head for evaluating the suction pressure. Caution must however be exercised when for system having marginal net positive suction head (NPSH) margins.

Figure 2.56 illustrates typical pump head and efficiency curves with overlaid system curves. Variability of the system curves is achieved by manipulating the pump

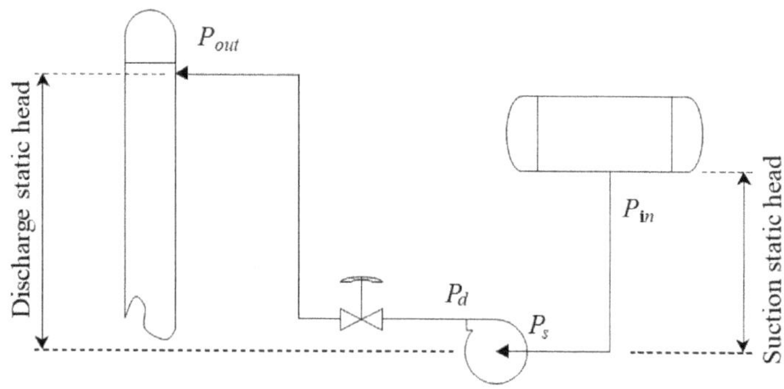

FIGURE 2.55 Example of a Pumping System.

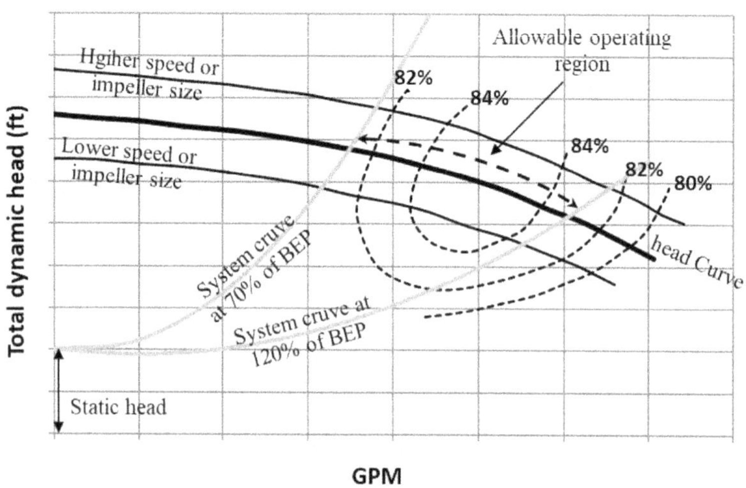

FIGURE 2.56 System Curve and Pump Head Curve.

discharge control valve. The actual operating point is defined by the intersection of the system curve with the pump head curve. Bloch [114] refers to the preferred or allowable operating region of a centrifugal pump to fall between 70% and 120% of the best efficiency point (BEP). The intersection of the system curves with the head curve illustrated in Figure 2.54 defines that allowable operating region. This is consistent with API-610, which also states that the rated flow shall be within 80–110% of the BEP. Bloch [114] states that operations too far away from the BEP cause liquid turbulence that may contribute to excessive vibration. The concept of allowable operating region is not fully understood by process engineers, especially in the context of debottlenecks where a pump can be assessed capable of delivering the required output as process engineers will likely consider the operating range between the minimum stable flow and the end of curve. Constant operation outside the allowable operating region will likely result in an increase of the meantime between failures, hence maintenance frequency. Two criteria define a pump minimum flow. First, the accumulation of heat will eventually cause the fluid within the pump to flash and cause significant damage. A pump should not operate for more than 5 seconds with the discharge valve close. Internal recirculation defines the second minimum flow criteria. Operating in this mode may cause excessive vibration eventually resulting in damage. The minimum stable flow is typically provided by the pump vendor. Minimum flow circuits are typically provided for protecting the pump. Those circuits may always be active when regulated by a flow orifice. Such configuration consumes capacity and limits the effective output. The use of a control valve on the minimum flow circuit ensures effective usage of the pump capacity.

The NPSH is another critical parameter for reliable centrifugal pump operation. Centrifugal pumps are designed to handle single-phase liquid flow; their performance and reliability are badly affected when vapors are present in the fluid. It is not unusual to pump liquids that are at or close to the boiling point (e.g., tower reflux, boiler feed water). In such application, liquid may flash into vapor inside the pump; this phenomenon is known as cavitation and could cause severe damage to the pump. The concept of NPSH is used for assessing the likelihood of cavitation by comparing the NPSH available to the NPSH required by the pump. The NPSH available quantifies the head above the vapor pressure of the fluid at the pump suction as described by Equation 2.183, where P_v is the vapor pressure of the fluid and P_{in} the pressure at the vapor/liquid interface of the suction drum, as illustrated in Figure 2.55. It is good practice to evaluate the NPSH available with the pump suction drum at its low–low level alarm point. For liquid at the boiling point, P_{in} at the vapor/liquid interface of the suction drum is equal to the vapor pressure P_v. If the friction losses are negligible, the NPSH available becomes equal to the static head ΔH.

$$NPSH_{Available} = \frac{1}{g}\left(\frac{P_{in} - P_v}{\rho} - h_l\right) + \Delta H \qquad (2.183)$$

The NPSH required is a characteristic of the pump and varies with speed and capacity. It is normally expressed in terms of feet of liquid absolute and is often plotted with the pump curve. Figure 2.57 illustrates an example of an NPSH plot where the NPSH margin is the difference between the NPSH available and required.

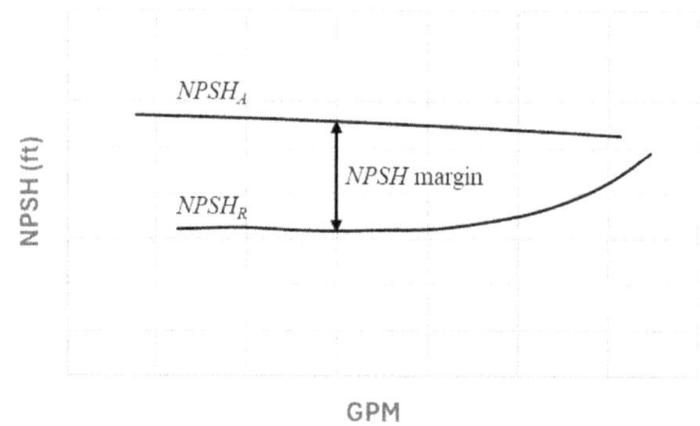

FIGURE 2.57 NPSH Required versus Available.

Pumping power. The mechanical work term in Equation 2.165 was assumed zero in establishing the flow equation based on the assumption that no mechanical work was taking place within the system considered (suction and discharge piping). However, mechanical work is transferred from the pump to the fluid, when using the pump inlet and outlet as boundary. Neglecting friction losses and elevation difference between the pump inlet and outlet Equation 2.165 becomes:

$$\int_{P1}^{P2} \frac{1}{\rho} dP + \hat{W} = 0 \tag{2.184}$$

In terms of power:

$$P_f = \hat{W}\dot{m} = -\dot{m}\frac{\Delta P}{\rho} \tag{2.185}$$

The power supplied to the fluid divided by the pump efficiency provides the power required to the pump which is typically expressed in BHP or watts:

$$P_b = \frac{P_f}{\eta} \tag{2.186}$$

where

P_f = Power delivered to the fluid
P_b = Power supplied to the pump
η = Pump efficiency
W = Mechanical work
m = Fluid flow rate
ΔP = Pump pressure differential ($P_s - P_d$ in Figure 2.53)

Affinity Laws. Pump curves are often supplied for a given speed and a set of pump impeller sizes. It is not unusual however to operate pumps at different speeds and/or consider replacement with different impeller sizes. Figure 2.56 illustrates how curves for different speeds or impeller size compare to the original head

TABLE 2.16
Affinity Laws

Impeller diameter (D) change	$Q_2 = \frac{D_2}{D_1} Q_1$
	$TDH_2 = \left(\frac{D_2}{D_1}\right)^2 TDH_1$
	$P_{b2} = \left(\frac{D_2}{D_1}\right)^3 P_{b1}$
	$NPSH_{R2} = NPSH_{R1} \left(\frac{D_2}{D_1}\right)^X$
Speed (RPM) change	$Q_2 = \frac{RPM_2}{RPM_1} Q_1$
	$TDH_2 = \left(\frac{RPM_2}{RPM_1}\right)^2 TDH_1$
	$P_{b2} = \left(\frac{RPM_2}{RPM_1}\right)^3 P_{b1}$
	$NPSH_{R2} = NPSH_{R1} \left(\frac{RPM_2}{RPM_1}\right)^Y$

curve. There are relationships, which help in predicting the head curve for different speeds and impeller sizes. Those equations are commonly called Affinity Laws or Fan Laws and are summarized in Table 2.16. The NPSH relationships are an approximation with the exponent x varying between −2.5 and 1.5, and y between 1.5 and 2.5 [105].

2.1.5.3 Compressors

Compressors are for gases what pumps are for liquids. There are three main compressor types used in the CPI, namely centrifugal, axial, and reciprocating. Covering all compressor types would exceed the scope of this book. This discussion will be limited to centrifugal compressors given their extensive use in the CPI. Similar to pumps, compressors reconcile pressure imbalance between process unit boundaries by increasing the pressure from a source to a destination. As for centrifugal pumps, it is accomplished by imparting kinetic energy to process gas by accelerating it through rotating impellers. As for pumps, compressors can be broken down into mechanical and process components. Process components are those in contact with the gas. The discipline cross boundary between mechanical and process engineers is associated with a different vocabulary. A typical ethylene plant crack gas compressor (CGC), as illustrated in Figure 2.58 [115], is used for facilitating the discussion. Process engineers refer to compression stages as a flange-to-flange section [116]. The compressor illustrated in Figure 2.58 has three bodies. A first body, which contains one process stage (first), a second body including two process stage (second and third), and a third body including two process stage (fourth and fifth). As illustrated in Figure 2.59, each of those stages is characterized by head and efficiency curves that vary with speed. For pumps, the head curve is defined by the pressure differential between the discharge and suction, expressed in head of a liquid and can be described by the following expression:

$$H = k_1 \int \overline{V}\, dP = k_1 \overline{V}\left(P_d - P_s\right) \tag{2.187}$$

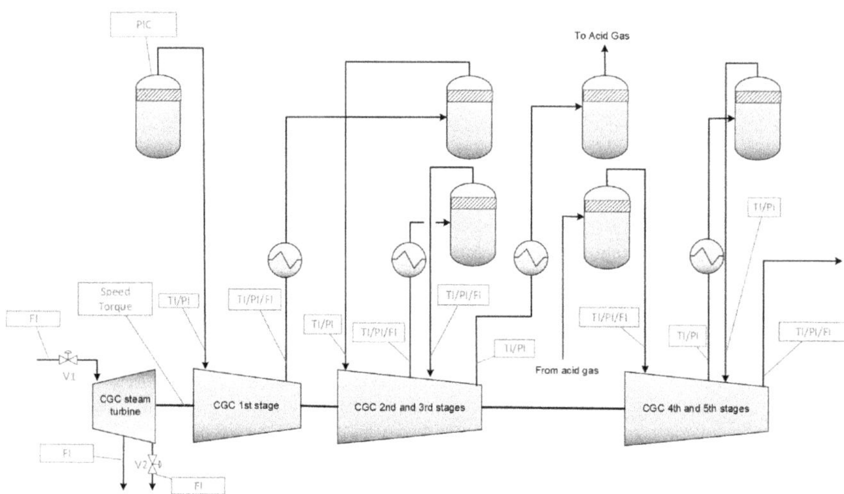

FIGURE 2.58 Typical Ethylene Plant Crack Gas Compressor [118].

Reproduced with Permission from Charge Gas Compressor Operation and Mechanical Consideration, A. Bernard, M. Pickard, 2025 AIChE Spring Meeting – Ethylene Producer Conference

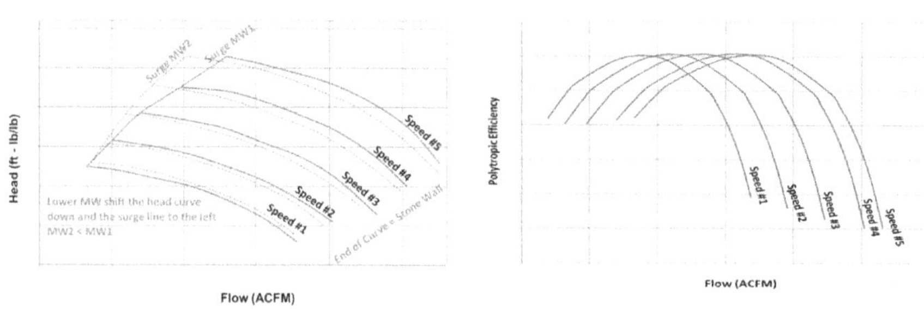

FIGURE 2.59 Compressor Polytropic Head and Efficiency Curves [118].

Reproduced with Permission from Charge Gas Compressor Operation and Mechanical Consideration, A. Bernard, M. Pickard, 2025 AIChE Spring Meeting – Ethylene Producer Conference

where $\bar{V}$ is the specific volume an k_1 is a constant for engineering units. For liquids, the specific volume is constant, which allows the simple resolution of the integral term to the right end side of Equation 2.187. For centrifugal compressors, the specific volume varies with pressure and results in a more complex relationship. The compression is assumed to be polytropic, in that it is described by the following expression:

$$P\bar{V}^n = Constant \tag{2.188}$$

where n is the polytropic exponent of compression. The PV relationship of Equation 2.188 can be introduced in the integral term of Equation 2.187, and after integration and rearrangement provides a relationship for the polytropic head:

$$H = \frac{Z_a R T_s}{MW} \frac{\left(\frac{P_d}{P_s}\right)^{\sigma} - 1}{\sigma} \tag{2.189}$$

With $\sigma = \left(\frac{k-1}{k\eta_p}\right)$, where k is the heat capacity ratio (C_p/C_v) and η_P the polytropic efficiency, which is a function of the discharge temperature. The power can be calculated by multiplying the gas mass rate by the polytropic head as described by Equation 2.190 when expressed in terms of gas horsepower:

$$Power\ (GHP) = \frac{Gas\ (lb/\min) \times Polytropic\ head(ft)}{33{,}000 \times \eta_p} \tag{2.190}$$

The compressor shaft horsepower is the gas horsepower divided by the mechanical efficiency. It is important to point out that the polytropic head curves provided by vendor are established for a given molecular weight and suction conditions. Deviations from those conditions can cause the curve to shift. For instance, as illustrated in Figure 2.59, a reduction in molecular weight causes the polytropic head curve to move down. It also causes the surge curve to move to the left. Significant deviation in suction temperature may also shift the polytropic head curve. Lapina [117] suggests validating the accuracy of the polytropic head curve against the condition for which it is being used by comparing a parameter θ to the reference conditions, where:

$$\theta = \left[(26.1MW)/(kZT)\right]^{1/2} \tag{2.191}$$

The polytropic head curve is assumed to provide good accuracy when the differences between θ and $\theta_{reference}$ is less than 10%.

Prior to site delivery, compressors are typically tested at the factory. For convenience and in many cases, for safety reasons, compressors are tested with test gas that has different thermodynamic properties than the intended process gas. Factory testing is described in ASME PTC-10 [118], which also provides methodology for extrapolating the results to the specified process conditions and thermodynamic properties of the process gas. The methodology as described by Bloch [119] and Godse [120] involves using the concept of equivalent tip speed, by multiplying the tip speed by the ratio of a reference gas acoustic velocity to the actual process gas acoustic velocity. It also involves conversion of the polytropic head curve in a reduced dimensionless form using flow and head coefficients, which are described by Equations 2.192 and 2.193:

$$Flow\ coefficient = \phi = \frac{700Q_s}{ND_s^3} \tag{2.192}$$

where Q is the compressor suction flow (ft^3/min), N is the rotational speed (RPM), D_s is the suction impeller diameter (in):

$$Head\ coefficient = \psi = \frac{H_p(1300)^2}{IN^2D_m^2} \tag{2.193}$$

where H_p is the polytropic head (ft-lb/lb), I is the number of impellers, and D_m is the average impeller diameter (inch).

The acoustic velocity is obtained from Equation 2.194:

$$Acc.\ Velocity = V_{ac} = \left(32.174 * 1{,}545Z * Z\,K * \left(T * 1.8 + 32\right) + 459.67\right)/MW \tag{2.194}$$

where Z is the compressibility factor, K is the heat capacity ratio, and T is the temperature in degree Celsius.

The above methodology has been used for re-mapping the polytropic head curve illustrated in Figure 2.59 for a lower gas molecular weight (MW2) than the original gas molecular weight (MW1). Understanding how operating conditions and gas thermodynamic properties influence the polytropic head curve is important as illustrated in Figure 2.59, a transition to a lower molecular weight may mislead the process engineer in concluding a negative head offset when assessing the performance of the compressor.

It must be noted that simulations will predict lower head and high volumetric flow for low-molecular-weight gas and higher head and lower volumetric flow for heavier gases. This is of particular importance as without polytropic head curve correction, it will under predict the discharge pressure for gases heavier than the design. This may result in pressure safety valve (PSV) set pressure exceedance. For light gases, failure to not accounting for curve correction can result in conditions exceeding choke flow.

Note that Affinity Laws described in Table 2.16 also apply to compressors. Bloch [119] states that Affinity Laws provide good accuracy on light gas involving small head such as what is typically experienced with air. However, less accuracy is expected on heavier gas and multistage compression.

Figure 2.58 illustrates some of the instruments that can be used for monitoring the performance of compressors. Although the polytropic head curve can be calculated from Equation 2.187, the use of process simulators is recommended as it calculates all the required thermodynamic properties at process conditions. Furthermore, the composition of the gas is expected to vary across a compression train similar to what is illustrated in Figure 2.58 due to the separation of condensed liquids in the multiple knockout drums. The gas composition from on-line analyzer our laboratory analyses is required. Inlet and outlet temperatures and pressures should be available along with either suction or discharge flow metering. Power can be measured by torque meters or by measuring electrical power on electrical motor or enthalpy balance on steam turbine [121]. Speed for variable speed compressors is required for evaluating performance of the compressor and can be measured by electronic revolution counter, mechanical revolution counter, tachometer, or stroboscope. A process simulator can be used for calculating the actual head curve by providing the gas flow, composition, inlet temperature, pressure, and outlet pressure. The polytropic efficiency is set by trial and error or through a design specification in a simulation for matching the discharge temperature. The actual polytropic head and efficiency can then be compared to those predicted by the vendor curve. The magnitude of the deviation between the actual and expected will provide an indication on the health of the compressor. Monitoring those parameters over time will show if the machine maintains a constant offset or if it is deteriorating.

Operating range. As for pumps, centrifugal compressors are limited at the low end by a condition referred as surge, which can be defined as the point of minimum stable flow [119]. It is defined by the surge line on the polytropic head curves illustrated in Figure 2.59. Operation below the surge point is associated with significant vibration that can cause serious mechanical damage. The vibration resulting from surge is explained by a flow reversal/forward flow cycle at cycle of about 20–80 ms [119]. Most compressors are protected from surge using false load or recycle circuits. Figure 2.60 illustrates different anti-surge strategies for the CGC

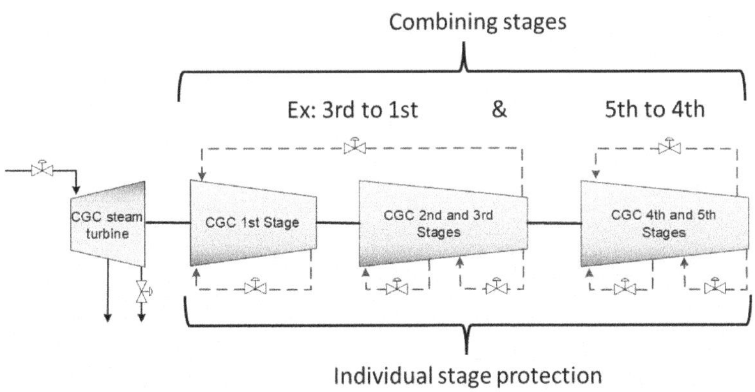

FIGURE 2.60 Surge Protection Strategy [118].

Reproduced with Permission from Charge Gas Compressor operation and Mechanical Consideration, A. Bernard, M. Pickard, 2025 AIChE Spring Meeting – Ethylene Producer Conference

used as example. The required recycle flow on a CGC typically involves relatively large pipes and control valves. As such, some will adopt strategies aimed at minimizing capital. Physical space constraint is another factor in plant retrofits that imposes design limitations. As such, the surge protection can be strategized to the limiting process stage. As shown in Figure 2.60, if satisfying the surge margin of the first stage also satisfies the second and the third, then there might be no reason to add additional circuits; same rational with the fourth and fifth stage. Therefore, this strategy would include only two false load circuits (i.e., third to first and fifth to fourth). Another important element, not illustrated in Figure 2.60, is to provide adequate cooling on the recycle streams to avoid building up heat compression. Compressors are generally protected from surge by anti-surge controllers. These controllers are intended to open the control valve regulating the false load circuit such that the operating point never reaches the surge point plus a margin. As shown in Figure 2.61, the surge line is represented by the minimum flow on the polytropic head curve. The polytropic head is not something that can be physically measured in the field. As such, process control strategy linearized the pressure ratio (i.e., discharge pressure/suction pressure) as a function of the square of the flow [122].

The end of the polytropic head curve characterizes the high operating end at a given speed. This high limit is referred to as stonewall (choke flow), a condition at which the gas reaches sonic velocity. Further decrease in head would not provide increases in volumetric flow. Bloch [119] states that choke is not particularly damaging to single-stage compressors but can aerodynamically affect the rotors and blades of multistage centrifugal compressors. In such situations, stonewall conditions should be avoided and may require protection.

The potential for choke flow brings an interesting aspect that can be missed in relation to surge protection. Going back to the third to first stage surge protection strategy illustrated in Figure 2.60, there is a potential for the second or third stage to reach stonewall, which would limit the ability of the third stage discharge

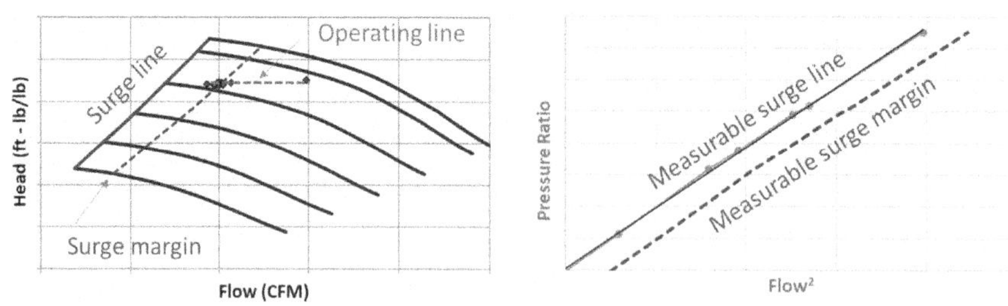

FIGURE 2.61 Surge Protection Controls [118].

Reproduced with Permission from Charge Gas Compressor Operation and Mechanical Consideration, A. Bernard, M. Pickard, 2025 AIChE Spring Meeting – Ethylene Producer Conference

in providing the flow required for meeting the surge margin of the first stage. This situation has been encountered in evaluating a CGC startup strategy on a methane (MW of 16) [123] on a compressor designed for a process gas having a molecular weight of 26. Simulation of the CGC on methane gas revealed that the predicted operating point would significantly exceed stonewall on the third stage with the first stage operating well below surge. Such situation would have caused an uncontrolled surge event. This re-emphasizes the importance of understanding the capability of the equipment. While the procedure documented in literature was sound, a third to first stage surge protection of the CGC was not adequate for implementing this procedure. On the other hand, individual process stage surge protection, as shown in Figure 2.60, would have been appropriate for implementing this procedure.

Nomenclature

A	area of the flame front cone – Equation 2.16
A	constant – Equation 2.18
A	heat transfer Area
A	exposed area defined by the volume – Equation 2.47
A	cross section
A_B	bubbling area in distillation tray
A_h	hole area in distillation tray
a_i	activity coefficient
AIT	autoignition temperature
B	constant – Equation 2.18
C_d	coefficient in clear liquid height equation 2.121
C_d	drag coefficient – Equation 2.159
C_{factor}	C-factor describing the vapor load in distillation
CFD	computational fluid dynamics
CFS	vapor flow, ft^3/s in distillation
C_i	concentration of component i
COT	coil outlet temperature
Cp	heat capacity
CPI	Chemical Process Industry
C_{st}	stoichiometric molar concentration of fuel in air
CSTR	continuous stirred tank reactor
Cv	valve flow coefficient
d	diameter
D	molar distillate rate
D_m	average impeller diameter – compressor
D_s	suction impeller diameter – compressor
e	equivalent thickness – Equation 2.47

F	overall exchange factor – Equation 2.23
F	feed in distillation
f	Darcy friction factor
FCC	fluid catalytic cracking
FD Fan	force draft fan
$f_{g,1}$	overall exchange factor – Equation 2.23
f_i	fugacity of component i
f_i^o	fugacity of component i at standard state
Fr	Froude number
F_s	superficial F-factor describing vapor load in distillation
FT-ICR-MS	Fourier transform-ion cyclotron resonance mass spectrometry
g	gravitational acceleration
g_c	force factor unit conversion – 32 lbm ft/s^2/lbf
GC-MS	gas chromatography-mass spectrometry
GHSV	gas hourly space velocity
h	fluid convection heat transfer coefficient
h_c	head of aerated liquid in distillation trays
h_d	dry pressure drop in distillation trays
H&MB	heat and material balance
HHV	high heating value
h_l	pressure drop through the aerated liquid in distillation trays
h_L	number of velocity head in hydraulic calculation
h_R	residual pressure drop in distillation trays
I	number of impeller – compressor
I	intensity of radiation – gamma scans
I_0	radiation intensity
ID Fan	induce draft fan
k	thermal conductivity
K	dry pressure drop coefficient in distillation trays
K	hydraulic resistance coefficient
k_0	pre-exponent factor in the Arrhenius equation
k_1 and K_2	rate constant in reaction kinetic
K_{eq}	equilibrium activity coefficient
K_i	K value in distillation
$K_{i,j}$	binary interaction parameter
KT	Kepner-Tregoe
L	liquid molar rate in distillation
L/D	reflux ratio
LFL	lower flammability limit
LHV	low heating value
LMTD	log mean temperature difference
m	coefficient – Equation 2.52
m_i	mass fraction of component i

MIE	minimum ignition energy
MW	molecular weight
n	constant – Equation 2.18
N	rotational speed - compressor (RPM)
n_{ip}	number of moles products
n_{ir}	number of moles reactants
NMR	nuclear magnetic resonance
NPSH	net positive suction head
OLE	object linking and embedding
P	pressure
P	power
P_c	critical pressure
PFO	pyrolysis fuel oil
P_i	partial pressure of component i
P_i^*	vapor pressure of component i
PLC	programmable logic computer
Pr	Prandtl number
P_r	reduce pressure
Q	heat duty
Q	volumetric flowrate in hydraulic calculation
q	volumetric rate
Q_R	radiant section useful heat duty
R	universal gas constant
Re	Reynolds number
RPG	raw pyrolysis gasoline
S_l	tubes exposed surface – Equation 2.23
SCR	selective catalytic reduction
SIL	safety integrity level
SIS	safety instrumentation systems
SNCR	selective noncatalytic reduction
S_u	burning velocity
T	temperature
T_c	critical temperature
TDH	total dynamic head
TMT	tube metal temperature
T_r	reduce temperature
u	gas velocity
U	overall heat transfer coefficient
UA	overall heat transfer coefficient x heat transfer area
u_b	tray hole vapor velocity in distillation tray
UFL	upper flammability limit
u_g^*	Wallis flow reversal criteria – Equation 2.130
UV	ultraviolet

v	volumetric gas flow – Equation 2.16
V	volume
v	velocity
V	vapor molar rate in distillation
VLE	vapor liquid equilibrium
VLOAD	parameter describing the vapor load in distillation
VOC	volatile organic compounds
W	mechanical work
x	medium thickness – gamma scan
x_i	mole fraction of component i
x_i	liquid mole fraction of component i in distillation
y_i	mole fraction of component i
y_i	vapor mole fraction of component i in distillation
Z	compressibility factor
ΔG	Gibbs free energy
ΔH	enthalpy
ΔH	height difference in hydraulic calculation
ΔHc	heat of combustion
ΔH_p	enthalpy of the products
ΔH_r	enthalpy of the reactants
ΔS	entropy

Greek Letters

θ	temperature of the tube outside wall
α	half angle of the flame – Equation 2.17
$\alpha_{i,j}$	relative volatility
ε	emissivity
η	pump and compressor efficiency
λ	latent heat
λ	stripping factor in distillation
μ	viscosity
μ	absorption coefficient – gamma scan
μ_{RI}	angle factor between refractories and receiving surface
ρ_L	liquid density
ρ_V	vapor density
σ	surface tension
Υ	activity coefficient
Φ	fugacity coefficient of vapor and liquid
Φ	flow coefficient – compressor
Ψ	head coefficient – compressor
ω	acentric factor

REFERENCES

[1] Wikipedia, "Control of Fire by Early Humans," *Wikipedia*, [Online]. Available: https://en.wikipedia.org/wiki/Control_of_fire_by_early_humans. [Accessed 5 January 2026].

[2] Wikipedia, "Industrial Revolution," *Wikipedia*. [Online]. [Accessed 5 January 2026].

[3] R. H. Perry and C. H. Chilton, *Chemical Engineering Handbook*, Fifth Edition, McGraw-Hill, 1973.

[4] ISO Standard 13442, *Natural Gas* – Standard Reference Conditions, First Edition, ISO, December 1996.

[5] API Standard 2000, Venting Atmospheric and Low-Pressure Storage Tanks, Seventh Edition, API, March 2014.

[6] ASTM D240–19, *Standard Test Method for Heat of Combustion of Liquid Hydrocarbon Fuels by Bomb Calorimeter*, ASTM.

[7] ASTM D4809–18, *Standard Test Method for Heat of Combustion of Liquids Hydrocarbon Fuels by Bomb Calorimeter (Precision Method),* ASTM.

[8] D. A. Crowl, "Minimize the Risks of Flammable Materials," *Chemical Engineering Progress*, vol. 108 no. 4, pp. 28–33, April 2012

[9] D. A. Crowl, *Understanding Explosion*, AIChE, 2003.

[10] ASTM, E681–09, *Standard Test Method for Concentration Limits of Flammability of Chemicals (Vapors and Gases),* ASTM.

[11] M. G. Zabetakis, Flammability Characteristics of Combustible Gases and Vapors, Bulletin 627 US Dept of the Interior, Bureau of Mines, 1965.

[12] S. Mannan, CSP, Lee's Loss Prevention in the Process Industries. Hazard Identification, Assessment and Control. Volume 1, 4th Edition, Elsevier Inc., 2012.

[13] G. L. Borman and K. W. Ragland, Combustion Engineering, McGraw-Hill, 1998.

[14] A. Ghosh, N. M. Munoz-Munoz, K. P. Chatelain, D. A. Lacoste, "Laminar Burning Velocity of Hydrogen, Methane, Ethane, Ethylene and Propane Flames at Near-Cryogenic Temperatures," *Applications in Energy and Combustion Science*, vol. 12, p. 100094, 2022.

[15] F. Yang, H. Zhang and X. Wang, Effects of Soret Diffusion on the Laminar Flame Speed of n-Butane-Air Mixtures, *Proceedings of the Combustion Institute* 33 (2011) 947–953.

[16] ASTM E659–15, *Standard Test Method for Autoignition Temperature of Chemicals,* ASTM.

[17] M. E. Ternes, "Trends in Corporate Sustainability," *Chemical Engineering Progress*, vol. 113 no. 8, pp. 63–67, August 2017.

[18] ASTM D92–11, *Standard Test Method for Flash and Fire Points by Cleveland Open Cup Tester,* ASTM.

[19] "Boom and Bust Coal. Tracking the Global Coal Plant Pipeline," Global Energy Monitor, CREA, E3G, Reclaim Finance, Sierra Club, SFOC, Kiko Network, CAN Europe, Bangladesh Groups, ACJCE, Chile Sustentable, 2023.

[20] API R. 535, Burners for Fired Heaters in General Refinery Services, Third Edition, API, May 2014.

[21] J. Charles E. Baukal, The John Zinc Hamworthy Combustion Handbook, Second Edition Volume 3, Applications, CRC Press, 2014.

[22] M. Bradford, R. Grover and P. Paul, "Controlling NOx Emissions Part 1," *Chemical Engineering Progress*, March 2022.

[23] M. Bradford, R. Grover and P. Paul, "Controlling NOx Emissions Part 2," *Chemical Engineering Progress*, vol. 118 no. 4, pp. 42–46, April 2022.

[24] E. Platvoet and C. Baukai, "Process Burners 10," *Chemical Engineering Progress*, vol. 109 no. 8, pp. 35–39, August 2013.
[25] C. E. Baukal, *The John Zinc Combustion Handbook*, CRC Press, 2001.
[26] C. Bonnet, Chapter 5 – Process Furnaces 4 Material and Equipments, Institut Français du Pétrole, Edition TECHNIP, 2000.
[27] J. Holman, *Heat Transfer*, Fifth Edition, McGraw-Hill, 1981.
[28] D. M. Himmelblau, *Basic Principles and Calculations in Chemical Engineering*, Fourth Edition, Prentice Hall, 1982.
[29] J. Charles E. Baukal, The John Zinc Hamworthy Combustion Handbook Second Edition Volume 1 Fundamentals, CRC Press, 2013.
[30] G. Towler and R. Sinnott, Chapter 19 – Heat Transfer Equipment, *Principles, Practices and Economics of Plant and Process Design*, Third Edition, Elsevier Ltd., 2022.
[31] V. B. Guthrie, *Petroleum Products Handbook*, First Edition, McGraw-Hill, 1960.
[32] N. Nordström and J. Cornish, "FTIR: A Flexible Tool for Industrial Gas Analysis," *Chemical Engineering Progress*, vol. 114 no. 6, pp. 50–54, June 2018.
[33] M. AlAbbad, R. Gautam, E. G. Romero, O. Chatakonda, J. W. Kloosterman, J. Middaugh, M. D. Agostini, *Characterization and Surrogate Formulation of Heavy Oil*, Elsevier, 2023.
[34] S. Mishra and R. Datta, *Embrittlement of Steel*, Elsevier Science Ltd., 2001.
[35] A. R. 538, Industrial Fired Boilers for General Refinery and Petrochemical Service, First Edition, API, October 2015, Reaffirmed April 2021.
[36] *Industrial Water Conditioning, Betz Handbook*, Betz Laboratories Inc., 1991.
[37] M. N. Congiundi and W. Stewart, "Severity Control of Pyrolysis Furnaces," in AIChE Spring Meeting, Ethylene Producer Conference, Houston, TX., 1991.
[38] C. Wallsgrove, "Cracking Furnace Fundamentals," in AIChE Spring Meeting, Ethylene Producer Conference, New Orleans, LA, 2002.
[39] H. N. Woebcke, "Process for Pyrolysis of Hydrocarbons". United States Patent 5,271,827, 21 December 1993.
[40] D. Gent, "Conversion of Flexible Cracker to Mostly Ethane Feed," in *AIChE Spring Meeting, Ethylene Producer Conference*, 2016, 2016.
[41] B. K. Sullivan, "Ethylene Cracking Heaters Troubleshooting," in AIChE Spring Meeting, Ethylene Producer Conference, Austin, TX., 2015.
[42] M. Györffy, M. Webb, E. Hendriks and H. Fraser, "Kubota Cracking Tube Technologies," in *AIChE Spring Meeting, Ethylene Producer Conference*, 2019.
[43] O. Hougen, K. Watson and R. Ragatz, *Chemical Process Principles Part II, Thermodynamics* 2nd Edition, John Wiley & Sons, 1959.
[44] "NIST Chemistry WebBook, SRD 69," *National Institute of Standards and Technology*. US Department of Commerce, [Online]. Available: https://webbook.nist.gov/chemistry/.
[45] A. K. Coker, *Ludwig's Applied Process Design for Chemical and Petrochemical Plants*, Elsevier, 2007.
[46] R. S. Huss, "Reactive Distillation for Methyl Aceate Production," *Computer and Chemical Engineering*, vol. 27, pp. 1855–1866, 2003.
[47] J. D. Roberts and M. C. Caserio, *Chimie Organique Moderne*, InterEditions, Paris, 1977.
[48] I. Z. Önsan and A. Avci, Multiphase Catalytic Reactors, Theory, Design, Manufacturing and Application, John Wiley & Sons, 2016.
[49] E. L. Mohundro, "Overview on C2 and C3 Selective Hydrogenation in Ethylene Plants," in AIChE Srping Meeting, Ethylene Producer Conference, New Orleans, LA, 2003.

[50] C. Griffiths and E. J. Angueira, "Overview of Acetylene Hydrogenation in Front End and Back End Acetylene Converters," in AIChE Spring Meeting, Ethylene Producer Conference, Virtual, 2020.
[51] B. Hadrich, "Identification of the Best Model and Parameters for T-Y-X Equilibrium Data of Ethanol-Water Mixture," in First International Conference and Exhibition on Chemical and Process Engineering, Tripoli, Libya, 2009.
[52] W. McCabe and E. Thiele, Graphical Design of Fractionating Columns, Industrial and Engineering Chemistry. 17, Pp 605, 1925.
[53] P. M. Mathias, "Visualizing the McCabe-Thiele Diagram," *Chemical Engineering Progress*, vol. 105 no. 12, pp. 36–44, December 2009.
[54] M. Lockett, *Distillation Tray Fundamentals*, Cambridge University Press, 1986.
[55] H. Z. Kister, Distillation Design, McGraw-Hill, 1992.
[56] M. Duss and R. Taylor, "Predict Distillation Tray Efficiency," *Chemical Engineering Progress*, vol. 114 no. 7, pp. 24–30, July 2018.
[57] R. Taylor and M. Duss, "A Tray Column Design Method Incorporating a New Simple Efficiency Model," in AIChE Spring Meeting, New Orleans, LA, 2019.
[58] H. Z. Kister, "Effects of Design on Tray Efficiency in Commercial Towers," Chemical Engineering Progress, January 2008.
[59] D. R. Summers, "A Brief History of Distillation Trays and Tray Correlations," in AIChE Spring Meeting, San Antonio, TX., 2022.
[60] R. Harrison, P. Todd, S. Rudge and D. Petrides, *Bioseparation Science and Engineering, Section 10.2.2 Vapor-Liquid Separation*, 2nd Edition, Oxford University Press, 2015.
[61] W. Stupin and H. Z. Kister, "System Limit The Ultimate Capacity of Fractionators," IChem.
[62] A. Bernard, "Tray Operating Window – Description and Applications," in AIChE Spring Meeting, San Antonio, TX, 2017.
[63] D. R. Summers, "Distillation Tray Design Through a Better Understanding of Operating Windows," in AIChE Annual Meeting, San Francisco, CA, 2016.
[64] H. Z. Kister, Distillation Operation, McGraw-Hill, 1990.
[65] D. R. Summers, "Facts at your Fingertips," *Chemical Engineering*, vol. 117 no. 6, p. 27, June 2010.
[66] D. R. Summers, "Sieve Tray Stability Factor Based on Efficiency Data," *Chemical Engineering*, vol. 128 no. 9, p. 31, September 2021.
[67] D. R. Summers and A. Sloley, "How to Handle Low Liquid Loadings," *Hydrocarbon Processing*, vol. 86 no. 1, p. 67, January 2007.
[68] H. Z. Kister, R. Dionne and W. Stupin, "Preventing Maldistribution in Multi-Pass Trays," *Chemical Engineering Progress*, vol. 106 no. 4, pp. 32–41, April 2010.
[69] D. R. Summers, "Designing Four-Pass Trays," *Chemical Engineering Progress*, vol. 106 no. 4, pp. 26–31, April 2010.
[70] W. Bolles, "Multipass Flow Distribution and Mass Transfer Efficiency for Distillation Plates," *AIChE Journal*, vol. 22 no. 1, pp. 153–158, January 1976.
[71] D. R. Summers and T. J. Cai, "Dry Tray Pressure Drop of Sieve Trays Revesited," *Chemical Engineering*, vol. 124 no. 8, pp. 38–41, August 2017.
[72] C. Colwell, "Clear Liquid Height and Froth Density on Sieve Trays," *Ind. Eng. Process Design and Development,* vol. 20, pp. 298–307, 1981.
[73] A. Bernard, "Distillation in Ethylene Plants – An Operators' Overview," in AIChE Spring Meeting, San Antonio, TX, 2022.
[75] H. Z. Kister, J. Scherffius, K. Afshar and E. Abkar, "Realistically Predict Capacity and Pressure Drop for Packed Columns," *Chemical Engineering Progress*, vol. 103 no. 7, pp. 28–38, July 2007.

[74] E. E. Ludwig, Chapter 9 – Packed Tower, Applied Process Design for Chemical & Petrochemical Plants, Elsevier, 1997.
[76] H. Z. Kister and D. Gill, "Flooding and Pressure Drop Prediction for Structured Packings," in *IChem Symp. Ser. 128*, pp A109-A123.
[77] A. W. Sloley, "Properly Design Thermosyphon Reboilers," *Chemical Engineering Progress*, vol. 93 no. 3, pp. 52–64, March 1997.
[78] E. V. Doorn and H. Z. Kister, "Stable Thermosiphon Reboiler Operation: With or Without a Valve in the Inlet Piping?," *Hydrocarbon Processing*, vol. 98 no. 2, pp. 73–76, February 2019.
[79] E. van Doorn and H. Z. Kister, "Stable Thermosiphon Reboiler Operation: with or without a Valve in the Inlet?," in *AIChE Spring Meeting*, New Orleans, LA., 2019.
[80] E. Van Doorn and H. Z. Kister, "Stable Thermosiphon Reboiler Operation: With or Without A Valve in the Inlet Piping?," *Hydrocarbon Processing*, vol. 98 no. 2, pp. 73–76, February 2019.
[81] A. Bernard, "Distillation Tower Statup Considerations in Ethylene Plants," in AIChE Spring Meeting, Virtual, 2021.
[84] ALPEMA, The Standards of the Brazed Aluminium Plate-Fin Heat Exchanger Manufacturers' Association, 4th Edition, The Brazed Aluminium Plate-Fin Heat Exchanger Manufacturers' Association (ALPEMA®), The Standards of the Brazed Aluminium Plate-Fin Heat Exchanger Manufacturers' Association, 4th edition, 2024, 2024.
[82] A. Bernard, "Distillation Tower Startup Following a Plant Revamp with Focus on Reboiler Operation," in AIChE Spring Meeting, Houston, TX., 2023.
[83] M. Hagan and V. Kruglov, "Understand Heat Flux Limitations on Reboiler Design," *Chemical Engineering Progress*, vol. 106 no. 11, pp. 24–31, November 2010.
[85] Z.-Y. Liu and M. Jobson, "The Effect of Operating Pressure on Distillation Column Throughput," *Computer & Chemical Engineering*, vol. 23, pp. S831–S834, June 1999.
[86] A. W. Sloley, "Effectively Control Column Pressure," *Chemical Engineering Progress*, vol. 97 no. 1, pp. 38–48, January 2001.
[87] H. Z. Kister and D. Hanson, "Control Column Pressure via Hot-Vapor Bypass," *Chemical Engineering Progress*, vol. 111 no. 2, pp. 35–45, February 2015.
[88] A. Bernard, "C2 Splitter Pressure Control Problems and Solutions," in *AIChE Spring Meeting, Kister Distillation Symposium*, 2025.
[89] B. Poling, J. Prausnitz and J. O'Connell, The Properties of Gases and Liquids, Fifth Edition, McGraw-Hill, 2001.
[90] A. Goodwin and J. Sengers, Chapter 4 – Cubic and Generalized van der Waals Equations of State, Applied Thermodynamics of Fluid, Royal Society of Chemistry (RSC, 2010.
[91] R. Agrawal, Y. Li, O. Santollani, M. Satyro and A. Vieler, "Uncovering the Realities of Simulation Part 1," *Chemical Engineering Progress*, vol. 117 no. 5, pp. 42–52, May 2001.
[92] H. Kister, "Can we Believe the Simulation Results?," *Chemical Engineering Progress*, vol. 98 no. 10, pp. 52–58, October 2002.
[93] AspenTech, Aspen Physical Property System, Physical Property Methods, Aspen Technology, Version V7, Aspen Technology, January 2009.
[94] Z. Braswell, B. Hanley and H. Ragsdale, "Reconsidering the Reliability of Chemical Process Simulators I: Implementation of the Murphree and Vaporization Efficiencies," *Industrial & Engineering Chemistry Research*, 2024.
[95] H. Z. Kister, Distillation Diagnostics An Engineer's Guidebook, Wiley, AIChE, 2025.
[96] H. Z. Kister, "Use Quantitative Gamma Scans to Troubleshoot Maldistribution on Trays," *Chemical Engineering Progress*, vol. 109 no. 2, pp. 33–42, February 2013.

[97] H. Z. Kister, "Apply Quantitative Gamma Scanning to High-Capacity Trays," *Chemical Engineering Progress*, vol. 109 no. 4, pp. 45–49, April 2013.
[98] H. Z. Kister, "New Frontiers in Quantitative Analysis of Packed Tower Gamma Scans," in AIChE Spring Meeting, Kister Distillation Symposium, Dallas, TX, 2025.
[99] ASTM D3370–18, *Standard Practices for Sampling Water from Flowing Process Streams,* ASTM.
[100] L. Machemer, "Sampling of Process Streams," Process Control & Automation, February 5th, 2020.
[101] L. Machemer, "Pressurized Piping: Sampling Steam and Water," *Chemical Engineering*, vol. 121 no. 1, pp. 42–47, January 2014.
[102] A. Bernard, "Debutanizer Bottoms Recycle for Enhancing Gasoline/Water Separation in the Quench System," in *AIChE Spring Meeting, Ethylene Producer Conference*, 2019.
[103] C. Hammond and E. Walker, "Ethylene Quench System Fouling," in *AIChE Spring Meeting, Ethylene Producer Conference*, 2006.
[104] B. R. Bird, W. E. Stewart and E. N. Lightfoot, Transpot Phenomena, John Wiley & Sons, 1960.
[105] S. Moran, "Pump Sizing: Bridging the Gap Between Theory and Practice," *Chemical Engineering Progress*, vol. 112 no. 12, pp. 38–44, December 2016.
[106] Crane, Flow of Fluids through Valves, Fittings, and Pipe, CRANE CO., 1986.
[107] T. W. Cochran, Simplifying Piping Network Analysis, *Chemical Engineering*, vol. 102 no. 4, pp. 114–119, April 1995.
[108] API, *Standard 521, Pressure-Relieving and Depressuring Systems*, Fifth Edition, Washington DC: American Petroleum Institute, 2007.
[109] O. A. Hougan, K. M. Watson and R. A. Ragatz, *Chemical Process Principles Part II* – Thermodynamics, John Wiley & Sons, 1959.
[110] F. Teng, P. Medina and M. Heigold, "Compressible Fluid Flow Calculation Methods," *Chemical Engineering*, vol. 121 no. 2, pp. 32–41, February 2014.
[111] A. Bernard, "Troubleshooting a Cooling Water System with a Rigorous Flow and Pressure Network Model," in *AIChE Spring Meeting, Ethylene Producer Conference*, New Orleans, LA, 2004.
[112] W. L. McCabe, J. C. Smith and P. Harriott, Unit Operation of Chemical Engineer, McGraw-Hill, 1985.
[113] J. D. Kumana and M. R. Suarez, "Analyzing the Performance of Pump Networks Part 1: Basic Theory," *Chemical Engineering Progress*, vol. 114 no. 1, pp. 34–42, January 2018.
[114] H. P. Bloch, Pump Wisdom Problem Solving for Operators and Specialists, John Wiley & Sons, 2011.
[115] A. Bernard and M. Pickard, "Charge Gas Compressor Operation and Mechanical Considerations," in *AIChE Spring Meeting – Ethylene Producer Conference*, Dallas, TX, 2025.
[116] S. Harvey, "Centrifugal Compressors in Ethylene Plants," *Chemical Engineering Progress*, vol. 113 no. 2, pp. 28–32, February 2017.
[117] R. P. Lapina, "Compressor – How Changes in Inlet Conditions Affect Efficiency," *Chemical Engineering*, vol. 97 no. 7, pp. 110–113, July 1990.
[118] ASME, *PTC 10–1997, Performance Test Code on COmpressors and Exhausters,* ASME.
[119] H. P. Bloch, *A Practical Guide to Compressor Technology*, McGraw-Hill, 1996.

[120] A. G. Godse, "Predict Compressor Performance at New Conditions," *Hydrocarbon Processing*, vol. 68 no. 6, pp. 77–79, June 1989.

[121] AIChE Committee, Equipment Testing Procedures, "Evaluating Centrifugal Compressor Performance," *Chemical Engineering Progress*, vol. 109 no. 7, pp. 35–39, July 2013.

[122] F. G. Shinskey, *Process Control Systems Third Edition* – Application, Design and Tuning, McGraw-Hill, 1988.

[123] G. Dunnells and K. Akhtar, "Minimum Flaring – Impact on COmpression System," in *AIChE Spring Meeting, Ethylene Producer Conference*, New Orleans, LA, 2004.

3 Troubleshooting – Consequences and Circumstances

3.1 CONSEQUENTIAL AND CIRCUMSTANTIAL TROUBLESHOOTING

The discussion in Chapter 2 was focused on functionality and potential deviation that may occur in unit operation typically found in the CPI. Dimensional questions, what, where, and why are associated with an object (i.e., unit operation) exhibiting a deviation. However, deviations experienced in an operating unit, while frequently being part of the root cause, are rarely the initiator for troubleshooting activities. The need for troubleshooting arises under different circumstances and the consequence of the problem will also influence the strategy and timelines. A rate constraint may affect the plant capacity but may not require a full outage, while environmental exceedance or process safety hazard may require an immediate plant shutdown. Troubleshooting performance deviation in an operating plant has the luxury of operating data and performance history. Commissioning a new facility, on the other hand, has no performance history that could provide a baseline. Change in plant behavior following a unit startup or following recovery from a process upset would focus on transient conditions and sequence of events rather than long-term performance history. Consequential troubleshooting relates to the dimensional question "what", not from the object causing the problem but from the perspective of consequences, which provide the framework for an initial problem definition. Circumstantial troubleshooting imposes different constraints with respect to time and decision-making. The classification and magnitude of the problem will dictate the strategy and urgency for finding a solution. This chapter discusses troubleshooting strategies based on the consequences, circumstances, and available data.

3.2 PROBLEM CLASSIFICATION

Process engineers involved in an operating unit work in a dynamic environment of constant change where today's priorities are different from those of yesterday and tomorrow. Effective process engineers must have the ability to identify priorities that matter the most. It is critical to understand that the rate of performance degradation does not exceed maintenance strategies such that it would not impede plant capacity, product quality, or cause environmental exceedance. The process engineer also plays a critical role in ensuring that the unit operates safely. Problems requiring

DOI: 10.1201/9781003449676-3

troubleshooting can therefore be classified in terms of observable consequences, such as performance degradation, capacity limitations, inability to make product sale specifications, environmental exceedance, and process safety.

3.2.1 Plant Performance Degradation

Performance degradation in a chemical plant is not unusual and often the result of tear and wear or fouling of process equipment. In some cases, non-optimal operational practices could translate into overall performance degradation. In an existing facility, degradation in performance is sometimes not recognized as a problem until it impacts target production volumes, product quality, environmental emissions, or process safety. The performance degradation rate of key equipment would often define the turnaround frequency of a unit. Compressors, pumps, heat exchangers, and towers operating in fouling service are typical equipment exhibiting performance degradation over time. When applicable and economical, some of those equipment are duplicated to avoid a full plant outage.

In many instances, early detection of performance degradation could minimize the full impact and in some cases with appropriate response restore performance. Early detection requires proactive behaviors normally accomplished through effective unit performance monitoring and key sets of performance metrics frequently reviewed. Proactive plants who have good performance monitoring programs tend to do better than those who are reactive. Process unit monitoring is discussed in more detail in Chapter 4.

The CEDs illustrated in Figure 1.4 in Chapter 1 breaks down problems' potential root causes into six categories. Those categories stimulate the thinking process as performance degradation symptoms may relate to different mechanisms. While early detection helps in acting in a timely fashion, understanding the mechanisms responsible for performance degradation will help in implementing effective remediation strategies intended to minimize performance degradation and maximize production asset availability.

People is the category in a CED that considers the human interface with the process. As stated in Chapter 1, the human factor is often overlooked due to fear of assigning blame to individuals. Human interaction with the process deserves attention as it brings an element of variability. Cars and drivers are used as an analogy to demonstrate the impact of human interaction relative to performance degradation. The performance of cars of identical designs and construction will likely deteriorate at different rates when operated by different drivers. Let's assume that the cars are performing the exact same tasks, moving from point A to B then B to A repeatedly. All the cars meet the basic objectives of getting their drivers to the destination. Therefore, normality might be defined as the ability of the car to bring the driver to its destination. Normality would also suggest that car maintenance occurs per the manufacturer's recommended schedule. However, drivers have different driving habits from speeds, breaking, and use of different systems within the car (i.e., heating, air conditioning, radio). As such, failure would likely occur at different rates. For some, failure might not be experienced as it would occur outside the maintenance recommended window, which can be defined as normality. For others, failure would occur prematurely, resulting in an unexpected trip to the garage for repair. In a chemical plant, process operators may follow the same procedures but may introduce an element of variability in the execution of their tasks, as some may be more aggressive than others in transitioning process variables. Compromises are often required for meeting the unit performance metrics, which in some cases could be detrimental to the long-term reliability of some process components. The method category in the CED is indirectly associated with people as it relates to operating procedures and strategies.

Troubleshooting process operators' behaviors is not an easy task and is rarely considered. However, failing in correcting behaviors that are detrimental to process equipment's reliability does not only impact the reliability of a process unit but also hinder process operator's development. Process engineers should pay attention when reviewing process trends to aggressive changes in process variable SPs, a behavior often referred as "Hammer Move" by process operators. Except for a few activities, process variables shall be ramped up or down gently to avoid disrupting the unit. Furthermore, frequent aggressive transition could cause thermal and/or mechanical stress that eventually results in failure by fatigue of process equipment's mechanical components. The use of operator training simulators (OTS), discussed in more detail later, is a great tool to exposed process operators to numerous conditions, with time spent reviewing and correcting improper actions when required. In any case, feedback should be provided in a constructive manner, and blame should be avoided.

Material such as change in feedstocks could likely cause significant shift in yields from the reaction section (furnaces or chemical reactors). Change in yields will shift the process loads in the recovery section and may create the illusion that some equipment experienced performance degradation or become prematurely constrained. For instance, an increase in streams' molecular weight would result in higher power demand in compressors, which could translate in capacity limitations. The higher molecular weight would also increase the pressure profile across a multistage compressor train and could mistakenly be associated with fouling. Change in distillation tower feed composition will result in a shift in internal hydraulic load, which may fall outside the capability of the internals. The fouling propensity of contaminants and impurities from feedstocks could cause premature fouling and may require a different protective chemistry and/or dosage.

In many instances, new feedstocks are assessed by lab analysis with a focus on contaminants known to affect the unit. New feedstocks are frequently introduced in the plant as experimental trial runs. Some of the impacts of the new feedstocks may manifest themselves quickly, while consequences from some reaction mechanisms may occur weeks or months later.

Measurement is a category where faulty instruments may cause operation outside the recommended range. As an example, operation of condensing steam turbines at lower exhaust pressure than design will cause higher level of moisture against wheels that may not have the mechanical characteristics or tolerance for it. While this operating mode may positively influence the energy balance, premature failure of the turbine would be expected. A faulty temperature transmitter in distillation tower used for regulating the heat stream to a reboiler may cause higher internal hydraulic loadings with a warmer temperature profile in the tower bottom, which would accelerate chemical reactions in fouling service. Both effects would lead to premature capacity bottleneck. In contrast, a high-temperature offset would however be picked up quickly as it would cause the tower bottom to go off specification.

Environmental conditions often involve atmospheric conditions, which influence cooling capabilities from cooling water systems and air exchangers (e.g., fin fans). Seasons (summer vs. winter), day versus night, and rainfall cause elements of variability in the performance of cooling towers and air coolers.

Radiant heat from sunlight may cause temperature rise or worse, thermal liquid expansion, which may result in overpressure on uninsulated stagnant volume or pipe dead legs. On the other hand, extreme cold weather could cause freezing. Both of those conditions could result in loss of process containments (LOPCs).

Equipment is often the recipient of the above deviations and the component that exhibits symptoms without necessarily often being the root cause of the problem.

It is important to distinguish between a gradual performance degradation over time and a sudden drop in performance. Expected gradual degradation would typically align with plant maintenance and turnaround schedule. As stated earlier, degradation in performance is normally expected and is not a problem if the plant meets its operational metrics and the rate of degradation remains aligned within maintenance and turnaround schedule. Higher than expected degradation rate requires attention to prevent costly unscheduled outages. Sudden changes in performance are often the result of a significant failure of the equipment or one of its components. Failure of seals and/or bearings in rotating equipment is not uncommon after a significant plant upset. Coking in pyrolysis furnaces is typical following a furnace trip and requires going through a decoke cycle prior to bring the furnace back on stream. Other equipment may also experience a change in performance after recovering from an abnormal transient. As an example, a section of a demister pad in a refrigeration system knockout drum lifted during an abnormal transient that caused the compressor antisurge valves to open quickly. The high velocity in the drum caused the demister section to lift, resulting in significant volume of liquid droplets traveling as side load to a multistage compressor. While the volume of liquid entrained provided quenching to the compressor side load, and reduced power, it constrained liquid refrigerant supply to downstream heat exchangers. This resulted in a capacity constraint, where fortunately the magnitude of liquid carryover was insufficient for causing damage to the compressor.

3.2.2 Capacity Limitations

Evaluating capacity limitations starts with the equipment and requires understanding of the design capabilities. Chapter 2 discussed the fundamental principles and operating capabilities of the major equipment typically found in the CPI. The ultimate plant capacity is defined by the process unit which first reaches its maximum design capability, often referred to as the bottleneck. That maximum capacity can be defined by throughput and/or run length. Maximizing the value of the asset is directly related to maximizing the productivity of the bottleneck. In manufacturing, the concept of overall equipment effectiveness (*OEE*) [1] is defined by the product of availability (*A*), performance (*P*), and quality (*Q*) as described by Equation 3.1:

$$OEE = A \; x \; P \; x \; Q \tag{3.1}$$

An effective plant design would be limited by the reaction section. The number of reactor units (furnaces or chemical reactors) is typically set, such that regeneration schedule maximizes the availability of the units. Constrained by recovery, while possible, would not be optimal as expensive assets such as furnaces or chemical reactors would be either off stream or at reduced rate. However, throughout a plant run length, units in the recovery section are expected to experience performance degradation as previously discussed. Optimal turnaround frequency would be set such that performance degradation would not severely impact plant profitability.

Fouling of process equipment is one of the main causes limiting plant run length. Constraints experienced downstream the reactor section, although not affecting the reactors themselves, indirectly impact their productivity as rates becomes limited by the recovery section. The fouling mechanisms are described in more detail in a later section. For the troubleshooter, it is critical to understand where fouling is taking place as it may not be related to the equipment showing the symptoms. For instance, a process engineer could easily conclude flooding on a distillation tower caused by fouling or mechanical debris as the root cause. Process variables such as high-pressure drops, gamma scans revealing flooding would support his conclusion. Taking an outage to validate and fix this problem would prove costly not counting that the process engineer would find the tower clean. The above analysis missed on a critical step in troubleshooting chemical processes. A proper H&MB around the tower would demonstrate that the feed conditions significantly deviate from the design basis. In this case, the feed is preheated prior to entering the tower. Before concluding on fouling, an H&MB would have demonstrated that the feed was entering the tower at a vapor fraction 20% less than specified by design. The shortage of heat had to be compensated for by the bottom reboiler, which increased hydraulic loads below the feed point beyond the tray capacity, hence explaining flooding. Chemical plants are often highly integrated, a shift in process loads could often mislead the troubleshooter. While troubleshooting may start at the symptomatic unit, it is critical for the troubleshooter to understand interactions/integrations with other process units when defining the boundary of the analyses. Development of an H&MB is also a critical component which will help identify the deviation. In the above case, the lack of attention and understanding of the feed preheater relation with the tower caused the capacity limitation. While this may fall under methods and equipment in the CED, it has a strong relation with the people category. Without assigning blame, this example highlights that proper troubleshooting would have avoided the embarrassment of a non-required outage. Communicating the learning of such events through training and operating procedures is also critical for preventing re-occurrence.

To a lesser extent than fouling, operation practices deviating from the design intent could shift process loads in units which would prematurely constraint plant capacity. Those deviations could be the results or response to external factors or design anomalies from which operation adapted but may result in shifting constraints if not properly evaluated. Those behaviors are also frequently referred to as acceptance of deviations.

3.2.3 Inability to Meet Product Stream Sale Specification

Depending on plant configuration, the inability to meet product sale specifications involves routing product to off-specification storage for re-processing or flaring. Without a quick resolution, a total plant outage becomes necessary as off-specification storage has limited capacity and extended flaring periods, besides wasting valuable molecules is environmentally not acceptable. This situation comes with the challenge of time and limited data for avoiding an outage. Before getting into any troubleshooting exercise, it is imperative to understand which parameter or components causes the specification deviation. This becomes the answer to the "What" dimensional question in a KT analyses and will help in figuring out "Where" or the origin of the problem whether internal or external to the process unit. The most common causes leading to product specification excursions or exceedances are discussed below.

Capacity limitations, as previously discussed, should impact throughput and production volumes. However, under certain circumstances the symptoms of capacity limitations may manifest themselves in terms of composition deviations. This primarily affects operating units conducting separation functions such as distillation towers, phase separators, and adsorption beds. It could also impact chemical reactors responsible for converting impurities into components that would not otherwise be separated by downstream units such hydrogenation of acetylenes in ethylene plants. Capacity limitations are commonly associated with the maximum throughput capability of process units. The lower operating end or turndown limitations of operating units are often overlooked but could also cause significant drop in efficiency, resulting in stream specification exceedance. The troubleshooter must therefore first correlate the plant throughput to process load imposed on the process units. This involves establishing H&MB along with equipment rating calculations. Validating operation within the operating window of the equipment would be a first step in eliminating capacity limitations as a root cause.

Leaks due to cross-contamination between process streams in heat exchangers or from improper valve position and/or blinding of intermittent utility streams could contaminate the product streams. Knowing the contaminants responsible for specification exceedance will lead to investigating every exchanger containing those contaminants throughout the flow path used by the product between the reactor effluent and the product stream battery limit. The likelihood of leakage can first be assessed by comparing stream pressures, which will help in reducing the list of potential suspects. Sensitivity analyses through material balance will help quantify the magnitude of a leak that would align with the content found in the final product. This exercise would further sort the list of potential suspects and could be facilitated with process simulations. In the case of contamination by utilities (e.g., nitrogen), all entry points shall be reviewed for proper valve and blinding. If blinding does not exist, passing valve remains a potential source until ruled out by testing or analyses.

Intermediate streams not meeting specifications have similarities with leakage as it allows impurities to flow forward to process units that have not been designed for their removal. As such, those impurities may follow the same path as the final product. A review of the process flowsheet and material balance should reduce the boundary of the evaluation to the process units responsible for removing the impurity/impurities responsible for the off-specification excursion.

Feedstocks and external recycles impurities could cause product off-specification if they upset the recovery section or if the recovery section has not been designed for such impurities. Methanol is frequently used in cryogenic services and pipelines for removing hydrate formation, which causes blockage. In ethylene cracking furnaces, 50–75% of the methanol contained in feedstock is converted to CO, which then follows the path of hydrogen in the recovery section. A small fraction of the hydrogen stream is used for the hydrogenation reaction. CO is known to moderate the catalytic hydrogenation reaction in acetylene reactors. A spike in CO could therefore affect the hydrogenation reaction, resulting in an acetylene breakthrough and off-specification ethylene product. Methanol is also used within an ethylene plant recovery section for the same purpose (i.e., removing hydrates in cryogenic service). Depending on the injection point, methanol tends to follow the path of the C3 products. When injected in the ethylene fractionator, it would follow the bottom ethane stream recycled to the cracking furnace, where it can create a spike of CO. If injected in the ethylene delivery system, it would impact the end users. Ethylene and C3 products become the feedstock to downstream facility in the value chain. It is therefore important to understand that any additives used for resolving an operational problem can become a source for off-specification excursion to downstream users. Many years ago, a plant converting ethylene to ethanol by direct hydration, a process no longer in use in North America, went off-specification every time the ethylene supplier pumped methanol in the ethylene delivery system as their recovery unit was not designed for methanol removal.

Unrealistic specifications may not align with the design basis of the unit. This would more likely apply to new facilities as stream specifications should be well established and supported by history. However, it is not unusual for marketing and sales to see a market and pricing opportunity by requesting a more stringent product specification. Meeting the tighter product specification could be feasible under some circumstances, which may mislead the team in making those commitments. For instance, was the tighter product specification achieved at rates lower than nameplate? If so, the new specification may negatively impact the ability of the plant in meeting the nameplate capacity. Feedstock variability may be another reason why the tighter specification was achieved. As an example, an ethylene facility with feedstock flexibility operated a propylene fractionator design for delivering a chemical grade propylene (92%) product. The tower had delivered a propylene composition of 96% for an extended period. Marketing saw an opportunity in negotiating a premium on the propylene composition. However, shortly after, the plant increased the volume of propane cracking. The propylene fractionator became limited due to higher reflux ratios resulting from the increase in the tower feed propane composition. The tower was therefore no longer capable of delivering the new 96% specification at the current plant rates.

As another example, the product specifications for a new ethanol plant were established by the marketing group. Eager to gain access to the market and beat competition, the specifications were established without consideration to the design basis and capability of the facility. This presented a significant challenge to operation, which had to significantly deviate from the intended design operating philosophy. Significant losses and false load operation resulted in wasting valuable molecules and significantly reducing the design capacity of the unit.

3.2.4 Environmental Exceedance

Environmental exceedances can be the result of an abnormal event resulting in the release of chemicals to the environment. Some of those incidents could be of minor or major consequences such as the Bhopal tragedy [2] in 1984 resulting in the death of several thousand people following the release of 28 tons of methyl isocyanate (MIC) in the atmosphere over a period of about two hours. A large fraction of major environmental incidents is closely tied to process safety practices. Some of the common themes found when reviewing incidents investigations often point to normalization of deviations and process safety culture.

On a more normal basis, environmental releases are frequently associated with losses and are an indirect measure of recovery effectiveness. As initially stated, chemical processes involve the transformation of feedstocks into a mixture of valuable chemicals. Nonetheless, many impurities of little to no value must also be separated from the product streams, which occurred through purges and/or vents. When feasible, some of those waste streams are burned in fired equipment due to their calorific value or release to flare for destruction. Contaminated water streams are treated in wastewater treatment units prior to being discharged in local sewers or natural water streams. Besides purges required for maintaining product quality, safety devices such as PSVs rupture discs, and high-pressure control valves may contribute to environmental releases. Molecular sieves and onsite catalyst bed regenerations are intermittent operations associated with flare releases. Fugitive emissions, defined [3] as unintentional leakage of gases from plant equipment, such as valves, piping flanges, rotating equipment's seals, and storage tanks, are another source of environmental releases. The characteristics of the above releases, composition, rates, and sometimes durations are normally regulated by local environmental agencies.

Flare operation, as previously discussed, must meet a minimum heating value for ensuring effective destruction of hydrocarbons. This is normally associated with a relatively small flame visible at the flare tip. Deviation from normality would first be observed by the size, luminosity, and behavior of the flame. The visual observation might in some cases be sufficient in correlating the incremental flare release to its origin. Other circumstances might not be so trivial as the incremental release may come from passing valves or other abnormalities. While incremental flare releases might still fall within emission allowable by regulation, they are symptomatic of abnormal losses, which, if not addressed, may trend upwards beyond acceptable levels. Beyond flame observation, correlating abnormal flare releases to their origin could be executed in three layers. The first layer involves reviewing process data that correlates directly with releases of material to flare. This involves trending the output of control valves, pressures upstream PSVs, and rupture disks with outlet stream tied to flare headers. Positive output on control valves and operating pressure excursion exceeding set pressures of PSVs or rupture disks would correlate with increased level of flaring. A second layer requires a flare gas analyzer and can be used for short listing potential sources based on the predominant components found in the flare gas. Ratios of 2–3 components may help pinpoint the units responsible for the release. A third layer, somewhat less reliable, trends process variables in the vicinity of flare sources. Using statistical process control (SPC) techniques, variables such as flows, pressure, or temperature exhibiting behavior outside control limits should be treated as suspicious. Increases in material balance offsets can also be used for identifying potential flare sources.

Wastewater variation in hydraulic and/or contaminants loadings may affect the ability of the wastewater treatment unit in meeting effluent specification for release in local sewer or natural waterways. Generally, releases to plant sewer systems are well understood and easily correlated to operating practices for which the wastewater treatment unit would have been designed. Unexpected releases would be the result of abnormal operation, potentially initiated from transient conditions or equipment failures. Tank overflow, drain valve left open, and seal failures are a few examples of abnormal releases. In most cases, releases to plant sewers are easily correlated to equipment or unit malfunctions. For some odd cases, it might be challenging to link a wastewater excursion to a specific unit. The quality of a wastewater effluent is generally quantified in terms of pH, suspended solids, turbidity, chemical oxygen demand (COD), biological oxygen demand (BOD), and/or total organic carbons (TOCs). From the perspective of organic material contained in the effluent, COD and TOC analyses are most likely preferred over BOD due to simplicity and quickness of the test. Nonetheless, all the parameters stated above do not correlate directly with components found in the process units. This challenge can be illustrated by an effluent excursion experienced by a specialty chemical plant, which has several units mostly using batch reactors and solvent recovery sections. Wastewater from each unit was discharged in the plant's common sewer and then released in the municipal sewer. The municipality operated a wastewater treatment unit on its own and had negotiated with the plant target levels of contamination. The plant had no wastewater treatment unit as each individual unit, if properly operated, had the capability of meeting the target levels agreed with the municipality. An unexplained exceedance of the plant effluent oil and grease and COD content caused a production disruption. One unit was suspected as the culprit based on previous history but without any process data supporting the current excursion. Process engineering suggested using an empty and available storage tank for collecting the effluent of that unit. This would allow continuous monitoring (analysis and flows) of the water release by the unit and rule it out if the water collected in the storage tank met specifications. While executing the above plan, the final effluent still showed signs of elevated oil and grease and COD. In the meantime, a review of activities conducted in other units revealed that a solution used for washing a vessel, while not expected to cause an environmental exceedance, would have contributed to dissolve and wash scale accumulated over the years in the plant sewer system. In this troubleshooting exercise, isolating the suspected unit by storing wastewater while helping was not necessarily what led to the root cause. This strategy was directed by the unit manager to avoid shutting down his unit while buying time for resolving the problem. A good review of activities prior to the environmental excursion and identification in what was different from previous practices was key in identifying and further investigating the nature of the wash as it was something new to the site.

As for losses to flare, correlating release to plant sewers can leverage on plant data, such as tank inventories, vessels levels, and valve outputs. Monitoring and reconciliation of plant and unit material balances also help in identifying losses and their potential origins.

3.2.5 Process Safety

Proactive process safety is intended for predicting scenarios leading to hazardous events and quantifying the risk of those scenarios. It is normally done through PHA, Hazard Operating Studies (HAZOP), or What If Analysis. As indicated in Chapter 1, process safety analysis has a lot of similarities with troubleshooting, where the root cause in troubleshooting becomes the initiating event leading to a hazard in process safety. The work process goes backward from the initiating event to the consequences rather than consequences to the most likely root cause. Those initiating events are the result of one or several deviations from the plant design intent. The quality of process safety analysis strongly depends upon the level of knowledge of those involved. One element of confusion frequently encountered with process safety methodologies is the distinction between double jeopardy and chain reactions or domino effects. Those effects could easily be missed without the participation of process experts having a deep understanding of the process unit evaluated and the integration it has with other units. Process safety hazard analyses methodologies are often conducted from P&ID and are broken down into small sections or nodes. Too often, for the sake of time, the scenarios are brainstormed but rarely thoroughly evaluated with likelihood and consequences established based on intuition. Although acceptable for simple systems, a deeper understanding of the sequence of events, transient heat and material balances, and equipment loads help, first in confirming the credibility of the scenario, then provides more clarity around the likelihood and consequences. Such deep analysis would also provide a better understanding in the design of safeguards and operational response. SIS and High Integrity Protection System (HIPS) design effectiveness strongly depend on a clear understanding of the mechanism and sequence of events triggering a hazardous condition. It is not unusual for SIS and HIPS systems to be designed based on intuition, which can potentially result in ineffective protection or nuisance trips.

Example 3.1: This illustrates the above discussion, where an analysis of the potential for cold temperature embrittlement in bimetallic demethanizers operating in ethylene plants [4] is being discussed. Bimetallic demethanizers were common in the 1960s and early 1970s ethylene plants, designed [5] with 3½ nickel alloy in the top section and killed carbon steel (KCS) in the bottom section. Figure 3.1 illustrates a hypothetical tower with the specification transition from 3½ nickel alloy to KCS occurring below the second feed point from the bottom. The right-hand side of Figure 3.1 shows the KCS section minimum design metal temperature (MDMT) and minimum allowable temperature (MAT). The MDMT is evaluated at the design pressure and independent from the operating pressure, while the MAT provides a relationship of the MAT for preventing cold brittle fracture as a function of pressure.

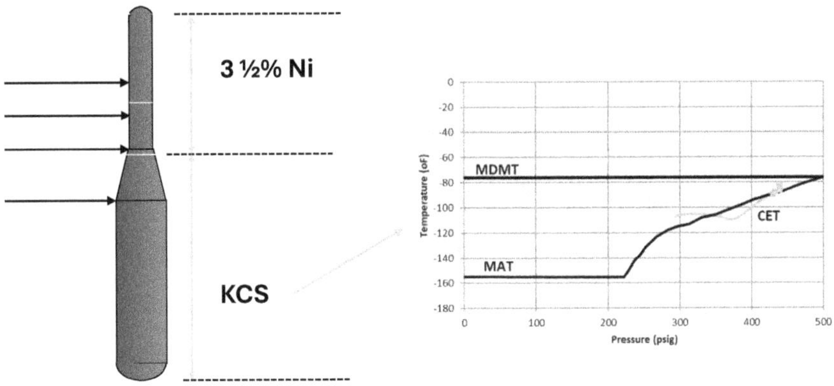

FIGURE 3.1 Bimetallic Demethanizer [4].

Reproduced with Permission from figure 3.5, Evaluating the Potential for Cold Brittle Fracture in Ethylene Plant through Dynamic Simulation, André Bernard, 2014 AIChE Spring Meeting – Ethylene Producer Conference

The use of the MDMT or MAT is described by API 579–1 [6, 7] for conducting a fitness for service evaluation, where using the MDMT would fall under a level 1 and the MAT a level 2 evaluation. The initiating event in this case is a sudden loss of reboiler duty caused by interruption of the heating stream flow rate (i.e., no flow). In a typical brainstorming session, the consequences would easily be identified as a potential cold temperature excursion below embrittlement on the bottom KCS section caused by the cold liquid traveling down the tower. The MDMT would likely be available from the vessel datasheet. However, the MAT must be established from API 579 methodology and might not always be available when conducting a safety review. Without requesting a deeper evaluation, the safety review team may recommend an SIS, which would stop feed to the tower and shut the reflux pump. This provides little input relative to the SIS configuration, such as what would trigger the SIS and would the SIS be effective in preventing a cold temperature excursion below embrittlement. A more effective approach would be to request a rigorous evaluation for this scenario. Establishing the MAT of the KCS section should be assigned to a qualified mechanical engineer and the transient conditions expected shall be evaluated by a process engineer. The use of dynamic simulations in this case would provide a non-destructive test ground for validating the hazard and testing appropriate response. The transient conditions obtained from dynamic simulations would provide the expected tower temperature pressure transient, referred as the critical exposed temperature (CET), at the KCS metal transition point. As shown in Figure 3.1, plotting the CET over the MAT validates the credibility of the scenario identified, where the CET drops below the MAT following the loss of reboil heat to the tower. In agreement with the review team, the aggressive timeline of the temperature drops points toward an SIS for protecting the tower. The conceptual functionality of an SIS that would stop the feed and reflux to the tower is illustrated in Figure 3.2. As shown, the trigger temperature of the SIS could not be a single value due to the temperature/pressure dependency

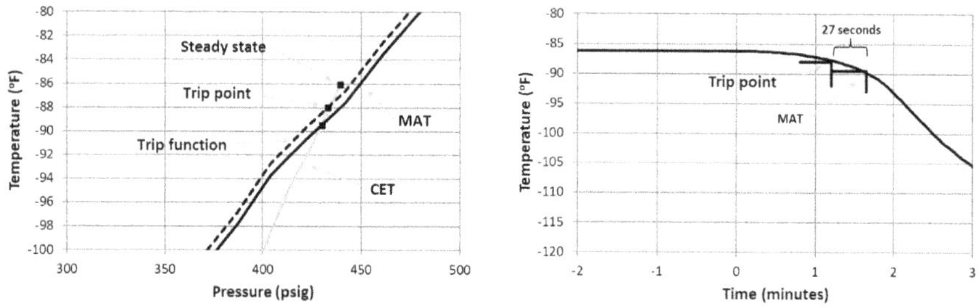

FIGURE 3.2 Conceptual Trip Functionality [4].

Reproduced with Permission from figure 3.13, Evaluating the Potential for Cold Brittle Fracture in Ethylene Plant through Dynamic Simulation, André Bernard, 2014 AIChE Spring Meeting – Ethylene Producer Conference

and the proximity of the operating temperature to the MAT. A temperature/pressure function parallel to the MAT is then proposed as the trigger to the SIS. Nonetheless, the close proximity of the temperatures provides little margin for setting the trip function trigger, offset by 1 °F from the MAT in Figure 3.2. The right-hand side of Figure 3.2 shows the aggressiveness of the temperature transient, with the operating temperature reaching the trip point 1.2 minutes after losing heat to the reboiler and 27 seconds away from dropping below the MAT without taking any actions. It can be concluded from the above discussion that an SIS would not be effective in protecting the vessel for the scenario evaluated and that replacing the vessel in this case would be more appropriate. While this is a fictional case, it can bring the question as of why was the operating point so close to the MAT? Was that an original design missed? The answer to this is that plants may undergo significant changes from their original design intent over the years. Changes in feedstocks, heat integration, and different operating parameters, to name a few, would likely expose process units to process conditions deviating from the original design. Before getting into the above situation, proactive process safety should assess the impact of those changes through a management of change (MOC) process. In 1992, the U.S. Occupational Safety and Health Administration (OSHA) announced the Process Safety Management of Highly Hazardous Chemicals regulations, which requires updating plant PHAs every 5 years, and includes provision for MOC and mechanical integrity (MI).

Example 3.2: This presents a review of the catastrophic incidents that happened at a pesticide plant in Bhopal India on December 3, 1984, which could have been avoided with good process safety management (PSM) systems. This incident has been widely discussed in literature and provides many lessons and learnings for preventing re-occurrence of such events. Among one of the lessons learned is how inadequately assessing operational changes can result in disabling the safety layers. The following discussion will be limited to operational changes resulting from a series of pumps' seal leaks. MIC was stored in two storage tanks with a third one available for diverting contaminated MIC on an emergency basis only. MIC is a highly reactive intermediate that was used for manufacturing various pesticides. It is a highly toxic substance that can be harmful or lethal if inhaled or absorbed through the skin. It reacts with contaminants such as rust and water exothermically. As shown

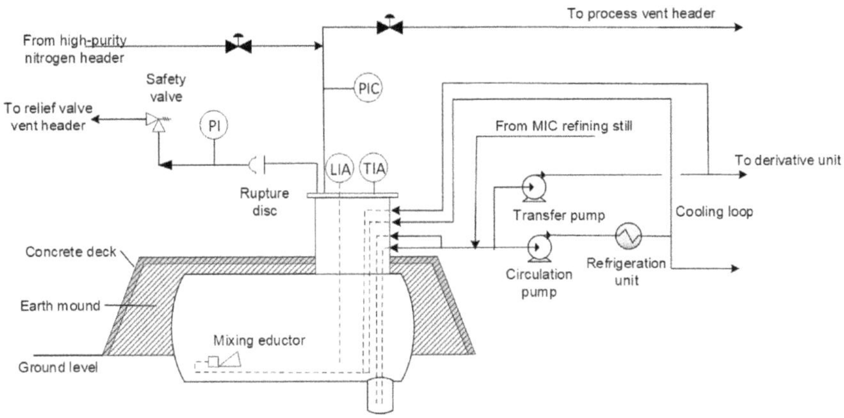

FIGURE 3.3 Bhopal MIC Storage Tank [8].

Reproduced with Permission from figure 3.2, Consider the Role of Safety Layers in the Bhopal Disaster, Ronald J. Willey, *Chemical Engineering Progress*, December 2014

in Figure 3.3, the tanks were equipped with two centrifugal pumps, one for transferring the MIC to a derivative unit where pesticides were produced and the second one ensured circulation of the MIC from the storage tank through a refrigeration unit then back to the storage tank. The refrigeration loop kept the MIC temperature in the storage tank to about 0 °C to prevent a thermal runaway in case of contamination. A high-temperature alarm set at 11 °C was installed on the tanks to warn process operators about an undesired reaction. Operators were expected to respond to the alarm by isolating the contamination source and/or reprocessing or destroying the contaminated MIC. The pumps experienced several seal leaks, creating MIC exposure risks to maintenance workers. Given the lethal nature of this chemical, operation established a procedure for transferring MIC to the derivative unit by pressurizing the tank with nitrogen to a pressure of 14 psig. For similar reason as above, the refrigeration system was shut down at one point due to potential MIC exposure from the circulation pump ensuring refrigeration of the MIC.

The above changes resulted in three deviations from the design intent. First, pressurizing the storage tank with nitrogen required isolating the tank from the pressure vent header (PVH). By doing so, air was able to migrate into the PVH and the relief valve vent header (RVVH). The PVH and RVVH were made of carbon steel and started to corrode, creating rust which catalyzes the formation of MIC trimer deposits. Water flushing in the PVH and RVVH were introduced as maintenance procedure for removing the MIC trimer deposits. This procedure is a second deviation from the original design, as under a continuous nitrogen environment corrosion would not have occurred nor the formation of MIC trimer deposits. Shutdown of the refrigeration loop is the third deviation. By shutting down the refrigeration system, the normal temperature of the storage tank rose from the design 0 °C to a range varying between 15 and 40 °C.

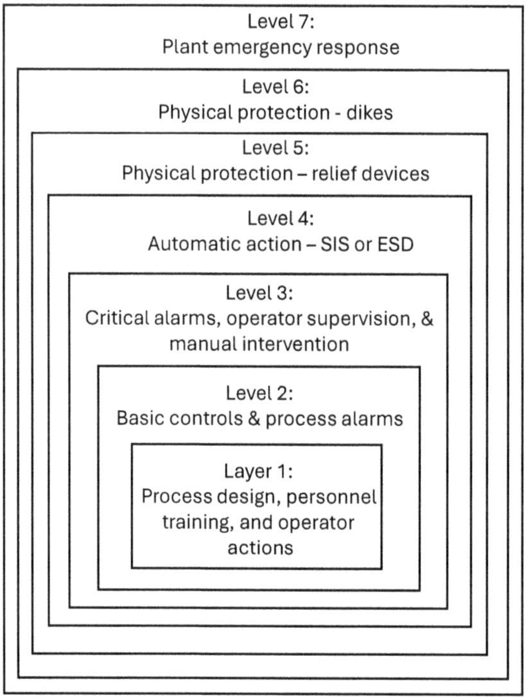

FIGURE 3.4 Layers of Protection [8].

Reproduced with Permission from figure 3.4, Consider the Role of Safety Layers in the Bhopal Disaster, Ronald J. Willey, *Chemical Engineering Progress*, December 2014

Figure 3.4 illustrates seven layers of protection that can be applied to a chemical process. The effectiveness of those layers of protection is described by the frequency of a consequence occurring from one or several abnormal conditions. In conducting a layer of protection analysis (LOPA), Equation 3.2 is used for correlating the frequency of the consequences (f_i) to the probability of an initiating event *IEFi* and the probability of failure of each independent layer of protection (*PFDi*):

$$f_i = IEF_i \times \prod_1^n PFD_i \tag{3.2}$$

Ronald J. Willey [8], using information available from the Bhopal incident, discussed the scenario of a major release of MIC vapor in the surrounding area resulting from the contamination of the storage tank by water. The initiating event of the Bhopal incident was attributed to about 500 kg of water entering the storage tanks. This created an exothermic reaction, which became a runaway due to the accumulation of heat. Based on an assumption that such an abnormal condition could occur once every 10 years ($IEF = 0.1\ yr^{-1}$), the probability of creating a vapor release of MIC can be calculated by evaluating the layers of protection. Table 3.1 compares the layers of protection as intended by design with the expected frequency of experiencing this

TABLE 3.1
Layers of Protection Probability of Failure Matrix [9, 10]

	Frequency	Initiating Event	Layer 1	Layer 2	Layer 3	Layer 4	Layer 5	Layer 6	Layer 7
	Consequences		**Design/ operating procedure**	**Basic controls**	**Alarm manual intervention**	**SIS**	**Relief devices**	**Dike**	**Plant emergency response**
Design	0.000001	0.1	0.1	0.1	0.1	1	0.1	1	0.1
After changes	**0.100000**	0.1	**1**	**1**	**1**	1	**1**	1	**1**
Change alarm SP	0.010000	0.1	**1**	**1**	0.1	1	**1**	1	**1**
Relief system	0.010000	0.1	**1**	**1**	1	1	0.1	1	**1**
Alarm and procedure	0.001000	0.1	0.1	**1**	0.1	1	**1**	1	**1**
Alarm/procedure/relief	0.000100	0.1	0.1	1	0.1	1	0.1	1	1

consequence once every million years ($fi = 1 \times 10^{-6}$) to the impact caused by the change in MIC transfer practice and shutdown of the refrigeration system. While literature reports other deviations related to design intent and operating procedures, the change in methodology for transferring MIC to the downstream unit created conditions promoting corrosion and the deposition of solids in the venting system. This was mitigated with a water wash procedure, which also deviates from the original design intent. Corrosion and the presence of water increased the risk of contamination and reaction with MIC compromising this layer of protection. From a control perspective, shutting down the refrigeration loop disabled the MIC storage tank temperature control, therefore its safety layer. The MIC storage temperature was always above the alarm point due to shutdown of the refrigeration system. Whether the alarm was always displaying a high status or if it was intentionally disabled, compromised this layer of protection. Because of corrosion issues originating from those changes in operational practices, the flare system was out of service, hence disabling safe operation of the relief system. All the layers of protection provided for this system were therefore disabled by implementing two operational changes with dramatic consequences.

Deviations from original design are not something unusual in chemical plants. The lesson learned from the Bhopal incident is that thorough implementation of MOC with proper evaluation of risk must be conducted when implementing deviations from design and established procedures. It can be argued that transferring the MIC by pressure differential and shutting down the circulation pump reduced exposure to maintenance workers. Understanding the impacts on the existing safety layers might have accomplished the same and reduced residual risk within similar levels by implementing a few additional modifications. Table 3.2 provides a typical risk ranking matrix commonly used in process safety that shows the consequences and likelihood relation relative to residual risk. In this scenario, the consequence drives the residual risk when assuming several fatalities within the plant and in the community. As shown in Table 3.1, re-enabling a single layer of protection like the SP of the high-temperature alarm to a new agreed high value assessed acceptable without refrigeration or reliable operation of the flare system would have reduced the frequency from 0.1 to 0.01 yr^{-1}. This would have reduced the residual risk from extreme to high. Re-enabling two of the four safety layers such as temperature alarm along with a procedure for destruction of the MIC on high temperature would have further reduced the frequency to 0.001 yr^{-1} and residual risk to moderate. Re-enabling the above two layers of protection could have been easily implemented. Adding a third layer of protection, by addressing the corrosion and formation of deposits, might have required more intrusive changes but would have further reduced the frequency to 0.0001 yr^{-1}.

It is worth noting that the aftermath of the Bhopal disaster has initiated a significant change in process safety culture. The above example also demonstrates that proactive process safety is an exercise that requires a multidisciplinary team primarily involving process safety engineers for facilitating the evaluation and quantification of risk and process engineers for identifying credible scenarios with their associated consequences. As shown in Example 3.1, developing layers of protection requires a good understanding of the process and its dynamic/transient behavior.

TABLE 3.2
PHA Risk Matrix

		Consequences				
		Limited	**Small**	**Medium**	**Large**	**Catastrophic**
		Consequences Based on Safety, Environmental and Safety Criteria Specific to Organization				
Likelihood	**Frequent** Expected to occur a few times in a year Frequency: > 10 yr^{-1} Probability: >10^0	High	High	Extreme	Extreme	Extreme
	Occasional May occur a few times in a year Frequency: 1–10 yr^{-1} Probability: 10^{-1}–10^0	Moderate	High	High	Extreme	Extreme
	Rare Has occurred in similar facilities Frequency: 0.1–1 yr^{-1} Probability: 10^{-2}–10^{-1}	Acceptable	Moderate	High	High	Extreme
	Very Rare Few records in similar facilities Frequency: 0.01–0.1 yr^{-1} Probability: 10^{-2}–10^{-3}	Acceptable	Acceptable	Moderate	High	High
	Conceivable Few historical records in the industry Frequency: 0.001–0.01 yr^{-1} Probability: 10^{-3}–10^{-4}	Acceptable	Acceptable	Acceptable	Moderate	High
	Improbable Has never occurred in the history of the industry Frequency: < 0.001 yr^{-1} Probability: <10^{-4}	Acceptable	Acceptable	Acceptable	Moderate	Moderate

Reactive process safety might be more aligned with typical troubleshooting thinking as it starts from a known consequence working backwards to the root cause. The term reactive is used to highlight that an abnormal situation has resulted in an incident that has already occurred. The investigation aims at understanding the root cause and the sequence of events for preventing re-occurrence and/or evaluating/ improving the response with the intent of minimizing the consequences. Process engineers with good understanding of the unit and troubleshooting skills are often the best qualified for such evaluations. Those incidents provide valuable data for understanding abnormal condition transients and effectiveness of SIS if involved in the incident. The following example illustrates a typical methodology followed for understanding an abnormal situation, from which the initiating event caused a chain reaction on other units.

Example 3.3: This involves a trip of a CGC in an ethylene plant that caused the two refrigeration compressors to trip shortly after [11]. This incident occurred a few months after commissioning a revamp project, which raised suspicion on changes implemented by the project as the root cause. Losing the refrigeration compressors is significant as it extends the unit downtime and re-start. It is also associated with the loss of valuable refrigerant to the flare system. A trip of the CGC without a loss of the refrigeration compressors allows a speedy recovery of the unit. The trip of the CGC itself is a different problem that has its own root cause. In this case, the interest is to understand why a trip of the CGC caused the refrigeration compressors to trip.

The problem statement can be formulated as "Why or What caused the 2 refrigeration compressors to trip on a trip of the crack gas compressor?" Unlike proactive process safety, the consequence is known, and the objective is to find the most probable root cause. This case involves a first CGC trip experience following a plant revamp. As such, historical data relating to previous CGC trips might be available and should be reviewed. However, caution must be exercised in drawing conclusions as design changes might have impacted the expected behavior of the system.

A data gathering step should follow the formulation of a problem statement. Note that the problem statement can evolve as more understanding is gained through the data gathering step. The data gathering step should include the following:

- Define the boundaries for the troubleshooting analysis. While this may change as more understanding is gained, the boundaries of the analysis should initially be focused on the system exhibiting the deviations or symptoms.
- A general schematic of the system investigated. A high-level schematic including all the process tags available from the plant data historian would avoid navigating through multiple PFDs and P&IDs. The schematic would also provide a starting point for a simulation model if deemed necessary. One aspect, which seems trivial but is often overlooked, is to understand the process objectives of the equipment being investigated. Without this basic understanding, it would be difficult to identify deviations from expected performance.
- Gathering of relevant raw process data trends in the time domain from the plant data historian. From the trends established time stamp of critical events, which will provide a starting point for establishing a sequence of events. Data from previous CGC trips should also be gathered for

comparison, understanding that differences might exist between the pre- and post-revamp configuration.

- Establish a pre-upset heat and material balance. The use of a process simulator would help in closing the gaps for missing instrumentations.
- Establish a sequence of events. The use of dynamic simulation would help reconciling assumptions relative to process variable transients and formalized a transient heat and material balance sequence. The transient process loads can then be compared against process equipment characteristics.
- Gather relevant process equipment design characteristics and compare the transient process loads established from transient heat and material balance to the operating window of the equipment.
- Depending on problems, the analyses can include hydraulic calculations of relevant circuits.

Further knowledge is gained throughout the data gathering steps, which may result in redefining the problem statements. The key questions in formulating the problem statement are: *Why is it a problem? What are the consequences?*

In this case, it is known that the ethylene refrigeration system cannot operate without propylene refrigerant as it provides condensing to the ethylene refrigeration compressor (ERC) discharge. The loss of propylene refrigeration would cause the discharge pressure of the ERC to rise, with the operating point transitioning to its surge margin. Upon which the anti-surge controllers would open recycle valves to protect the compressor from surge. Without the heat sink provided by propylene refrigerant, the recycle stream will cause a rise in temperature due to accumulation of the heat of compression. The compressor will therefore trip on high temperature through an SIS or by manual intervention from an operator if a trip mechanism does not exist. However, not tripping that compressor would result in significant mechanical damage if the high-temperature excursion does not get addressed.

System boundaries, schematic, and system objectives. From the above discussion, the boundary of the troubleshooting exercise initially focuses on the propylene refrigeration system, from which a high-level schematic is illustrated in Figure 3.5. The propylene refrigeration compressor (PRC) is a four-stage machine providing four levels of refrigeration. The propylene refrigerant temperature is set by the pressure profile of the compressor/knockout drum configuration. At each level, refrigeration is provided to process users by evaporation of liquid propylene in heat exchangers. Vaporized propylene is then recompressed by the PRC as it gets routed to the first, second, and third stage suction. The fourth stage suction of this compressor is extracting propylene by design to satisfy the heat and material balance. Propylene vapor from the third and fourth stage knockout drums are condensed by providing heat to reboilers. To complete the refrigeration cycle, propylene vapor from the PRC discharge is de-superheated through air coolers and condensed against cooling water in heat exchangers. Condensed propylene is collected in an accumulator drum, from where it supplies liquid to different refrigeration levels by cascading from one level to the next. Flash vapor from each level is either condensed by process users or returned to the compressor for re-compression. Liquid inventory in the second, third, and fourth stage drums is ensured by level controllers importing from the adjacent

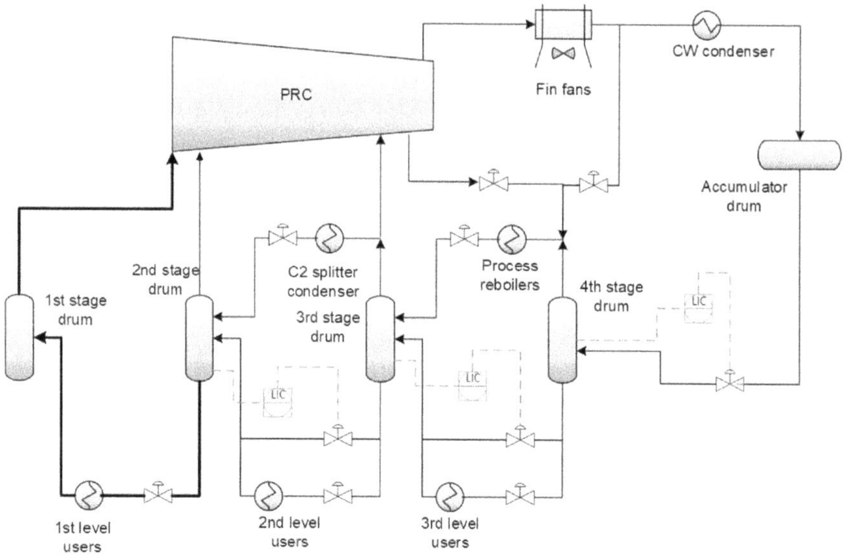

FIGURE 3.5 Propylene Refrigeration System.

higher-pressure drum. The first stage drum is a dry drum collecting propylene vapors from the first level users.

High-level sequence of events. Trends of the three compressors' speed provide the timeline of each machine trip as observed by their respective speed ramp down trend to zero. Those trends reveal that the PRC tripped within 5 minutes after the CGC. The ERC followed the PRC within about a minute. Trends of the propylene refrigeration system knockout drum level reveal that a high-level excursion in the second stage drum energized an interlock protecting the compressor from high drum level. The above steps related to system schematics, boundaries, and objectives are frequently skipped by operation as they consider themselves knowledgeable of the area for which they have responsibilities. As such, the operation group concluded that the high level experienced in the second stage drum was the result of controller tuning not being fast enough in responding to an aggressive transient condition. Another theory was that changes implemented by the project team might have been responsible for the PRC trip. Consequently, a process engineer involved in the project was tasked to investigate the trip. Being external to the operation group, the process engineer followed a methodology aligning with the above description. After collecting information from operation, establishing the initial troubleshooting investigation boundaries, and fully understanding the intent of the system, pre- and post-trip H&MB were conducted.

H&MBs. Plant data from the data historian and process simulation are used for establishing the pre-trip H&MB. The propylene flow rates are not measured for most of the heat exchangers. Process simulations are therefore instrumental in evaluating the heat balance, where the propylene heat duties get calculated from the associated process side. Table 3.3 compares the propylene refrigeration system heat balance

TABLE 3.3
Propylene Refrigeration System Heat Balance by Level

	Pre CGC Trip Heat Duty (%)	Post CGC Trip Heat Duty (%)
Third Level		
• C2R desuperheater	11.9	Assumed same
• CGC chiller	14.7	0
• Distillation condensers	73.4	0
Total	100	11.9
Second Level		
• Distillation condensers	32.5	0
• C2R condenser	67.5	Assumed same
Total	100	67.5
First Level		
• C2 splitter condenser	75.7	0
• Process gas chiller	19.5	0
• C2R subcooler	4.7	Assumed same
Total	100	4.7
Fourth Level Reboilers	100	0
Third Level C2 Splitter Reboiler	100	Assumed same

before and after trip of the CGC. The heat duties are expressed on a relative basis, with the total level residual duty post CGC trips expressed as a percentage of the pre-trip conditions. At this point, the post trip heat duties are based on assumptions, supported by associated process variables extracted from the data historian.

The post-trip heat balance reveals a significant reduction in total heat duties from the first level users estimated at 4.7% of the pre-trip value. The first stage drum is a dry drum; in other words, liquid from the second stage drum must vaporize as it travels through the first level users for ensuring a continuous flow path to the compressor suction. The propylene demand for the first level users is regulated by level controllers. In the absence of vaporization, the level controller closes the valve, and by doing so closes the flow path of liquid propylene from the second stage drum to the compressor suction. This flow path is shown by the bold line between the second stage drum and compressor suction in Figure 3.5. The ethylene fractionator plays a major role in the propylene liquid balance surrounding the second stage drum. The mass of propylene vaporized by the ethylene fractionator overhead condenser is more than twice the mass of propylene condensed by the tower reboiler at steady state. Losing the ethylene fractionator condenser duty therefore results in a significant liquid imbalance around the second stage drum. The heat duty on the ethylene fractionator reboiler was assumed the same as the pre-trip conditions, while plant data suggest a total loss of condensing duty. This observation raised questions as

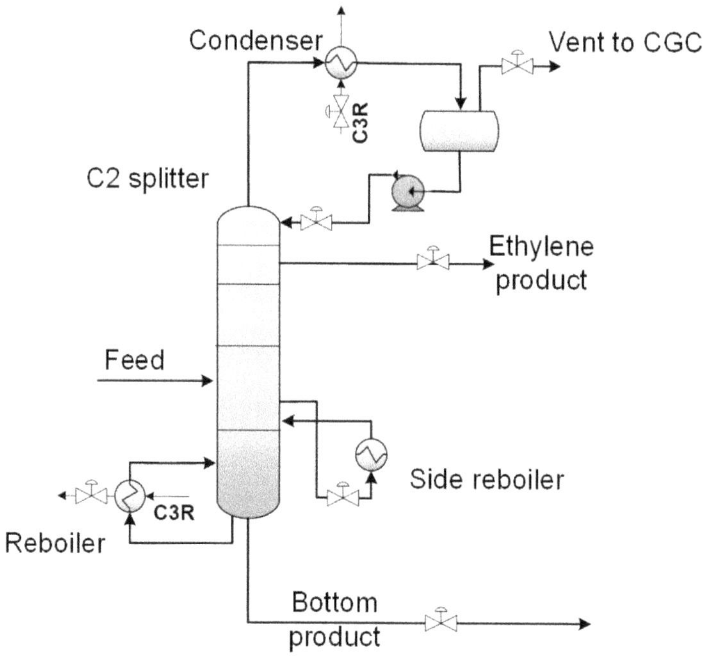

FIGURE 3.6 Ethylene Fractionator.

to why the ethylene fractionator still has heat duty on the reboiler but none on the overhead condenser. This requires extending the boundary of the evaluation to the ethylene fractionator tower, as illustrated in Figure 3.6. The feed enters the tower as a slightly superheated vapor and as such provides some load to the overhead condenser. A pasteurization section prevents accumulation of non-condensable, which are vented from the reflux drum. The vent stream rich in ethylene is recycled to the CGC. The reflux is regulated by a level to flow cascade control loop. The ethylene product is withdrawn a few trays below at a rate aligning with the tower material balance. The bottom product rate is regulated by the tower sump level. The tower is equipped with a side reboiler where warm process gas from the CGC discharge provides heat. Propylene (identified as C3R in Figure 3.6) vapor from the third stage drum provides heat to the bottom reboiler, and propylene liquid from the second stage drum provides cooling to the overhead condenser. A CGC trip causes interruption in process gas forward flow. This is associated with a feed interruption to the ethylene fractionator and a loss of heat duty on the side reboiler. The side reboiler heat duty corresponds to about 40% of the total tower reboil duty. While the bottom reboiler still provides heat to the tower, the overall heat duty drops by more than 40% when accounting for the contribution of the vapor feed. The reflux being controlled by a level to flow cascade control loop does not react immediately to this aggressive reduction in heat load to the tower, and as such starves the overhead condenser by collapsing the vapor traveling up the tower.

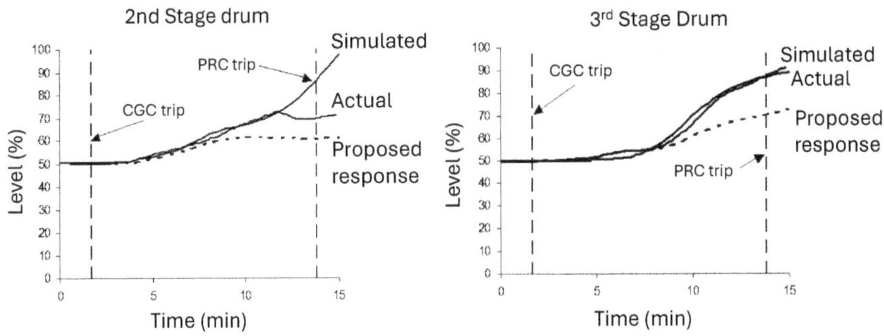

FIGURE 3.7 Propylene Refrigeration System Second and Third Stage Drum Level Transient.

Historical data were reviewed from previous trip of the CGC, considering that the revamp project had implemented some changes. However, similar behavior and interaction between the propylene refrigeration system and the ethylene fractionator would likely occur in the pre-revamp plant configuration. As expected, review of previous CGC trips confirmed the chain reaction to the PRC and ERC compressors. The only difference noted was variability with regard to which drum was experiencing the high level, second or third stage drum.

Discussion with process operators reveals that under those circumstances the focus was on attempting to drain excess liquids from the drums in order to avoid tripping the PRC, thus treating the symptoms rather than their root causes. In fairness to process operators, without effective direction and training for responding to such an event, the likelihood of doing the right thing is extremely low given the time available for decision-making. The imbalance caused by the ethylene fractionator was not well understood and, as such, not given the appropriate attention.

Dynamic simulations, besides helping in understanding the sequence of events and mechanisms leading to a trip, provide a non-destructive test ground for establishing an effective response. Figure 3.7 compares plant data to simulation and a proposed simulated response for the second and third stage drum. In this event, operators had been draining the second stage drum as observed by the deviation when comparing simulation and plant data. The PRC ended up tripping on high third stage drum level, where both simulation and plant data align.

Problem statement refinement. The trip of the CGC may have triggered conditions promoting imbalance in the propylene refrigeration system. However, the problem is not the CGC itself but rather the inability to re-balance the system under those transient conditions. The problem statement can therefore be re-formulated as *High PRC Second and Third Stage Drum Level.*

Using the KT dimensional questions:

What: High second and/or third stage PRC drum level.
Why Is It a Problem: Can trip the compressor on high drum level
When: Following a CGC trip.

Where: PRC second and/or third stage drum
Probable Cause: Heat and material imbalance causing liquid accumulation in the second and third stage drum.

The solution to this problem is to set up a sequence of action intended to re-establish balance in the propylene refrigeration system. To do so, opening the flow path of propylene from the second stage drum to the compressor suction is critical and can be achieved by ensuring heat duty on the ethylene fractionator overhead condenser. On a trip of the CGC, the tower should be transitioned to total reflux operation, in which condenser and reboiler duties will balance each other out. To alleviate the aggressive loss of feed and side reboiler, the reflux flow should be reduced to align with the heat supplied from the bottom reboiler. This would require the board operator to switch the reflux control from remote cascade (level to flow) to manual mode. Output settings would have to be specified by process engineering along with the targeted propylene flow to the reboiler. Process operators typically like transitioning process variables by manipulating SPs. However, such aggressive events required quick transition, making proportional-integral-derivative (PID) controllers tuned for normal operation ineffective. The above sequence can be executed quickly by a board operator and can avoid or minimize the need for a field operator to drain the second and third stage drums. This strategy can be tested on the dynamic simulations used for evaluating the upset. As shown in Figure 3.7, simulation of the proposed response suggests that the accumulation of liquids remain below the high-level trip in the second and third stage drums, thus preventing trip of the PRC after a CGC trip.

This example illustrates the importance of conducting a rigorous investigation following a significant process upset. Reactive troubleshooting of process upsets is frequently not being given sufficient attention, as the consequences are believed no longer relevant. However, re-occurrence of those events, if not handled properly, may cause significant downtime, damage, and/or potentially safety hazard. As demonstrated in this example, proper troubleshooting goes beyond observation of raw process data from trends in the time domain and gut feel. The initial review conducted by the operation group concluded on controller tuning as the root cause and suspicions on changes implemented by the revamp project. Those conclusions, nevertheless, do not lead to any solutions. PID controllers used for normal operation cannot be tuned for aggressive transient conditions experienced during an upset. Even if they were, there would not be any basis for the tuning parameters, and no possibility of proving effectiveness until the next upset. A deeper understanding of the system and transient H&MB is often necessary for identifying the most probable root cause. This understanding provides the framework for developing effective responses to similar abnormal conditions, which are to be communicated to process operators through operating procedures and training. In this example, the trip of the CGC resulted in a loss of forward flow, which created a heat and material imbalance on the propylene refrigeration system. Operation of the ethylene fractionator was found to be a key contributor to this imbalance but also provided handles for restoring the balance. Those findings are quite different than the initial conclusion drawn by the operation group and would not have been found by limiting the evaluation to a review of process variables trends in the time domain.

3.3 INHERENT CHARACTERISTICS OF CHEMICAL PROCESSES AFFECTING PERFORMANCE

The most common causes affecting the effectiveness of an operating unit and its run length are related to fouling and reliability of rotating equipment and reactors [12], with fouling being a subset of the factors affecting the reliability and run length of the latter two. Other major factors are related to failure by fatigue from repetitive cycles of mechanical and thermal stresses.

Those characteristics are described as inherent to a chemical process due to the presence of impurities in feedstocks, undesired side reactions, and transient conditions either planned or unplanned. Awareness of those characteristics enforces applications of strategies intended for minimizing their consequences. Those characteristics are a subset of causes initiating the problems described in the previous section.

3.3.1 FOULING

Fouling is the result of accumulation of undesired material within a process unit. It is one of the biggest causes of equipment performance degradation and one of the main criteria for limiting an operating unit run length. Fouling can also be associated with corrosion as deposition of fouling material on metal surfaces can cause localized environment promoting corrosion, referred as under-deposit corrosion.

Fouling can be the result of material already present in the streams supplying the unit. For example, suspended solids contained in cooling water may foul heat exchangers. Scaling may occur on metal surface at high temperature when in contact with fluid containing salts exhibiting solubilities that are inversely proportional to temperature. Fouling can also be the result of reactive components contained in the process fluids, which under favorable conditions will form oligomers and/or polymers that will accumulate in the process units. Many of the fouling reactions occurring in process units are promoted by temperature and residence time. The presence of rust, peroxides, and/or oxygen is also known to catalyze/initialize fouling reactions. Heat exchangers, towers, pumps, and compressors are units typically affected by fouling. The fouling mechanism of steam cracking furnaces is related to coking, which has already been discussed in Chapter 2. Piping systems can also experience fouling, where the development of a scale layer on the pipe surface will increase the hydraulic resistance of the system due to reduction of the effective pipe diameter and change in surface roughness.

Heat exchangers. Symptoms of fouling depend on the operating unit affected. For heat exchangers, a distinction must be made between solid chunk constraining flow to the exchanger to a fouling layer building up on the heat transfer surface. For instance, chunk particles, covering part of the tube sheet, will restrict the flow and be characterized by high-pressure drops. While the total flow might be sustained, its distribution will be affected and result in a reduction of the effective heat transfer area. In such cases, however, the resistance to heat transfer remains unaffected inside the exchanger. On the other hand, scale or polymer deposition on the tube surface, besides increasing pressure drop, adds resistance to heat transfer. From temperatures and flows, the overall heat transfer coefficient can be calculated and plotted over time using the following expression. It should be noted that Equation 3.2 combines the overall heat transfer coefficient U with the heat transfer area A. The U value can be calculated on its own if A is known. However, for convenience, as A is a constant the behavior of the U value would still be reflected on a time plot when expressed as UA:

$$UA = \frac{Q}{LMTD} \tag{3.3}$$

where:

$$LMTD = \frac{(T_H - t_H) - (T_c - t_c)}{Ln\left[(T_H - t_H)/(T_c - t_c)\right]} \tag{3.4}$$

with TH and Tc being the hot-side hot and cold temperatures and tH and tc the cold-side hot and cold temperatures. In design practice, correction factors apply to the LMTD calculations based on exchangers' design configuration. However, for the purpose of a high-level evaluation, using the uncorrected LMTD provides valuable information relative to the performance of a heat exchanger over time.

Figure 3.8 illustrates the *UA* value and process flow rate of a heat exchanger, with the plot on the left side showing normal operation and the plot on the right a significant loss of performance. From the discussion in Chapter 2, it is known that the overall heat transfer coefficient U is flow dependent. This is illustrated by the process flow following the *UA* trend, with flow increases reflecting rise in *UA* value. The plot on the right illustrates a significant drop in *UA* value at one point in time, where the flow rate/*UA* value correlation no longer fits the behavior illustrated on the left plot. Note that the loss in *UA* value is also preceded by a significant drop in process flow rate followed by a recovery. If the exchanger does not have pressure gauges on the inlet and outlet, the pressure drop is unknown and cannot be used for diagnosing fouling and/or hydraulic restrictions. Note that in the absence of pressure indication, the increase in pressure drop can be validated by the output of control valves associated with the operation of the exchanger. Higher output for similar flows would confirm the incremental pressure drop. The plot on the right side of Figure 3.8 also suggests that fouling occurs inside the exchanger as it negatively affects the heat transfer coefficient when comparing the *UA* value for flow similar to what can be observed on left side plot.

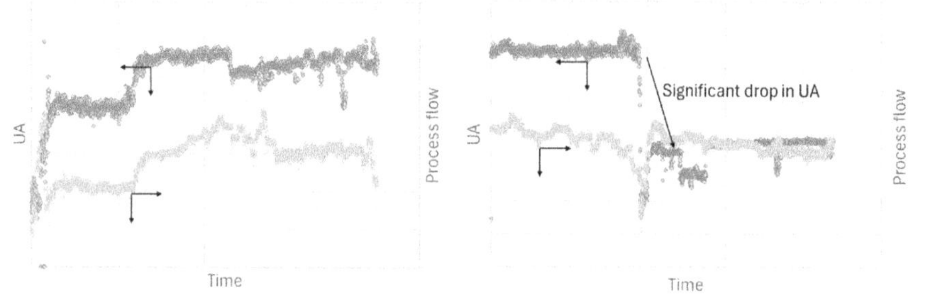

FIGURE 3.8 Impact of Flows and Fouling on Heat Exchanger Overall Heat Transfer Coefficient (*UA*).

Besides the fouling propensity of the fluid, operating conditions also play a role in the fouling mechanisms of heat exchangers. Low rates result in low velocity and increase the likelihood of particles settling inside the exchanger. Conversely, high velocity minimizes settling and provides shear forces on the tube surface, which helps in preventing deposition of scale or polymeric material. High TMT due to large temperature differences between the hot and the cold stream promotes fouling reactions and deposition of salts or film of non-volatile material due to the impact on solubility or vaporization at the tube surface.

Distillation trays. Fouling also affects the capacity and efficiency of distillation towers. Deposition and adherence of solids on the active area reduce the hole diameter and affect the vapor and liquid flow path across the tray. Figure 3.9 illustrates the impact of a 1/16″ polymer layer adhering to the circumference of ½″ hole sieve deck, which reduces the effective diameter by 1/8″. The picture shown on the right side of Figure 3.9 suggests this to be relatively mild fouling. Nonetheless, this corresponds to 25% reduction in effective hole diameter or 43% reduction in the overall hole area of the tray panels. Assuming the same operating line (i.e., same fractionation efficiency), this would result in a 16% reduction in capacity due to the jet flooding limit shifting down. This assessment assumed that fouling is equally distributed across the tray and that efficiency is unaffected, which, based on the picture on the right side of Figure 3.9, does not seem to be a bad assumption. While of little importance, the reduction in hole area causes the weeping limit of the tray to shift down as well.

Pictures of the trays illustrated in Figure 3.10 are examples of more severe fouling. The picture on the left side of Figure 3.10 illustrates a fouling layer covering more than 80% of the holes. This tower had to be brought down due to very-high-pressure drops and flooding excursion that severely impacted the tower capacity. The picture on the right side of Figure 3.10 shows significant accumulation of material on the trays. While some of this material might have been in suspension during operation, some of the polymer contributed to reducing the hole area, but unlike the tray illustrated in Figure 3.9, the deposition is more significant and more localized. This suggests the development of stagnant zones as material was accumulating on the trays and would explain why the fouling material did not settle uniformly when the tower was brought down. Besides reducing the jet flood limit, the localized accumulation of material affects the vapor and liquid flow path. Tray efficiencies are highly influenced by the flow path length used by the liquid across the active area of the tray.

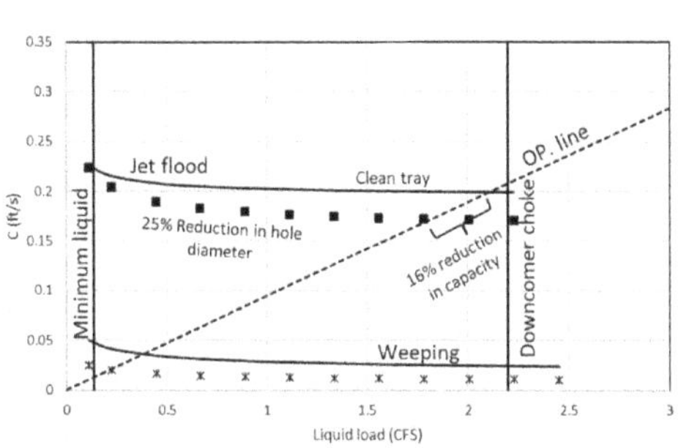

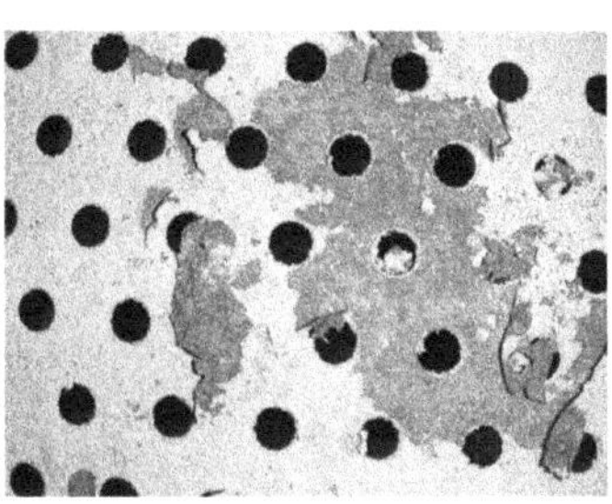

FIGURE 3.9 Impact of Fouling on Jet Flooding and Weeping.

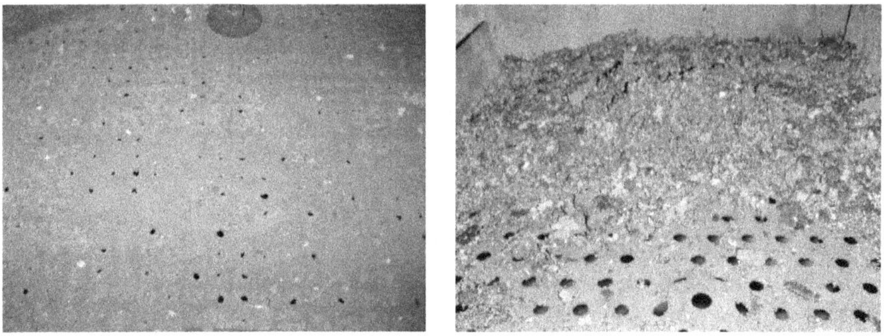

FIGURE 3.10 Severe Fouling.

This assumes that the liquid contacts the rising vapor as it travels from the inlet to the outlet of the tray. Blocked holes caused the vapor to flow preferentially through the remaining opened holes, creating liquid stagnant regions. The liquid flow path is strongly dependent on the hole distribution, with designers often using slotted holes or push valves for minimizing stagnant areas on the tray and maximizing the flow path length followed by the liquid. The crest over the tray outlet weir is obtained from the Francis Weir formula:

$$h_{ow} = 0.48F_w\left(\frac{q}{L_w}\right)^{2/3} \tag{3.5}$$

where Fw is a correction factor accounting for distortion in flow path as it approaches the weir. It is a function of the ratio of liquid load (in gpm) to the weir length in ft. q is the liquid load in gpm and Lw is the weir length in inches. The crest over the weir should therefore be constant across the weir length regardless of the level of aeration. Therefore, tray areas covered with solids should be characterized by low aeration or clear liquid, which will have little, if any, contact with vapor as it travels over the weir than to the tray below after flowing through the downcomer. The effective L/V ratios will therefore be different, and the effective flow path length reduced, thus affecting the overall tray efficiency. While there are exceptions, fouling reactions are known to typically occur in the liquid phase. Fouling reactions are promoted by the presence of reactive components, temperature, and residence time. As such, the stagnant pool of liquid developing over the fouled tray section contributes to promoting those conditions as well as making settling easier.

Troubleshooting fouling on tray towers. The outcomes of fouling on tray towers are characterized by an increased pressure drop, reduced capacity and efficiency [13]. Plotting the pressure drops as a function of feed rates, top tray vapor rate (typically the sum of the reflux and distillate), and/or heat to the reboiler should provide an indication on the fouling progress of the tower. Deviation in the plots when compared to the beginning of the plant run would be symptomatic of fouling. Plotting the reflux ratio over time will also give an indication on potential loss of efficiency, keeping in mind that the reflux ratio is also dependent upon the feed quality and composition as well as the distillate purity. Evaluating the progress of fouling through gamma scans should extend beyond a single scan line. Multiple scan line may reveal stagnant areas and poor distribution. Early detection of fouling may provide the opportunity to reverse or minimize the impact by adjusting operating conditions, attempting online cleaning and/or using inhibitors or dispersants.

Pack towers. Trays can be designed for providing more robustness toward fouling than random and structured packing. While grids provide robustness to fouling and falls under the structured packing family, it is more applicable to heat transfer applications. Although used in some applications, packing should not be used in fouling service. Unlike trays and while not the intent, packing provides an effective mechanical matrix for stopping and containing solid material inside a tower. Nonetheless, if used in fouling service, random packing has the advantage of being washable with high-pressure water outside the tower during outages. Fouled structure packing cannot be washed and if used in fouling service shall be replaced as required. The risk of pyrophoric fire is another problem associated with packing in fouling services. The fouling material accumulated in a pack bed can ignite when exposed to air during tower outages. Special precautions must be taken when opening packed towers during shutdowns. As such, proactive process safety troubleshooting should be part of shutdown preparation with assessment of the risk of pyrophoric fires and specific instructions for risk mitigation and response. Regardless of service, it is good practice to assume that the risk of pyrophoric fire is present. Some of those precautions should include but are not limited to:

- Availability of fire protection equipment.
- Keeping the packing wet by circulating water.
- Monitoring temperature, as a rise in temperature may indicate the start of a reaction.
- If a reaction is detected, close all manways and keep water flowing to the tower. This would starve the pyrophoric reaction from oxygen.
- Keep the packing wet as it is being removed and stored for further washing or disposal.

The quench section of ethylene plants is a good example of the use of packing in fouling service and is explored in more detail through the following example.

Example 3.4: Figure 3.11 illustrates the quench section of a typical ethylene liquid cracker, where the combined crack gas stream from the furnaces flows through a primary fractionator and quench water tower before supplying the CGC. Packing offers lower pressure drops than trays and helps reduce the cracking furnaces' coil outlet pressure, which is beneficial to cracking yields. However, due to the fouling propensity of fuel oil and raw pyrolysis gasoline produced from cracking, operation of the primary fractionator frequently becomes the unit limiting the plant run length, thus increasing turnaround frequency. The main objective of the primary fractionator is to condense fuel oil from the combined furnace streams and fractionate gasoline to required specification. This is accomplished through two pump-around circuits. The first one involves quench oil, which is collected in the tower sump and cooled

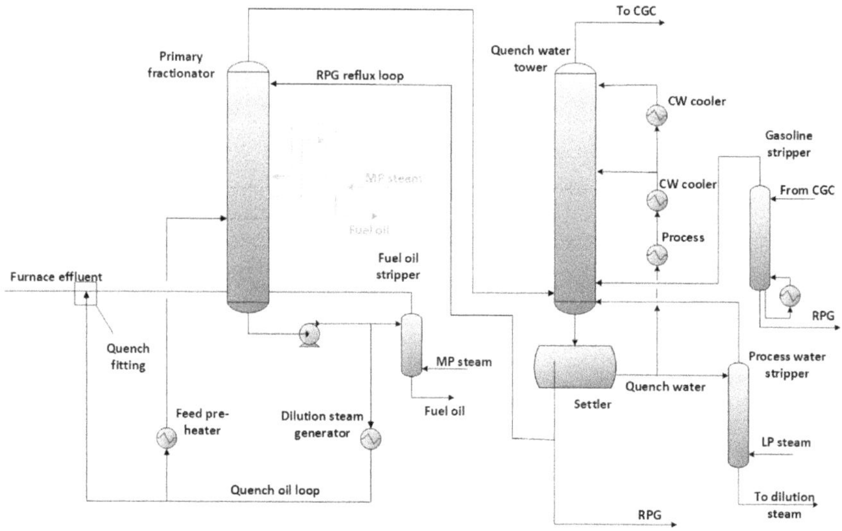

FIGURE 3.11 Ethylene Plant Quench Section [13].

Reproduced with Permission from figure 3.5, Strategies for Mitigating Fouling in Ethylene Plants, André Bernard, *Chemical Engineering Progress*, March 2025

down by boiling water in the dilution steam generators and providing preheat to the furnaces feed. The cool quench oil is then used as mid-reflux for condensing fuel oil in the lower section of the tower. Some plants use quench oil to further reduce the temperature of the furnace effluent exiting the primary quench exchangers by direct contact in individual furnace quench fittings. A draw of quench oil is routed to a fuel oil stripper for closing the fuel oil balance around the tower. The fuel oil stripper is intended to increase the middle fuel oil content of the quench oil stream, which is known for providing good heat transfer properties and lower fouling propensity. The second pump-around involves the circulation of gasoline collected in the settler of the quench water tower, which reflux the top of the primary fractionator. The gasoline reflux rate is set for condensing the light fuel oil and pushing it down the primary fractionator such that the gasoline content collected in the quench water tower meets specifications. It is important to note that the only exit point for the light fuel oil boiling range is through the fuel oil stripper. This consequently reduces the temperature of quench oil collected in the tower sump, hence reducing the temperature driving force for boiling water in the dilution steam generators. Furthermore, it makes the quench oil stream lighter and results in the migration of higher volume of light fuel oil in the top section of the tower due to the stripping effect of crack gas when in contact with the quench oil mid-reflux. Higher gasoline reflux must then be used for countering this effect. Ethylene licensors have mitigated this consequence by drawing a mixture of light and middle fuel oil boiling range above the mid reflux point and process that material to a side stripper as illustrated in light gray in Figure 3.11. This provides an exit point for light fuel oil higher up in the tower, resulting in a reduction in gasoline reflux and higher quench oil temperature. This design alteration, besides optimizing the operation of the primary fractionator, provides some mitigation to fouling by reducing the residence time of gasoline components in the top section and by selectively recycling more middle fuel oil components from the bottom fuel oil stripper.

The gasoline refluxing the primary fractionator contains reactive components such as indene and styrene, which are known to react by free radical polymerization. Quench oil contains polynuclear aromatic (PNA) hydrocarbons, which are artificial asphaltenes produced by cracking. The high quench oil recirculation rate causes the PNA content and residence time to increase. In high concentrations, PNAs [14] are known to stack together causing high quench oil viscosity and precipitate when reaching a critical mass. The primary fractionator temperature profile ranges from about 120 °C in the top to temperature which can reach 200 °C in the bottom depending on systems. Those high temperatures are favorable for free radical polymerization reactions and high oil viscosity due to the stacking of PNAs. In terms of residence time, the high reflux and quench oil circulation rate cause fouling precursors, such as indene and styrene, to have residence times of about 1–4 hours. The quench oil loop being even more significant than the gasoline reflux has an estimated residence time of about 5 days [15] for heavy fuel oil components. Note that residence times are feedstock specific and tend to increase as plants migrate toward lighter feedstocks.

High temperatures and high residence times are conditions favorable for promoting fouling reactions. Minimizing the tower temperature profile would not eliminate fouling but would reduce the reaction rate, hence potentially extending the run length of the tower. However, this comes at a cost given the heat integration of this system with dilution steam generation and furnace feed preheating. Reducing residence time would require cutting on the reflux rate, which would cause exceedance in raw pyrolysis gasoline (RPG) quality. This would also be associated with a rise in temperature in the top section and may offset the impact of reducing the residence time. Similarly, cutting the quench oil circulation will hurt the heat integration and cause a warmer temperature profile or require a higher gasoline reflux rate.

The accumulation of fouling material in the top bed of primary fractionators has been reported to form an inverted cone of solid material [16]. As illustrated in Figure 3.12, random packing beds are characterized by a liquid film flowing at the boundary of the tower shell and packing. Fouling reactions typically occur in

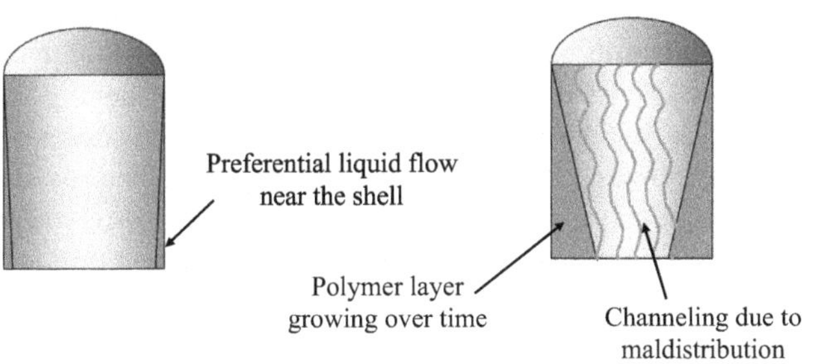

FIGURE 3.12 Fouling Progression in Random Packing Bed.

the liquid phase. The reaction rate also increases with temperature. As such, it is expected that polymer will initially form in the liquid layer present near the vessel wall. The thickness of the polymer layer would be expected to increase while moving downward in the tower, given the rise in temperature. Primary fractionators are relatively fat towers, with diameters of about 30 ft and over. This makes it more challenging for distributor design and sensitivity to levelness. Furthermore, fouling accumulation may also affect the effectiveness of the distributor. Poor distribution will promote channeling, hence impacting efficiency and creating further zone promoting polymerization and accumulation of solids. The growth of polymer over time will reduce the effective cross section of the tower and eventually limit plant capacity, at which point a unit outage is required for removing, cleaning, and/or replacing the packing.

Troubleshooting pack tower. As per the problem classifications previously discussed, this might not be a problem until it affects the plant capacity, the ability to meet product specifications, environmental exceedance, and/or process safety issues. However, fouling is a slow mechanism that can go undetected for a long time, such that once symptoms of performance degradation become apparent, it is irreversible and will likely result in an unscheduled outage. Proactive monitoring helps understanding that the fouling rate remains in alignment with the expected unit run length. With packing operating in fouling service, this involves frequent gamma scans of the tower (once or twice per year). To help interpretation of gamma scans, a dry scan should be conducted with the dry packing in the tower before starting up the unit. A dry scan provides a baseline for correlating the liquid hold-up and/or accumulation of solid material. As for trays, monitoring pressure drops as a function of feed rates or vapor loading can also be done. However, the accuracy of the measurements depends on the sensitivity and calibration of the pressure transmitters as packing pressure drops would typically range in inches of liquids. Trending the overall H&MB would also help in noticing performance degradation. For instance, rise in required reflux rates and shift in tower temperature profiles would suggest changes in efficiency. Early detection of fouling rate exceeding expectation could be mitigated by adjusting operating conditions and/or injecting chemicals known to solubilize and/or disperse solids accumulating in the tower. The use of inhibitors can be done when the fouling reaction mechanism is known, and it occurs in the tower. In the above example, this may apply to reaction involving indene and styrene but would be ineffective with PNAs as they are a product of cracking and already present in the stream feeding the tower.

Design and turnaround practices also have influence on fouling. Carbon steel can rust, and rust is known to catalyze fouling reactions. Opening vessels during turnaround expose the metal surface to moist air, which causes corrosion of the surfaces. Furthermore, grinding inside vessels generates small carbon steel particles, which will likely rust and provide the perfect catalyst for initiating fouling reactions. It is therefore critical to communicate to a shutdown team the importance of keeping the vessel clean and free of any rust particles before closing it. Passivating the metal surface is also part of best practice for re-commissioning a process vessel after a turnaround. From a design perspective, specifying stainless steel internals in fouling service helps in reducing the rate of fouling reactions. Stainless steel was also reported [14] as providing less adherence for solids to grow on the internal components of process vessels, hence making solids more likely to washout.

In fouling service, tray design should provide larger hole diameter or bigger valves with larger lift than on clean service. Movable valves should also be avoided as they might stick in a close position. Downcomers should be designed to avoid stagnant volumes and should be designed for low residence time. With packing, if the objective of the tower is providing heat transfer, then grid with large open area should be preferred over random or structure packing. For mass transfer applications, if trays cannot be used for whatever reason, then random packing would be better than structured packing. Stainless steel is typically specified for a different reason than fouling but provides advantages over carbon steel in fouling services relative to reduction in fouling reaction rates and reduction in adherence.

Compressors. Depending on the nature of the gases being processed, compressors can experience fouling. The heat of compression causes high temperatures, which promotes fouling reaction. The metallic components in some cases have been reported as catalyzing some of those fouling reactions. In known fouling service, the reaction rate and the potential accumulation of solids within the compressor can be mitigated by limiting the heat of compression through small process compatible liquid injection. By doing so, the heat of vaporization of the liquid stream contributes to limiting the temperature rise experienced by the gas. Further to limiting the temperature, compatible process wash oil injection contributes to provide a liquid film characterized by a high degree of solvency, which prevents solid deposition on the compressor impellers. The use of coating on compressor impeller is another mean of limiting fouling. Compressor performance degradation is typically characterized by deviation from the head curve and lower efficiency. Reduction in polytropic efficiencies is typically characterized by an increase in discharge temperature.

Characterization of fouling reactions in process equipment relies on the combination of empirical and analytical evaluations. Effective troubleshooting of fouling services requires an understanding of the components involved and the reaction mechanisms. When sampling is possible, the fouling propensity of a process fluid can be studied at laboratory scale. A hot liquid process simulator (HLPS) [14] has been used for studying the fouling propensity of the gasoline pump-around circuit described in Example 3.4. An HLPS is a laboratory-scale system, which exposes a small volume of process fluids to a temperature cycle, by circulating the fluid in close circuit between a storage vessel and a heater tube. A filter at the outlet of the heater tube is used for collecting solids generated by a suspected fouling reaction. The rate of the fouling reaction is indirectly correlated to the pressure drop across the filter over time. The experiment can be conducted for different temperatures and tube metallurgy. This provides high-level fouling reaction kinetics as well as dependency to metallurgy. Observation of the tube surface after completion of the experimentation cycle provides qualitative information relative to the adherence of fouling material on the tested metallurgy. Solids collected in the filter can be characterized by laboratory analyses, which provides key information relative to the components involved and the reaction mechanism. The HLPS offers the advantage of continuous system, which, if set correctly, can approach plant conditions. Fouling reaction mechanism can also be evaluated in batch systems, where process fluids are exposed to a constant temperature over a defined period of time. Samples are taken for analyses at known time intervals during the length of the experiment. A similar system has been used for studying the fouling propensity of quench oil in primary fractionator [17]. Given the complexity of the system and components involved, the fouling reaction rate was correlated to the rise of viscosity over time. Unlike the HLPS, the batch system has no capabilities for assessing the impact of the metallurgy on the fouling rate and potential adherence of the material on metal surfaces.

The above laboratory-scale experiments are applicable to fluids that can be sampled and conserved at near atmospheric conditions. In many cases, the operating conditions significantly differ due to high pressure and/or extremely cold or high temperature. Under such circumstances, samples of fouling material are limited to what can be collected from filters or other area of accumulation that can be sampled with the plant on stream. However, the best opportunities for gathering data relative to fouling are during plant turnaround, when samples of fouling material can be collected from the process vessel where they have settled. Rossana et al. [18] described visual observations that may provide clues on the origin and fouling mechanism in process equipment. The location of fouling materials, for instance, if showing up on the underside of distillation trays and/or layer of fouling material peeling off the shell surface above the froth, can be an indication of fouling occurring in the vapor phase. In such a case, use of a different chemical treatment than use in the liquid phase would be required. The coloration and texture of the fouling material provide other clues. Colors of aldol polymer vary with molecular weight, from yellow, orange, red to black. The texture of the polymer, from sticky, gluey, or solid, provides other clues. Laboratory analyses would provide a more definite characterization, which should allow tracing back the most likely reaction mechanism.

Eliminating or minimizing fouling can be done by removing the contaminants responsible for the fouling reactions or using chemicals for inhibiting the reactions or dispersing the solids resulting from the fouling reactions. The proper methodology relies on a good understanding of what is causing the problem, therefore why time spent on understanding the mechanism, using the above strategies, will pave the way to an effective program.

3.3.2 Catalyst Deterioration

As stated in Chapter 2, catalyst stability is defined by its ability to retain its activity when exposed to changes in physical and chemical properties. A loss of activity is associated with a reduction in selectivity and eventually capacity. Losing selectivity means a reduction in feedstock conversion toward the desired products. A drop in the reactor effluent product concentration increases the load on the separation unit. Depending on the magnitude, the reactor or separation unit could eventually reach a capacity constraint. In some applications, catalytic reactors are used for converting impurities into something that can be easily separated. A premature or unexpected loss of activity in such reactors can result in off-specification products. For example, the loss of catalyst activity in hydrogenation reactors used in ethylene plants may first be observed by a reduction in selectivity and could eventually result in acetylene slippage causing an off-specification product excursion.

Catalyst deactivations occur over time due to unwanted reactions resulting in the formation of oligomers and coke which can block active sites. Catalyst sintering can also occur over time and result in a reduction of the surface to volume ratio of the catalyst. Heavy metals, such as lead, mercury, and arsenic, are considered poison to catalyst as they permanently bond with the active sites. The presence of heavy metals typically comes from the feedstocks or through external streams. If present in feedstocks and/or recycle streams, they must be removed upstream of catalytic reactors. The most common removal techniques involve adsorption in molecular sieve media.

3.3.3 Mechanical Failure

Rotating equipment. Most mechanical failures are frequently associated with rotating equipment, with seal failure topping the list [19]. From a process safety perspective, a seal failure triggers an LOPC, which, depending on the nature of the fluid, may result in environmental release and potential fire and explosion. Failures can also be caused by worn-out bearings or significant loss of performance where excessive wear of mechanical components results in internal recycles that significantly impacts the capacity of the unit. In either case, the equipment must be shut down for repair. Depending on criticality and spare availability, this can cause a total unit outage. The cause of those failures can be divided into mechanical and/or process deviations. The mechanical aspect is related to equipment design, material selection, and maintenance practices. For instance, improper clearances, shaft misalignment, and/or improper lubrication are related to maintenance activities. From a design perspective, improper mounting base/foundation and/or inadequate piping system can result in excessive vibration causing failure of mechanical components. Furthermore, material incompatibility with the process fluids will negatively affect reliability of the equipment. The Bhopal incident previously discussed is an example where poor design and material selection resulted in poor pump reliability involving frequent seal replacement. At the design stage, it is imperative that process and mechanical engineers work together to ensure that the equipment meets the process objectives and ensure compatibility of the mechanical components with the process characteristics. Besides potential failure of mechanical components, process engineers often overlook the leak rate from seals. Although small and compatible with the process fluid, the leak rate may cause off-specification product. A case study has been documented in literature [20] where the use of dry gas seal on a new compressor caused off-specification products due to impurities contained within the seal gas.

Operation of rotating equipment is strongly related to process conditions, which, if operated outside the reliable boundary of the equipment, can result in premature failure. For centrifugal pumps, cavitation is probably one of the primary mechanisms leading to premature failure. Cavitation can be the result of phase changes or the presence of entrained gases in the pump. Phase change inside the pump is known to cause significant vibration resulting in failure of the mechanical components. Entrainments of gases may cause pitting on the impeller and affect the pump capacity but is not as detrimental as phase change inside the pump. Boiler feed water, steam condensate, and liquid reflux to distillation towers are applications where liquids enter a pump suction from a vessel with its content at or near the boiling point.

Sufficient static head must be provided between the drum vapor/liquid interface and the pump such that the suction pressure exceeds the vapor pressure of the fluid. As described in Chapter 2, in such systems, cavitation is avoided by providing sufficient NPSH margin, which is an indirect measure of the pressure margin required from the vapor pressure of the fluid. While pumps handling fluids near boiling are more susceptible to cavitation, other systems operating far from boiling could also be affected depending on the pump type and system configuration. For example, cooling water pumps with suction lift configuration could experience cavitation. In cooling water systems, there are typically multiple pumps operating in parallel to supply a large network of heat exchangers. Depending on system and pump characteristics, there might be a minimum number of pump requirements for avoiding cavitation. Below that minimum, the system curve causes the operating point to move to the right of the pump head curve with the associated NPSH required rising quite significantly and potentially exceeding the available.

Shutting off a valve downstream a centrifugal pump would cause the pressure to rise to the shutoff head but will also cause the temperature to rise as the energy supply to the pump is converted into heat. There is therefore a minimum flow required for preventing the fluid inside the pump from reaching its boiling point, which can be defined as the minimum thermodynamic flow. Below the minimum thermodynamic flow, flashing and high temperatures will cause pump damage. However, there is also a minimum stable flow, which is higher than the minimum thermodynamic flow. Below the minimum stable flow, suction recirculation will likely be experienced, which is a condition that may promote noise, vibration, and erosion. Centrifugal pumps are typically equipped with a minimum flow circuit to avoid operation below the minimum stable flow. Some minimum flow circuits are automated with a control valve; others are configured with a flow orifice. The disadvantage of a flow orifice is that it consumes pump capacity and energy. Operations may elect to close the minimum flow circuit when running out of pump capacity, thereby compromising pump protection.

Literature [21] suggests that the mean time between repairs and life cycle of a centrifugal pump can be improved by operation at lower speed and by limiting the operating range from 70% to 120% from the best efficiency point (BEP) which is defined as the allowable operating range (AOR). This does not mean that a pump cannot be operated reliably beyond the BEP. A pump can be operated through the range of head curve supplied by the manufacturer between the minimum stable flow and to the end of the head curve. Continuous operation outside the AOR, will likely result in increased frequency for maintenance. Infrequent operation beyond the AOR due to abnormal plant conditions may not impact the mean time between failure. Those criteria must be taken into consideration when evaluating pump capacity and sparing philosophy.

There are similarities with compressors, which also have a reliable operating range. Minimum flow on compressors is defined by the surge point and the maximum capacity by stone wall (i.e., gas reaching sonic velocity).

All rotating equipment (i.e., pumps, compressors, turbines, and electric motors) are characterized by critical speed, which is defined as the speed that is equivalent to the natural frequency of vibration. Rotating equipment have multiple critical speeds that must be avoided as they result in significant vibrations that will cause failure of the mechanical components of the equipment. Rotating equipment must transition through at least one critical speed when brought up from a shutdown state. To avoid problems, it must accelerate quickly through the critical speed such that there is insufficient time for resonance frequencies to amplify vibrations. Knowledge of the critical speeds is important, as it must be avoided when operating the equipment.

Both pumps and compressors are affected by external vibration caused by water hammer or vibration induced by two-phase flow regimes.

Static equipment. For some, mechanical failures are limited to rotating equipment. However, static equipment is also subject to mechanical failure. Internals in knockout drums and distillation towers can get damaged by pressure surges resulting from plant upsets. Movable valves on distillation tray deck are known to fail by fatigue due to extended operation around the turndown point causing them to shatter. Of higher consequences are LOPC typically occurring in the form of leaks due to mechanical or thermal stress. Those transient conditions typically occur during transient conditions experienced during startups, shutdowns, and upsets. Except for brittle fracture, mechanical failure due to mechanical and thermal stress occurs after a number of cycles. Brittle fracture, on the other hand, is more severe as it does not give any warning. It is typically referred to as break before leak. Operation and process engineers must recognize the circumstances leading to those failures and have systems and procedures in place to reduce the risk to acceptable level.

Corrosion is another form of mechanical failure that can result in LOPC or cross-contamination in heat exchangers. Corrosion can be caused by incompatible material of construction of static equipment, inadequate pH controls of aqueous streams, and/or corrosion under deposit in fouling services.

3.4 CIRCUMSTANTIAL TROUBLESHOOTING

Troubleshooting strategies are influenced by the circumstances under which they are executed. Those circumstances distinguish themselves by the state of the operating unit, whether operating at steady state or in transient conditions such as startup. They can also include an element of novelty when dealing with a new facility or revamp of an existing one. Those circumstances affect the troubleshooting strategy from the perspective of available data and people experience. The time factor is another element which is driven by both consequences and circumstances. As previously discussed, off-specification products, environmental exceedance, and process safety risk can limit the time available for bringing the plant back to normality. From a circumstance's perspective, a plant upset imposes troubleshooting and decision-making in real time. This section discusses different circumstances and troubleshooting strategies based on available data.

3.4.1 Existing Operating Units

The reason for covering existing operating units under circumstantial troubleshooting is that the strategy and data availability differ from revamp and new facilities. While the problem classification described in the previous section applies to any operating units, whether existing or new, an unmodified existing facility has the advantage of available historical plant data. Differences in process variable behavior over time may highlight deviations in unit performance. However, variability must be distinguished from deviations as it may be driven by feedstocks flexibility, intentional changes in reactor yields for satisfying an optimum product slate, and/or variation in rates imposed by sales and inventories. Suspected performance deviations can first be assessed by comparing historical data sets reflecting similar operating conditions. Unreconciled differences with emphasis on the problem classification discussed in the previous section would validate the presence of deviations, which may require troubleshooting.

Operational activities occurring in an existing facility can be divided into three main categories: steady-state operation, startup, and shutdown. Startups and shutdowns are infrequent activities of short duration but, if not properly executed, could impact proper operation of the unit once at steady state. A subdivision of startup and shutdown includes regeneration of catalysts, adsorbents, and decoke cycles of steam cracking furnaces, which occurs while the overall plant is in operation. Those cyclic operations involve transitions that have the potential of creating an off-specification excursion. Poor regeneration sequence of catalysts and/or decoke of steam cracking furnaces can reduce the run length and performance of those units, hence negatively impacting the plant capacity.

3.4.1.1 Steady-State Operation

Steady-state operation is characterized by stable operation with, in theory, no transient behavior. However, in reality, transient conditions occur and are the result of several activities frequently executed during normal operation of the unit. Maintenance activities, for instance, may require bringing equipment down for cleaning or normal maintenance due to expected tear and wear. Variability in feedstocks and reactors' yields could be altered intentionally for promoting coproducts or as the result of catalyst deterioration. Effective response to those transient conditions involves human interactions in conjunction with the control systems. The most common causes affecting the ability of an operating unit in meeting its expected performance during steady state have been discussed in previous sections. The advantage of troubleshooting the steady-state operation of an existing facility is the availability of a large volume of historical data. The troubleshooting methodology remains similar as examples previously discussed. H&MBs and process loads applying to the different unit operations can be compared to similar conditions previously experienced. Deviations in behaviors from past experience will provide clues relative to the potential root causes.

3.4.1.2 Shutdown

Planned shutdowns typically follow established procedures and are less likely to cause harm to the unit if properly executed. Nonetheless, they introduce transient conditions and an element of variability that could explain problems that may arise once re-starting the plant. Not to mention that planned shutdowns are normally done for executing unit turnaround where numerous process vessels, pumps, and compressors are open for cleaning or repair. Maintenance activities themselves introduce risks during execution of the work and may also affect the expected performance of process equipment once re-starting the plant.

The impact of scheduled turnaround is one element by itself that, if not properly planned and executed, could have significant impact on plant performance, reliability, and turnaround frequency. Let alone unplanned outages resulting from improper maintenance activities. A chemical plant does not generate money when it is down for maintenance. Fixed costs and the burden of additional men hours required for executing shutdown work put a lot of pressure on management for shortening duration and extending turnaround cycles. However, these objectives are somewhat contradictory to each other and must be optimized as they could significantly impact plant reliability. Turnaround successes are frequently measured against scope, schedule, and budget. Although important, they neglect the most important aspect of the turnaround objective, which is to give the plant reliable operation at capacity for the duration of its targeted run length. This metric is indirectly related to scope definition, where, if properly defined and executed, would meet that objective. In an existing plant, the scope definition should leverage on plant history. Noting however that changes may influence the rate and magnitude of performance degradation. For instance, transitions in feedstocks, processing of new external streams, changes in operating conditions, and/or changes in chemical additives can be associated with new problems never encountered. This may result in unexpected fouling material accumulations in location historically found clean. Findings of those new problems during turnaround require deviation from the original scope. Omission to fix those problems can seriously constrain the plant after re-start or impact reliability, which may result in a costly unplanned outage.

Typical pitfalls associated with the planning and execution of turnarounds are associated with experience and knowledge. Scope definition is done in the early stage of the planning phase. As stated earlier, an existing facility can leverage on the scope and findings from previous turnarounds. Scope addition may include execution of capital projects requiring a plant outage for tie-in or alteration to existing equipment. Inspection reports from previous turnaround should highlight equipment approaching its end of life with decision on replacement. New problems experienced since the last turnaround may also require new additional scope. Those often manifest themselves throughout the run length with the appearance of symptoms rarely in alignment with the turnaround planning schedule. Nonetheless, ignoring them may seriously impact the plant reliability on its next run. This is an area where performance monitoring can help in identifying and ranking the criticality of the anomalies that can seriously affect reliability. Process engineers must therefore be proactive in monitoring the performance of their unit and be in close contact with the turnaround team. Replacement versus reuse is often questioned on vessel internals. Without consulting the process engineer, a turnaround planner may wrongly assume that some internals may just be inspected or cleaned, while in fact they would need replacement. While this is not an absolute rule and may depend on design, planning on removing internals such as mist eliminators, and structured packing should be associated with replacement as the geometry and integrity of those internals may be altered when removed regardless of their cleanliness. Depending on design, geometry, and robustness, some random packing can be removed clean and reused. Reusing crushed and deformed packing should however be avoided. Distillation trays can be washed and reused. Cleaning is more efficient when executed outside the tower. Cleaning foul trays inside a tower may seem faster, but it is more difficult and is associated with safety concerns. Furthermore, it can move fouling material to undesired locations. It is critical to ensure that the tray components do not get damaged during handling, washing, and re-installations.

The increased work density [22] (men hours/outage day) during a turnaround is typically managed using external contractors. Some external expertise may also be required for specialty items. The experience and skill set of external resources are an unknown and it would be a mistake to assume that the work would be completed per expectations without some quality assurance. Contractors may have good intentions, but they do not understand the process and as for all those involved in the execution of the turnaround are under time pressure. Technical support from site representatives is therefore paramount for proper executions. Some of the problems commonly observed that can seriously impact a smooth plant re-start and reliable operation are associated with cleanliness of the work. Protecting cavities and open pipes from falling debris is critical. Small tools, grinding, and welding residues can accumulate in location that could block or damage downstream equipment, such as pumps and control valves. The classic hardhat or raincoat left in tower downcomer might be obvious. Nonetheless, less obvious objects such as small tools and plastic bags are not always evident and can be missed. In fouling service, grinding dust and welding slag, besides causing potential blockage, may catalyze chemical reactions. Carbon steel grinding dust will likely oxidize into rust, which is known to catalyze free radical polymerization reactions in some applications.

Improper re-assembly of vessel internals is another problem requiring special attention. Re-installing distillation trays after cleaning is as critical as installing new trays. It takes only one bad tray for creating a pinch point that will seriously impede on the tower capacity. Experience and skill set of the boiler maker's tasks with installation is a variable that can be mitigated by setting up a tray as mockup on the ground nearby the tower. Depending on the tray complexity, each tray can be assembled on the ground prior to moving the parts inside the tower, such that the installation crew can familiarize themselves, ask questions, and ensure that all parts are accounted for. Frequent inspections as the tray gets re-installed would capture defect quickly and avoid re-works. Those inspections should be coordinated with the installation crews to minimize work disruption. Process engineers should get involved in those inspections to understand and visualize how the equipment is laid out and works. It is not unusual for young engineers to struggle relative to what needs to be checked. Reviewing the drawings and setting up an inspection check sheets of key measurements and observations is a proactive way of making inspections more effective. It also provides a record of inspection that can become useful in future troubleshooting. Taking pictures is also important as it provides a mean for observing elements that may later be scrutinized but were missed during the inspection. Inspections should focus on measurements that are relevant to proper operation of the internals, for instance, ensuring that nuts and bolts are tight, tray panels are properly oriented, and that no gaps above tolerance are present between hardware components. Special attention should be given to anything that may create pinch points. Ensure that openings defining the flow path of liquids and vapor are not obstructed. For example, the downcomer clearance on trays in an extractive distillation tower supplied with live steam was found inconsistent due to tray panels that were bent. The deformation was caused by several steam upsets that had exposed the tray panels to uplift forces. Since those anomalies were found during the tower inspection and were not part of the scope, they were reported but not fixed. The unit had to come down shortly after startup due to premature flooding of those trays.

The last man in a tower is typically a boiler maker tasked with closing the tray manways. Using a checklist with everything he brings inside the tower is good practice that provides a safeguard relative to potentially leaving something behind.

Fouling materials tend to settle in process vessel when shutting down. Cleanliness and proper re-assembly of process equipment is the most common problem associated with turnaround.

Unplanned shutdown can be forced by the trip of critical equipment or done in a controlled manner due to premature deterioration requiring an outage for repair. Shutdown forced by a trip can be detrimental as they are often uncontrolled and impose aggressive transient conditions on the equipment. Tripping cracking furnaces, for instance, is associated with thermal stress on the mechanical components. Furthermore, depending on the thickness of the coke layer deposit on the radian coil surface, additional mechanical stress can occur due to the different expansion coefficients between coke and the radiant tube material. Those stresses can result in tube failures. A sudden loss of forward flow and the impact on the transient H&MB can cause mechanical and thermal stresses on equipment. Process material may migrate to areas that could become problematic on re-start. Sudden loss of process functionality due to the trip of a critical component is an event that requires quick action by process operators and does not provide the luxury of time for properly responding to it. Process operators' instincts are to keep the unit on stream, which may, under some circumstances, contradict with the appropriate path required for bringing the unit to a safe park. This emphasizes the need for training and procedures. Semi-automated trip responses are an effective means for guiding process operators in responding to upset scenarios. This, however, assumes that the nature of the upset and the appropriate response is well understood. Response to upset involves real-time troubleshooting, which strongly relies on process operator experience and knowledge. As discussed earlier, this level of knowledge is acquired through post events/reactive troubleshooting.

Plant re-start following upsets is also critical as if not properly executed may harm some equipment. For instance, migration of fluids and compositional shift resulting from an upset can cause significant problems on a re-start. It is good practice to drain tower sumps, knockouts drums, and low points prior to re-start. This practice will minimize fouling, undesired chemical reactions, and potential liquid carryover that may damage downstream equipment. In some heavy fouling services, a pre-startup wash with solvents may prevent problems related to fouling and undesired reactions. Although a clean service, alcohol (methanol or propanol) wash is a practice recommended on cold boxes of ethylene plants before it warms up, due to the potential accumulation of NO_x (NO, NO_2, N_2O_3) and its reaction with diolefins, which creates NO_x gums that are unstable and can detonate. For tray tower, fouling material contained inside the tower will settle on the tray deck and inside downcomers. Circulation of a compatible solvent prior to re-start may prevent those solids from blocking downcomer exits and the tray open area. Tower re-start will also influence the impact of fouling. While vapor velocity across the open area of trays may clear the solids from the holes, the high temperatures may cook the fouling material on the tray or promote further polymerization of the residue. The liquid flow, on the other hand, may scour the solids' residue away from the open area but also contribute to entraining the solids through the holes under severe weeping conditions, which may contribute to blocking those holes. Draining tower sumps and reboilers will prevent fouling and potential reboiler stall on re-start. A tower slump resulting from an unexpected shutdown is frequently associated with a loss of pressure, which makes the tower sump content heavier and more likely to pinch on temperature with the reboiler heating medium.

Particular attention should be given to potential accumulation of water in undesired locations before re-starting the plant. The presence of water in hot oil service can cause pressure surges in tower [23], which may destroy the internals from the pressure surge caused by the expansion of water when shifting from liquid to vapor. If an unplanned outage required vessel entry, it is very likely that the vessel would go through a steam out for decontamination as well as for O_2 freeing before startup. The Process Safety Beacon section of *Chemical Engineering Progress* (CEP) of April 2017 [24] reports such incidents, where condensate accumulation in a process vessel following a steam out was intended to be drained out before transfer of hot oil. Due to a block valve incorrectly set to a close position, water could not be drained out of the vessel. The supply of hot oil caused the water to flash with the associated rapid expansion causing the vessel to rupture. The oil released from the vessel rupture ignited causing an explosion.

Plant trips can also be associated with cold temperature excursions. In which case, the vessel wall temperature of each vessel at risk of cold brittle fracture should be assessed before executing a plant re-start.

As discussed earlier, there are many elements that may impact the performance and safety of the process when navigating through an upset and unplanned shutdown. The priority is to bring the plant to a safe state and ready for a safe re-start. Under those circumstances, operation is typically under time pressure with decision-making relying on experience and preparedness to such conditions. Those elements should be addressed as proactive process safety and be part of plant PHAs and be documented in operating procedures. Post-event process should also be reviewed for capturing any new learnings.

Post shutdown operation performance should meet a baseline consistent with operation following previous turnaround unless significant alteration has been implemented. Unfortunately, performance evaluations following a turnaround are rarely done; hence, there are hardly any records from previous turnarounds relative to plant performance. Troubleshooting is done reactively as operational problems are encountered. Post shutdown performance evaluation is one way of evaluating the success of a turnaround, but also to identify equipment for which performance may not have been reestablished to expectation. Post turnaround performance evaluation falls under unit monitoring which is discussed in more detail in Chapter 4.

Assessing unit performance following a re-start caused by an upset is also of particular importance for identifying any step changes in equipment performance.

3.4.1.3 Startup

In most cases, total plant startups are infrequent activities following a unit outage as plants typically aimed at minimizing downtime. Nonetheless, scheduled outages are required due to regulatory inspection and maintenance requirements. Upsets causing plant shut down, although not desired and infrequent, do occur and require re-starting the unit from a safe park mode. As such, process operators and engineers get little exposure to unit startup. Due to frequent role rotation in some organizations, it is not unusual that unit process engineers may not even have experienced a unit startup. The lack of experience and nervousness of the operating crews may result in extended startup periods that are costly, not counting the risk of damaging process units through the transition. Unlike steady-state operation, troubleshooting and decision-making during a startup transition are done under tight time constraints. This creates situations where improvisation can lead to undesired states, which besides affecting the normal course of startup may result in process safety incidents. Effective startup execution and troubleshooting of unexpected events strongly depend on intuition and knowledge, which is enhanced by the level of preparation. An existing plant has the luxury of historical data from previous startup. Post-startup evaluation, while not always done, is good practice, where learning from issues experienced during the previous startup transition can be further studied. This can be classified as a reactive troubleshooting exercise intended to find the cause of problems experienced during previous startup transitions and recommend alteration to the startup procedure addressing the gaps identified and preventing reoccurrence. Total plant startups typically follow a turnaround, where the operating group including process operators and engineers are extremely busy. Startup is the least of their concern during those busy times, and when coupled with fatigue accumulated throughout the turnaround period add to the challenge for good startup preparation. Preparation and training the operating crews before the plant turnaround, while better than nothing is not as effective when compared to a rehearsal within days of the actual startup. However, given the availability of process operators and the pressure to start the plant once closing the last vessel, it often becomes the only option. Dedicating a technical startup team to direct/coach the board operators throughout the startup execution may fill the gap when early training is not an option. OTS has proved effective [25] relative to rehearsing startup procedures and associated refinement has improvement observed from previous startup evaluations. OTS also provides an ideal training tool for board operator training.

While good preparation would minimize consequences, problems are to be expected during the execution of a startup sequence. The following provides a description of the most common problems occurring during a startup sequence.

Process safety. Starting up a plant from atmospheric to normal steady-state conditions involves a transition associated with sequential displacement of air, nitrogen followed by hydrocarbon inventory, and pressurization. Considering that a total plant startup typically follows a turnaround also means that process vessels have just been closed, hence increasing the potential of leakage due to inappropriate torqueing procedures of flanges, vessel manways, and other bolted process boundaries. Use of appropriate gaskets can also be a cause for leaks. Process temperature transient conditions if too aggressive can induce thermal stress on process equipment that may result in leaks or even cause damage. Startups are associated with a higher risk of cold temperature excursion that may exceed the MAT of process vessels. Warm or near ambient hydrocarbon gas should be used for displacing nitrogen from process vessels prior to pressurization. Introduction of light liquid hydrocarbons in a nitrogen environment is known to cause a significant drop in temperature due to the initial hydrocarbon partial pressure imbalance. High-pressure differential between vessels caused cold temperature excursion by the Joule–Thompson effect. The thermal stress caused by those temperature excursions besides potentially causing leaks can also result in cold brittle fracture of process vessels if the pressure/temperature goes below the MAT of the equipment. Process leaks can be avoided through leak testing and detection throughout the startup sequence. The potential for cold brittle fracture can be avoided by first understanding the cause and aligning the startup procedure for preventing those events. Monitoring process vessels' temperature and pressure relative to their associated MAT curves throughout the startup sequence provides a proactive detection mechanism. The appropriate response to a vessel experiencing a potential MAT exceedance would be to reduce the pressure immediately. The drop in pressure will further drop the temperature, but the stress on the vessel will be reduced, providing the temperature/pressure remains above the MAT. The cause of the cold temperature excursion should be understood and the vessel skin temperature warmed up prior to repressurized the vessel. In the absence of any means for warming up the vessel, time becomes the only way. Pressurization shall not occur until the vessel skin temperature provides sufficient margin from the MAT.

Potential for over-pressure is also higher during a startup transition due to higher volumes of hydrocarbons released to the flare system. Until process units meet their targeted specifications, off-specification streams are released to flare at rates exceeding normal operation. If the release rates exceed the capability of vent valves or if any kind of obstruction limits the vent rate, then the pressure of the vessel may climb and exceed the set pressure of safety valves protecting the vessel. Popping up safety valves is considered a safety incident, and contingencies shall be in place to prevent such occurrence. This can be done by finding an alternate venting location or pulling heat if the overpressure occurs on a distillation tower. Another cause of over-pressure involves improper field valves alignment and proper blinding, de-blinding configuration. Those should be addressed through pre-startup checklists.

Components like ethylene may experience decomposition if exposed to high pressure and temperature. While this hazard is more predominant in the commissioning of ethylene pipelines, decompositions have been known to occur in other services. For instance, ethylene product pumps in ethylene plants raise the pressure of the ethylene collected from a product drum to supercritical conditions before supplying pipelines. During commissioning, the pumping system is isolated from the pipeline. If nitrogen is not purged out of the piping system between the pump and pipeline, startup of the ethylene pump would compress nitrogen. The presence of diatomic gases like nitrogen is known to have a higher compression temperature due to their higher Cp/Cv ratio. The high temperature associated with high pressure can then cause ethylene to decompose. Reaction runaway that might be experienced in acetylene hydrogenation reactors in ethylene plants can result in ethylene decomposition as well. While the normal operating pressure of those reactors may not be favorable to ethylene decomposition, a runaway reaction will cause a significant rise in temperature. The pressure of the vessel would likely increase, and the runaway reaction may transition to ethylene decomposition. Response to ethylene decomposition is not always intuitive. This first reaction might be to stop the pump and the ethylene flow. However, maintaining the flow and reducing the pressure will provide quenching and prevent the piping system from over-pressure. Literature provides similar incidents associated with commissioning of product meters [26]. Response to reaction runaways might be more intuitive, where removal of the reactants and nitrogen purge will kill the reaction.

Symptoms surrounding the risks described earlier must be well understood by the operation staff as well as the appropriate response. Without upfront knowledge, there is not sufficient time when encountering those conditions for appropriate diagnostic and response. Startup preparation is another example requiring proactive troubleshooting for identifying potential problems, their most likely root causes, and the appropriate response.

Oxygen freeing and dry outs are pre-startup activities that, if not properly executed, could damage process equipment. Depending on plant complexity and process requirements, the startup sequence may differ but would typically involve freeing the unit of oxygen. This is frequently done with steam or nitrogen. Nitrogen is used in units where water would be undesirable such as cryogenic units commonly found in ethylene plants. Besides freeing the oxygen content, hot nitrogen would be used for drying out cryogenic sections, where water would otherwise freeze and cause hydraulic restrictions and potential equipment damage. Progress of unit dry outs is monitored through dew point measurements at specific locations.

The surfaces of the flue gas side of fired equipment are normally made of refractory material. While down under ambient temperature, moisture from air gets adsorbed in the pores of refractory material. Rapid increase in temperature would cause water adsorbed in the refractory to vaporize. The expansion of water when transitioning from liquid to vapor would exert pressures within the pore of the refractory and may initiate cracks. This would affect the integrity of the refractory and increase heat leakage. The equipment manufacturer would typically specify a warmup rate that would allow drying out the refractory and minimizing thermal stress on metallic components. The warmup rate would typically refer to the temperature ramp rate as a reference. The warmup rate is managed by sequencing burners light up and firing rates.

Startup transitions are associated with several challenges from instrumentation often malfunctioning, unstable control loops, unusual thermal gradients, and, but not limited to, operation below equipment turndown.

Instrumentation and process control are frequent problems experienced on startup. It is not unusual for process variables during the transition of an operating unit through the different stages of a startup sequence, to be outside the range of the instruments. This is something often missed and not recognized. Many OTS systems do not even mimic this issue. Depending on DCS configurations, values below the instrument ranges may display the lowest value of the range, which may misguide the board operator. A near-miss incident has been reported in literature [27] where the startup procedure stated a minimum pressure requirement for the introduction of liquid hydrocarbon in process vessel. The pressure value read from DCS was well above the permissive, while the actual pressure was significantly lower than the minimum pressure requirements. This resulted in a cold temperature excursion, which remained above the MAT of the vessel due to the low pressure. Another case [25] involves an anti-surge system protecting a compressor that kept the minimum flow valve in the fully open position. The turbine driving the compressor maxed out on power and the speed started to drift down toward a critical speed, which would have caused the compressor to trip. The startup crew had rehearsed the startup on an OTS and process engineering calculations suggested significant margin from the compressor surge line. The board operator was directed to switch the anti-surge system on manual and start pinching back on the anti-surge valve output. Process control diagnosed that the problem had been caused by the temperature transmitters associated with the anti-surge system. The actual temperature was above the transmitter range and displayed a bad value. Under those circumstances, the anti-surge system was configured to protect the compressor by keeping the anti-surge valve in a fully open position.

Pressures and temperatures are typically not problematic when operated within the range of the transmitters. Level and flow transmitters tend to be more problematic. Many level and flow transmitters are based on pressure differential, with the transmitter interfaced with the process through tubing. On startup, it is not unusual for debris or unintended fluid to be present in the tubing causing erroneous values.

Process operators tend to operate controllers on automatic mode as soon as they can. Their logic is that they no longer need to pay attention to the process variables that are automatically controlled. Unfortunately, startups involve numerous transient conditions that are often below or near turndown of process equipment. The analogy can be made to flying an airplane near stall speed. The oscillatory behavior of PID controllers can cause instability, which can be eliminated by operating control loops on manual. It is critical to identify the controllers that initiate oscillatory behavior as they affect the stability of other controllers. As an example, a refrigeration compressor started to exhibit oscillatory behavior when operated without process loads. Five controllers were associated with the operation of the compressor, one for regulating the suction pressure, two for controlling the surge margins by regulating recycle flows, and two for controlling interstage temperatures by regulating liquid refrigerant injection for preventing the accumulation of the heat of compression. The temperature controller was in cascade mode with remote SP provided by a tabular function, which released a SP based on the suction pressure and a margin from the refrigerant dew point. Variation in suction pressure caused variation in temperature SPs. The variation in pressure also influenced the surge margin calculation resulting in oscillatory behavior. All controllers were switched to manual mode, with output ensuring sufficient surge margin and temperature preventing accumulation of the heat of compression. Once stable, the suction pressure controller was switched back on automatic and operated that way until stability was confirmed. The temperature controllers were switched to automatic (not cascade) at SP closer to the dew point corresponding to the suction pressure. For the time being, the anti-surge control remained on manual with controller outputs providing extra margin from surge. The surge controllers were brought back to automatic mode once sufficient process loads were introduced, which brought more stability.

3.4.2 New Facility

Unlike an existing plant, new facilities have no history. The baseline relative to expected performance is limited to the design H&MBs and equipment data sheets. Commissioning a new plant frequently involves transition in rates that do not reflect the design H&MB. Feedstock characteristics and reactor yields may differ from the design basis. Many of those deviations are linked to design methodologies and assumptions. The accuracy of the design is strongly related to the experience of the design team. Although there are always some differences such as geographical location and feedstocks quality, the performance of plants built on proven technology is less likely to exhibit significant deviations from design, especially when designed by experienced licensors who have gained knowledge from previous projects and follow a proven design methodology and established guidelines. Design gaps are more likely with engineering, procurement, and construction (EPC) firms mandated to do design work on processes for which they have little or no experience. Involvement of different engineering contractors in the different phase of a project can result in design gaps due to different practices and guidelines. Equipment vendors create another interface that may also cause variability in design parameters and guidelines if not closely monitored for consistency of design data.

A study phase would typically precede any design effort, where a new plant concept driven by an economic analysis suggests approximate range of feedstocks, product slate, and volumes. Technology may or may not be selected at that stage. Process design alternatives are evaluated for framing the economic viability of the concept. Completion of the study phase delivers more clarity on the owner's objectives and design basis relative to feedstocks, product slate, and production capacity. Impacts on the new plant ability to meet those objectives are relatively minor at this stage unless imposed by a set of constraints that may affect the smooth operations of the new facility. For example, constraining storage between units may cause unnecessary outage of downstream units during off-specification excursions. Feedstock delivery and quality, if not well understood and poorly assumed, may have a significant impact on the new plant performance. Overstating the concentration of key reactants in the feedstocks may seriously impact on the design, hence threatening the success in achieving the owner's objectives. Many assumptions arising from the study phase will provide the foundation for the design basis. A good understanding of the new facility alignment within the supply chain is also critical. For instance, if the new facility is to process a residual stream from an upstream plant, then could the quality of that stream be influenced by changes arising over time from feedstocks and product specifications on the upstream plant? Is the technology under consideration well known with proven design data and guidelines or is this new technology requiring pilot testing for providing design data required for scale-up? Those elements must be clearly defined before starting the design phase. One step in troubleshooting a new facility would be to review all those assumptions and objectives, and identify any deviations that may impact the expected performance.

The design of a new facility, as for any major project, is broken down in multiple phases. Cost estimation and reviews are conducted at the end of each phase to ensure alignment with the project business proposition. Approval and funding are awarded by the owner if the project business proposition remains aligned with the owner's objectives. The following discussion, however, focuses on the process design work as troubleshooting involves deviation from expected process performance rather than economics. The only economic aspect linked to troubleshooting is related to operating cost deviations caused by poor performance of the new plant. For example, not meeting the targeted capacity besides affecting loss revenues from sale opportunities results in an increase of fixed cost per production unit. The effectiveness of the process is independent from market conditions related to feedstocks and products.

Figure 3.13 provides a high-level schematic of project workflow from Pre-Front End Engineering (Pre-Feed) to detailed design. In Pre-Feed, alternative technologies and process flowsheets believed to meet the owner's objectives are being evaluated. Following Pre-Feed, the project moves into Front End Engineering Design (Feed) where conceptual process design of the selected alternative is being done. There is a bit of overlap between detail design and Feed. However, once in detail design, P&IDs should be frozen and the process changes be kept to a minimum as they could be detrimental to the project schedule and execution. As illustrated in Figure 3.13, process simulations are at the center of many activities during Pre-Feed

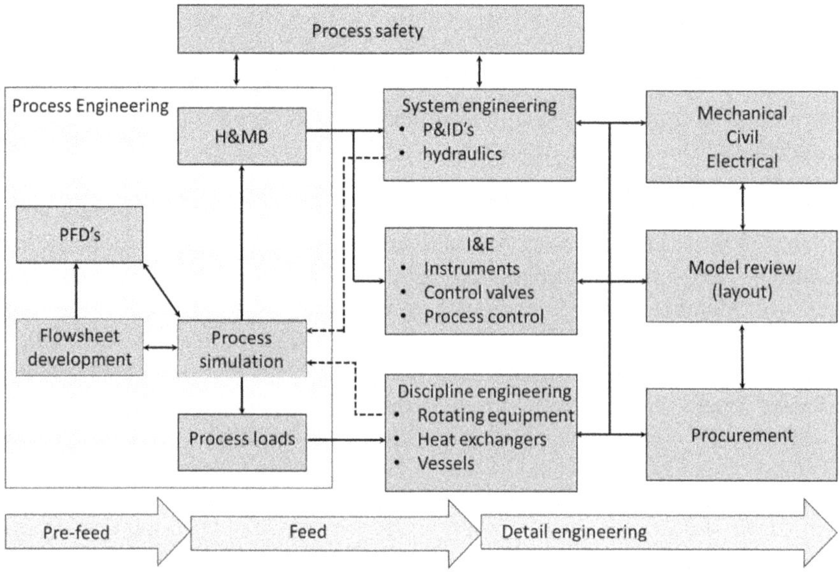

FIGURE 3.13 Project Workflow [28].

Reproduced with Permission from figure 3.11, Ethane Feed Vaporizers Startup Difficulties, André Bernard, 2024 AIChE Spring Meeting – Ethylene Producer Conference

and Feed stages, as they provide inputs for scaling equipment and evaluating hydraulic circuits. The process simulations provide streams by streams flow, temperature, pressure, composition, and properties. They also provide process loads required for designing major process equipment, such as vessels, towers, heat exchangers, pumps, and compressors. The accuracy of the simulation is therefore paramount for properly scaling the equipment. Many simulation inputs originate from a design basis document. A design basis document should be developed in the early stage of the project and be updated throughout the progression of the project. Such document states many assumptions from feedstock composition to utilities process conditions. In troubleshooting a new plant, process engineers would typically look at equipment datasheets. It is a good starting point. However, as stated earlier, the process numbers in the datasheet should originate from the simulation. If the actual process loads significantly differ from the datasheet, the reason for the difference should be investigated. For example, it is not unusual that equipment might be designed for meeting the conditions imposed by an off-design case. Because of process integration, the process loads on some equipment could be higher during a startup transition than at steady state. For those exceptions, it is common practice to state normal, maximum, and minimum loads in the datasheets. The next step, when a significant gap exists on the performance of a process equipment relative to its datasheets, would be to review the design basis and simulations, which are described in more detail below. Sometimes, this may require going back in the project history as any change after FEED may not always be communicated between disciplines. Although there should be processes and procedures for handling design changes, they do not always involve a full recycle back to process calculations and equipment specifications. This situation is more likely when transitioning to different people and engineering contractors as a project moves from one phase to the next. Those transitions are associated with a discontinuity in workflow between different organizations, where recycling back to the process simulations may not always be easy and may be short-circuited for keeping the project on schedule.

As an example, a licenser specified heat integration between a steam reboiler and steam generator. Both exchangers had kettle reboilers where steam condensing from the tube side of the reboiler was intended to flow by gravity to the shell of a nearby steam generator. To do so, the steam generator had to be at a lower elevation than the reboiler to ensure gravitational liquid flow. A different contractor was selected for procurement and construction. Suspecting an error in the drawings relative to the elevation of both exchangers, the contractor elected to review the drawings and put the exchanger at the same elevation. The discontinuity between the different contractors made it harder to review that decision along with the pressure of keeping construction on schedule. This change significantly affected the energy balance of the unit.

In another example, the transition from a licensor to an engineering contractor, as the project moved into detail engineering, resulted in an under designed heat exchanger. The process engineer misunderstood the exchanger process load sheet released by the licensor, where the pressure drop stated was based on an assumed pressure profile between operating units. Given that piping layouts were to be developed during detailed engineering, the licensor could only provide the desired pressure profile between operating units. As such, the pressure drop in the exchanger process load sheet included allowance for the piping system, static head, and the exchanger itself. The process engineers wrongly assumed that the pressure drop was limited to the exchanger and specified it as the allowable in specifications released to vendors. The exchanger design proposed by vendors, although they met the thermal criteria, was more compact with high-pressure drop, which disrupted the pressure profile specified by the licensor. Fortunately, in this case, the problem was detected before construction but after fabrication of the exchanger when review of detail hydraulics revealed significant deviations from the licensor pressure profile. The exchanger was re-designed with the appropriate allowable pressure drop. It might be easy to blame the process engineer in this case. However, different licensors and engineering contractors use different design methodologies, which can easily create confusion when interfacing different contractors. The core expertise and know-how of engineering firms are typically dedicated to front end engineering. Resources due to incremental workload in detail engineering are frequently backfilled with temporary external hiring. Minimizing those inconsistencies during transition requires the involvement of a project team that is involved for the entire life cycle of a project so they could provide guidance and ensure continuity during transition to different contractors.

Design basis provides many of the assumptions used for configuring process simulations. It should provide feedstock characterizations, the list of components, and definition of pseudo-components to be used in the process simulations. It should also state product streams' quality specifications and target production rates. Definitions of utilities specifications, such as water, cooling water, steam, and electrical power supply, should also be provided. Conditions affected by geography such as raw water availability and quality, atmospheric conditions, including ambient temperature, pressure, and relative humidity, would normally be included in the document.

Feedstock characterization is an important element to validate as changes in composition or presence of impurities may significantly affect reaction yields as well as separation. The design basis should also state the list of components to be used in the process simulations. Feedstocks would normally be sampled and analyzed. If not, that should be one of the first activities for ensuring consistency with the design basis and process simulations. While feedstock contaminants and impurities are not always or fully addressed in a design basis and simulations, search for contaminants that are symptomatic of the problems experienced by the new facility should be conducted. Loss of catalyst activity in catalytic reactors would be one example where catalyst poison might be present in the feedstocks.

Simulations. It is common practice to limit the number of components in a design simulation for practicality and ease of convergence. A rigorous analysis of feedstocks may have numerous components for which a large number of them are insignificant relative to the expected performance of the unit. Many of the less significant components are frequently lumped together into a single component believed to mimic the behavior of the group. Large hydrocarbons are sometimes grouped together and modeled as pseudo components that are characterized by an average boiling point, specific gravity, and molecular weight. The precision of the simulations depends on several factors with feedstock characterization be one of them. Omission of key components, although in small concentrations, could have significant impacts on the actual performance of a unit. This can result in reactor yield deviations due to catalyst poisoning or unaccounted reaction mechanisms. The absence of some key components may affect the actual reflux ratios over those simulated. As an example, methyl acetylene and propadiene were lumped together as methyl acetylene in the simulation of a depropanizer. Although their combined weight fraction was relatively small at 1.2% in the feed stream, lumping those components together caused the simulation to under predict the reflux by about 8% and resulted in a production bottleneck. As a first step, a validation of the overall material and energy balance would provide a level of confidence relative to the ability of the list of components used in the simulation in predicting the plant performance. If not, the process unit exhibiting the largest deviations from the simulations should be identified. Sensitivity analysis can then be conducted on that specific unit relative to different components, which have been explicitly excluded from the simulation. Missing components should be added to the simulation if they are found to have significant impact on the process loads. If there is no sensitivity, then the troubleshooting should investigate other parameters.

As discussed in previous sections, the selection of proper thermodynamics and property methods is paramount to the accuracy of simulations. Experienced licensors would typically use proprietary internal property packages, which have proven effective over the years. Given the proprietary nature of their work, the simulation itself is rarely shared with the clients and/or other contractors. The output of the simulation is communicated through tabular stream by stream material balances, which includes transport and thermal properties. Process loads are provided for major equipment such as towers, rotating equipment, and heat exchangers. Although this information is useful, it reflects the design case and may not reflect problems that may arise as the new unit is being brought on stream and is still transitioning toward design rates. Thus, unless the licensor is actively involved and can provide simulation support through commissioning and performance testing, the owner needs to develop its own simulation tools for troubleshooting. Good practice would be to have the owner process engineering team to develop simulations during the design stage. Matching the licensor simulations provides a few benefits in the project execution. First, the licensor simulation results can be used as benchmark for calibrating the owner in-house simulation. Second, by developing simulations, the process engineering team is more engaged, and rather than just verifying tables of data and schematics, gain more knowledge of the process. Simulation development becomes a tool for identifying gaps, thus understanding the criticality of process variables and enhancing healthy technical discussions between owner and licensor. Understanding that the licensor may not share proprietary data but may very well provide guidance to owner process engineers for simulation configurations may help closing process gaps. Third, as discussed in previous section, simulation tools are essential for troubleshooting process equipment, such as distillation towers, compressors, and others. Developing a library of simulations will not only help execution of the project but will also keep the simulations as troubleshooting and/or monitoring tools, once the project is closed and the licensor is no longer immediately available.

The above discussion refers to recognized licensors dealing with mature technologies, but what about less experienced contractors or owners undertaking part or the entire design work? The simulation remains critical, but the knowhow may not match years of experience gained by recognized licensors. Process simulations softwares have significantly evolved over the years and are more accessible. Managers with little to no experience in process simulation may perceive them as easy to use with their configuration more suited for junior engineers. Although graphical user interface (GUI) has made them easier, nonetheless, proper configuration requires engineers who have strong understanding of thermodynamic methods and properties as well as strong process engineering background. Major pitfalls in process simulations are around improper thermodynamic method selection which affects the entire flowsheet heat and material balance due to improper vapor/liquid (VLE) predictions. Furthermore, process simulator comes configure with default method for calculating the thermal and transport properties of mixtures. Those properties are calculated based on mixing rules, which may differ based on the nature of the components involved (e.g., polar vs. non-polar). Thus, significant plant deviations in heat and material balance from the design simulation may suggest inadequate thermodynamic methods selection. Note however, the simulation boundary stream composition should be validated with laboratory analysis prior to reaching the conclusion. When suspecting mismatch in thermodynamic method selection, it is recommended to reduce the scope of the simulation to a limited number of process units (e.g., a single distillation tower). In many cases, the selected methods might be appropriate, but the simulator databank may require more accurate input. These would include binary interaction parameters, pure components critical properties, and acentric factors. While design and operations process engineers are not expected to be thermodynamic experts, they should be able to recognize the gaps between plant data and simulation that can be attributed to thermodynamic predictions. The next step would be to consult with experts and/or gather equilibrium and pure components data from literature and compare them with the simulator databank.

While the VLE data in the simulation may provide an accurate H&MB, improper mixing rules can result in significant gaps in transport and thermal properties. Consequently, this may result in improper design of process equipment. Properties such as vapor/liquid densities, viscosities, and liquid surface tensions are involved in flooding correlations used for sizing distillation equipment. Heat exchanger design relies on transport properties such as densities and viscosities along with thermal properties such as thermal conductivity and heat capacity. Besides molecular weight and densities, compressor design requires heat capacity ratios and compressibility factors. Many of the above properties are also required for hydraulic evaluation of piping, valves, and fittings. Validation of those properties can be done by consulting subject matter experts, gathering properties published in literature, and/or measuring some of those properties in the laboratory.

Commissioning a new facility is a troubleshooting exercise by itself. Systems are typically commissioned in sequence. The sequence is established based on the level of interdependency between systems. In a new facility, electrical power would be commissioned first as it does not depend on any other upstream system other than power supplied from the grid. Power is required by many subsystems and as such must be energized first. Next, utility systems would typically be sequenced in the following order: plant and instrument air, water supply and treatment, cooling water and steam generation. Both power and utility systems would be assessed for their stability, reliability, and capacity before moving forward to the next steps. At this point, the operating unit would be pre-commissioned with activities that are dependent on process, which could involve O_2 freeing, dry outs, leak checks, and other processes. Introducing hydrocarbon feed may involve many parallel activities depending on the level of integration between equipment and operating units. During a startup, there is a transition between inventorying hydrocarbons and purging inert such as nitrogen to pressurizing the unit, stabilizing flows, and bringing all units to specifications. Good practice would involve aiming for an initial steady state that meets stable turndown of the unit. Rate can then be increased by step with plant performance evaluated at each step.

Process operators commissioning a new unit have no previous experience or feel for expected behavior and need guidance and coaching relative to operating conditions. An input/output table established based on step increase in rate may prove useful. Inputs in such table are defined as process variables for which operators have control like SPs. Output process variables are outcomes expected from the inputs. This kind of table not only is useful for process operators but also provides process engineers with a reference relative to how the plant is performing relative to established targets. Those tables are typically built from process simulation runs done for each step increment in rate. Significant deviations from expectations can be identified and resolved prior to proceeding with another rate increase. Such systematic rate progression is more likely to uncover problems as they appear rather than being hidden by the consequences they may cause on other systems. For example, an underperforming feed preheater on a distillation tower would shift load to the bottom reboiler resulting in an increase in vapor liquid circulation between the tower bottom and feed point. By first identifying the exchanger problem at lower rates and then using the input/output process variable table, the commissioning team may elect to explore and fix the problem of the exchanger prior to increasing rate. If they do elect to increase rate, they are already aware of the exchanger problem. As such, if the tower experiences flooding, then they could relate the flood excursion to the exchanger rather than questioning the performance of the tower internals. Contrarily, aggressive rate increase to target capacity may result in numerous problems all at once, which would complicate troubleshooting and identification of the root causes.

OTS are a combination of hardware and software interfacing to an operating station identical to those used by process operators in a control room. They have gained popularity over the years with their use extending beyond training. The operating station is the human/control system/plant interface. In an OTS, as illustrated in Figure 3.14, the plant, transmitters, and actuators are replaced by a process simulation model. In some cases, the DCS and PLCs are emulated in the OTS system. The DCS or DCS emulator is identical to the plant such that DCS configuration can be tested in the OTS ahead of time, that is, if the development of the OTS is properly scheduled in the project execution. In a new plant, availability and functionality of the OTS 6 months to a year before commissioning and startup would provide significant benefits.

An OTS system provides an excellent tool for proactive troubleshooting, meaning that it may help identifying potential problems and developing corrective actions and/or strategies to overcome those problems if they occur during commissioning. Troubleshooting a new plant during commissioning and startup involves a significant amount of real-time troubleshooting. Effectiveness of real-time troubleshooting

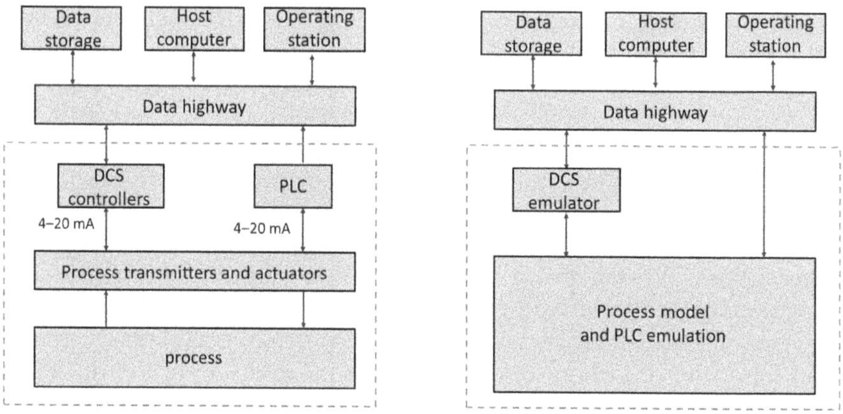

FIGURE 3.14 Process/OTS – Control System Architecture [25].

Reproduced with Permission from figure 3.1, Planning an Olefins Unit Startup with an Operating Training Simulator as Part of a Plant Revamp Commissioning Strategy, André Bernard, G. Ridell 20243 AIChE Spring Meeting – Ethylene Producer Conference

relies strongly on knowledge, experience, and intuition. Although members of the operating and technical staff might have had previous experience in similar facility, any new plant would have differences in configuration and dynamic behavior. Furthermore, many team members may not have any or little experience with similar processes. An OTS provides a non-destructive test ground for enhancing the knowledge and intuitions as well as raising the experience level.

OTS are typically developed by specialized companies. They are typically built from design H&MB, P&IDs, and equipment specifications. The involvement of the owner can be limited to scope definition, factory acceptance testing (FAT), and site acceptance testing (SAT). Some of these activities could also be delegated to engineering contractor overseeing the design. However, assigning a technical team in the early stage of an OTS development would provide significant benefits. First, the technical team would be exposed to modeling difficulties that typically have a strong connection with plant problems. As such, the team will gain significant knowledge of the process. During model integration to the DCS systems, they will become familiar with the console graphics and be exposed to numerous tag numbers associated with indicators and controllers. FAT and SATs typically involve an overview of steady-state conditions, a unit startup, and shutdown. The licensor would typically provide a startup and shutdown procedure as part of the operating manual. In some cases, the owner may bear that responsibility. Regardless, besides experiencing modeling deficiencies, simulating the startup sequence in particular will likely highlight weaknesses and/or inconsistency that may require re-adjusting the sequence. However, the objective of the FAT and SAT is not to fine-tune the startup sequence but rather to validate the functionality of the model; issues with the startup and/or shutdown sequence should be documented as needing further development once the simulator is approved and available. Bernard and Riddell [26] highlighted the importance of understanding modeling strategies aimed at building robustness, which cause deviations in actual behavior. Strong collaboration between the modeler and the technical team is critical for assessing if some behaviors are related to model limitations if it is to become an actual problem. In the cases reported by Bernard and Riddell, issues with anti-surge controls on a refrigeration compressor were causing high discharge temperature and difficulties in stabilizing the compressor during the initial stage of startup. Many rehearsals on the OTS raise the knowledge surrounding the problem with the development of strategies for overcoming those problems if they were to occur. That level of preparedness proved effective during the actual startup of the compressor as similar symptoms were observed with a rapid transition to a contingency strategy established during the OTS rehearsals. In this case, this proved effective for real-time troubleshooting and prevented significant delays in startups as well as preventing equipment damage from undesired trip or exposure of mechanical components to high temperature. In another example, extra caution was applied during the commissioning of a distillation tower due to aggressive pressure surge experienced on the OTS. While the level of caution may have caused small delays, transient operating conditions were adjusted after validating that the tower had a significant slower response than that experienced with the OTS. Good communication and collaboration with the modeler allowed the technical team to quickly conclude on the reason why the model dynamic was different.

Written operating procedures are sequential by nature, while in fact many activities need to occur in parallel. Rehearsal on training simulator will help understanding the timelines of the different activities. Process operators' responsibility and startup tasks can then be better defined such that those parallel activities get executed at the appropriate time and prevent unnecessary loss of valuable chemicals to flare or waste.

The technical team discussed earlier should be composed of process operators having experience with similar operating unit and process engineers. Their responsibilities should be geared toward the OTS development, validation, or development of startup and shutdown procedures and training of the new process operators.

Process operators in a new plant would need to familiarize themselves with the DCS graphics, the numerous control loops, and safety systems. Studying P&IDs is part of it, but training in an OTS system significantly enhances the effectiveness of training. Once the procedure has been fully established by the technical team, startup rehearsal will help minimize poor decision-making and provide operators with an opportunity to start building mental models and intuition around expected plant behavior. All this will therefore provide the operation and technical team with real-time troubleshooting skill necessary for commissioning a new plant.

MOC. The importance of management of change was previously discussed under proactive process safety. The early days of operation in a new facility are associated with learnings but also with performance deviation from design. Operation and the technical team must resist the urge to make changes that are not thoroughly reviewed as they could introduce hazards or initiate other problems. The plant should not be used as a test ground; proper process engineering shall be conducted when considering changes. Changes perceived as benign and quickly providing the illusion of benefits may shift the problem to a downstream unit. Recycle streams, for instance, are notorious for creating potential accumulation of contaminants within the process. In some cases, it may take days or weeks before observing negative consequences. Because of the time lag, it may not be intuitive to relate the consequences to changes made days or weeks earlier. However, assessing the proposed change through process simulation would provide more insight and quantification on the potential accumulation. It would also provide estimate on expected changes of process variables. The data can then be used in the MOC process for assessing the impact of potential changes relative to process hazards.

At this stage, it is also important to document troubleshooting initiatives, process evaluations, and MOC forms, as this will start building a knowledge base library for the new plant. The MOC process should also ensure updates of key documentation such as P&IDs and operating procedures.

3.4.3 Existing Facility Revamp

Revamp of existing facilities is frequently driven by increased capacity. However, other reasons may drive the need for revamping a facility such as change in feedstocks, environmental, and/or efficiency improvements. Depending on the magnitude of changes involved, revamp projects can be simple or challenging from a design perspective. They do not have the same degrees of freedom associated with the design of a new facility and must address numerous constraints imposed by plot space, compatibility of technologies, and existing infrastructure. Changes in codes and regulations also add on complexity as modified systems must often be upgraded to the most recent codes and regulations. Unlike new plants, a revamp project involves experience process operators. However, this can also introduce biases as to how the revamp unit must be operated. As such, training is critical and the use of OTS can greatly enhance the performance of operation. From a troubleshooting perspective, revamp projects combine the elements described earlier for existing and new facilities. Some elements unique to revamp projects are described below.

Scope of a revamp project, in many cases, does not include the entire operating unit and associated infrastructure. The rationale for excluding units or areas from the scope is not always rigorously evaluated. In some cases, they can be based on intuition such as when a reduction in process loads is expected. In other cases, the design margins and historical performance may suggest sufficient capacity. One common mistake when making those assumptions is to ignore turndown issues. An operating unit may have sufficient capacity, but turndown limitation may require false load through recycles, which may impact the energy balance and cause dilution of some streams. As an example, high dilution of reactants due to recycles required for satisfying the minimum flow to a chemical reactor will result in a reduction of the heat of reaction. There is a threshold, where the residual heat of reaction will no longer be capable of sustaining the reaction. Off-specification excursions will then occur on reactor used for converting impurities, such as hydrogenation reactors. Operations may mistakenly attribute the root cause to catalyst activity, whereas, in reality, the operating conditions are thermodynamically not favorable for the reaction to proceed. Operations in false loading distillation towers at higher reflux ratio to satisfy tray turndown is another example. This operating mode will impact reboilers and condenser duties. While the tower and its ancillary have the capacity for the false load operating mode, the impact on utility and/or refrigeration systems may not be fully accounted and could result in bottlenecks. It is therefore important to understand how out-of-scope equipment are integrated with the units that are within the scope boundaries.

Nomenclature

A	availability
AOR	allowable operating region
BEP	best efficiency point
BOD	biological oxygen demand
CED	causes and effects diagram
CGC	crack gas compressor
COD	chemical oxygen demand
CPI	Chemical Process Industry
DCS	distributed control system
EPC	engineering procurement and construction
ERC	ethylene refrigeration compressor
FAT	factory acceptance test
FEED	front end engineering design
f_i	frequency of a consequences
F_w	correction factor in Equation 3.5
GUI	graphical user interface
H&MB	heat and material balance
HAZOP	hazard operating studies
HIPS	high integrity protection system
HLPS	hot liquid process simulator
h_{ow}	crest over the tray outlet weir
IEFi	probability of an initiating event
KCS	killed carbon steel
KT	Kepner–Tregoe
L/V	liquid/vapor molar ratio in fractionation
LMTD	log mean temperature difference
LOPA	layer of protection analysis
LOPC	loss of process containments
MAT	minimum allowable temperature
MDMT	minimum design metal temperature
MIC	methyl isocyanate
MOC	management of change
NPSH	net positive suction head
OEE	overall equipment effectiveness
OSHA	U.S. Occupational Safety and Health Administration
OTS	operator training simulator

P	performance
P&ID	pipe and instrumentation diagram
PFD	process flow diagram
PFDi	probability of failure of each independent layer of protection
PHA	process hazard analysis
PID	proportional-intergral-derivative
PLC	programmable logic computer
PNA	polynuclear aromatic
PRC	propylene refrigeration compressor
PSM	process safety management
PSV	pressure safety valve
PVH	pressure vent header
Q	quality
Q	liquid load in GPM
RPG	raw pyrolysis gasoline
RVVH	relief valve vent header
SAT	site acceptance test
SIS	safety instrumentation system
SPC	statistical process control
t_c	cold temperature of cold fluid
T_C	cold temperature of hot fluid
t_h	hot temperature of cold fluid
T_H	hot temperature of hot fluid
TOC	total organic carbon
U	overall heat transfer coefficient
UA	overall heat transfer coefficient × heat transfer area
VLE	vapor–liquid equilibrium

REFERENCES

[1] P. T. Otis and D. Hampson, "Improve Production Scheduling to Increase Energy Efficiency," *Chemical Engineering Progress*, vol. 113 no. 3, pp. 45–51, March 2017.

[2] K. Bloch and B. K. Vaughen, "The Bhopal Gases Tragedy – Part I: Process Safety Culture," *Chemical Engineering Progress*, vol. 120 no. 9, pp. 23–28, September 2024.

[3] K. Sotoodeh, *Prevention of Valve Fugitive Emissions in the Oil and Gas Industry*, Elsevier Inc., 2021.

[4] A. Bernard, "André Bernard, Evaluating the Potential for Cold Brittle Fracture in Ethylene Plant through Dynamic Simulation," in *AIChE Spring Meeting, Ethylene Producer Conference*, Austin, TX.

[5] J. Messinger and M. Findlay, "Cold Safety Risk Assessment of Existing Bimetallic Demethanizers," in *AIChE Spring Meeting, Ethylene Producer Conference*, Houston, TX.

[6] R. E. King, "Auto-Refrigeration/Brittle Fracture Analysis of Existing Olefins Plants – Identification of Potential Excursions," in AIChE Spring Meeting, Ethylene Producer Conference, New Orleans, LA, 2004.

[7] A. 5.–1. FFS-1, Fitness-For-Service, API, June 5,2007.
[8] R. J. Willey, "Consider the Role of Safety Layers in the Bhopal Disaster," *Chemical Engineering Progress*, vol. 110 no. 12, pp. 22–27, December 2014.
[9] R. Deshotels and R. D. Zimmerman, Cost Effective Risk Assessment for Process Design, McGraw-Hill, 1995.
[10] M. Oliverio, "Risk Matrix for Canada," in *PSMD Symposium*, November 2021.
[11] A. Bernard, "Analyses of a Propylene Compressor Trip with Dynamic Simulation," AIChE Spring Meeting, Ethylene Producer Conference, Houston, TX, 2006.
[12] C. Cagnolatti, "Top Seven Causes for Lost Olefins Production," *Hydrocarbon Processing*, vol. 94 no. 4, pp. 31–36, April 2015.
[13] A. Bernard, "Strategies for Mitigating Fouling in Ethylene Plants," *Chemical Engineering Progress*, vol. 121 no. 3, pp. 32–38, March 2025.
[14] A. Bernard, T. M. Pickett and B. M. Manek, "Flux Oil Import to Quench System Risk, Ethylene Producer Conference," in AIChE Spring Meeting, Ethylene Producer Conference, Chicago, IL, 2011.
[15] A. Bernard, "Evaluation of Fouling Precursors Residence Time in Distillation Tower," in AIChE Spring Meeting, Ethylene Producer Conference, San Antonio, TX, 2013.
[16] M. e. a. Sprague, "Performance Evaluation and Fouling Mitigation in A Gasoline Fractionator,," in AIChE Spring Meeting, Ethylene Producer Conference, Orlando, FL, 2006.
[17] A. Bernard, F. Fahiminia and G. Kim, "Challenges of Primary Fractionators on NGL Feed," in AIChE Spring Meeting, Ethylene Producer Conference, San Antonio, TX, 2013.
[18] D. M. Rossana, J. Burgin and A. Khanlari, "Visual Clues Gives Insight into Ethylene Service Fouling," *Chemical Engineering Progress*, vol. 116 no. 5, pp. 63–69, May 2020.
[19] J. E. Olsen and B. G. Lewis, "Predictive Maintenance Basics for Process Safety Engineers," *Chemical Engineering Progress*, vol. 119 no. 5, pp. 26–33, May 2023.
[20] F. C. Subcommittee, "Feedstock Contaminant War Stories CO_2 , Ammonia,," in AIChE Spring Meeting, Ethylene Producer Conference, Houston, TX, 2007.
[21] H. P. Bloch and A. R. Budris, *Pump User's Handbook Life Extension* 4th Edition, The Fairmont Press, Inc, 2014.
[22] J. B. Nichols, "Recent Trends in Global Olefins," in AIChE Spring Meeting, Ethylene Producer Conference, New Orleans, LA, 2024.
[23] H. Z. Kister, *Distillation Troubleshooting*, John Wiley & Sons, 2006.
[24] P. S. Beacon, "Is Your Vessel Really Empty," *Chemical Engineering Progress*, vol. 113 no. 4, pp. 20, April 2017.
[25] A. Bernard and G. Riddell, "Planning an Olefins Unit Startup with an Operating Training Simulator as Part of a Plant Revamp Commissioning Strategy," in AIChE Spring Meeting, Ethylene Producer Conference, Houston, TX, 2023.
[26] S. T. Coleman, "Ethylene Decomposition: A Tutorial," in AIChE Spring Meeting, Ethylene Producer Conference, New Orleans, 2019.
[27] M. Kennedy, "Autorefrigeration Near-Miss at Equistar Chemicals, A Lyondell Company," in AIChE Spring Meeting - Ethylene Producer Conference, Atlanta, GA, 2005.
[28] A. Bernard, "Ethane Feed Vaporizers Startup Difficulties," in *AIChE Spring Meeting* – Ethylene Producer Conference, New Orleans, 2024.

4 Troubleshooting – The Data, the Tools, and Monitoring

4.1 THE DATA, THE TOOLS, AND MONITORING

Chapter 2 discussed the fundamental principles and operational characteristics of major equipment involved in chemical processes. Those concepts were applied in consequential and circumstantial troubleshooting discussed in Chapter 3. In both chapters, effective evaluation of process equipment relies on accurate documentation and process data. The work process discussed in Chapter 1 described methodology for structuring the data in a way that makes it effective for identifying the most likely route causes. This chapter discusses the type and source of data use for troubleshooting. It also touches on how effective use of data in performance monitoring minimizes reoccurrence of undesired deviations by enabling proactive behaviors, as well as how good monitoring program may speed up troubleshooting when needed.

4.1.1 The Data and the Tools

4.1.1.1 Engineering Documentations

Engineering documents, such as P&IDs, equipment data sheets, and mechanical drawings, to name a few, lay out the foundation for effective troubleshooting. Without those documents, the troubleshooting scope may expand to reverse engineering the process by re-sketching block flow diagram (BFD), physical measurements, and performances testing of equipment. Although it may be argued that historical data alone may suffice for troubleshooting a recent problem, the absence of proper engineering documentation makes it difficult to visualize the process and impossible to establish design normality. What may be perceived as normal is sometimes the result of years of malpractice, commonly called acceptance of deviations. As an example, troubleshooting the performance deterioration of a suspected coalescer in a specialty chemical plant started with a unit shutdown and internal inspection of the coalescer. The coalescing element had totally deteriorated into small pieces explaining the poor performance. However, this deterioration did not occur overnight and would not explain recent quality issues with the final product. A process engineer was then assigned to conduct a process study on the unit with the intent of finding the most likely root cause of the recent quality problems. This case study occurred in the early 1990s where two major transitions in the CPI were occurring. First transition was the early days of digitalization where maintenance records transitioned from paper

 DOI: 10.1201/9781003449676-4

to electronic database. Second transition was the quality assurance era driven by companies seeking ISO 9000 certifications. In fear of failing the ISO 9000 certification audit, this plant had destroyed many paper records including maintenance and original unit operating manuals. This complicated the task of the process engineer, who established the last replacement of the coalescing element to over 30 years after discussion with experienced process operators. Without the original operating manual and many of the equipment specifications, the process engineer with help of process operators and plant walked through sketched out a BFD. The plant had recently transitioned from analog control panel to DCS. However, data collection to a data historian was not yet configured. As such, material balances on the BFD were based on data collected by process operators on recent log sheets complemented with laboratory analyses. Due to absence of many equipment data sheets and specifications, operating points relative to the design capacity of equipment were evaluated using vendor data matching equipment nameplates and physical dimensions. Given the age of the plant, proxy approximations were often required. A Gantt chart of the process sequence was put together as the chemical reactor in this unit was operated in batch mode. The finding of the process studies concluded that many of the operating units were operated beyond their design capabilities. Although not affecting the quality of the final product, this mode of operation was associated with larger losses of valued chemicals. The recent quality degradation was attributed to the accumulation of solid impurities over years of operation in a few separator vessels. Cleaning of those vessels at a reasonable maintenance frequency would have prevented the recent quality issue experienced by the plant. The Gantt chart illustrated that proper sequencing of operation would improve productivity while operating within the design capability of the equipment.

The above discussion demonstrates the importance of good engineering documentation and availability of maintenance record for executing troubleshooting activities or conducting process studies. Over 90% of the above process study was spent creating information that would have normally been available from engineering documents. The absence and maintenance of engineering documents in this case also blurred the real design normality of the unit. The lack of management of change practices resulted in acceptance of deviations over the years. In the above example, this had significant consequences on quality and productivity.

Although the above example may be out of date with current business practices, other circumstances may challenge the ability of an organization in keeping their documentations up to date. Plants may change ownership; re-organization may affect the effectiveness of MOC practices, and digitalization of documentations may not always occur smoothly. The following discussion describes engineering documents that are essential for troubleshooting and comments on circumstances that may require validation of those documents.

P&IDs are documents that are essential to operation, troubleshooting, and process safety. They provide a substantial amount of information, from the process flowsheet to instrumentation in place, controls, pipe sizes, and specifications. They also provide high-level information on major equipment, such as vessel dimensions and design capacity on heat exchangers, pumps, and compressors. Due to the extent of information displayed, they are referred to as good communication tools between disciplines [1]. P&ID symbols, tagging, and layouts are standardized, as documented in ANSI/ISA 5.1 and PIP PIC001.

P&IDs are essential for conducting PHA, making their accuracy critical. As such, depending on locations, process safety regulations may require P&IDs to be current and accurate. This emphasizes the need for effective MOC processes, given the number of changes occurring in a chemical plant over years. An investigation report from the U.S. Chemical Safety and Hazard Investigation Board (CSB) [2, 3] discussed the implication of inaccurate P&IDs on the explosion followed by a release of flammable hydrocarbons caused by thermal expansion of liquids trapped in a close section. In this case, omission of valves, check valve, and strainer misled the PHA teams and those conducting MOCs during subsequent changes. As previously discussed, PHAs can be considered a form of proactive troubleshooting. Post-event troubleshooting may not be subject to the same consequences as the incident investigated has already occurred. Nonetheless, they may mislead the troubleshooting team if the P&IDs are not accurate. In conducting PHA and/or as part of MOC practices, it is advisable to conduct field checks for validating the P&IDs. Plants should also have processes in place for operations and unit process engineers to report discrepancy in P&IDs, given their knowledge of the unit and frequent use of P&IDs in executing their functions.

PFDs and H&MBs are generally produced as part of the original plant design. Although not as necessary as the P&IDs, they can provide a good starting point in troubleshooting due to their simplicity. The design PFDs are typically associated with a stream-by-stream H&MB, which besides providing components balances, also should include stream properties. The H&MB data are extremely useful for conducting process calculations as most of the process variables are typically not measured in the plant (e.g., transport and thermal properties). However, judgment is required as departure from the design operating conditions may cause significant deviations from the H&MB reported properties. In such cases, process simulation, which is discussed in more detail later, can be used as an alternative.

BFDs are not considered engineering documents but as discussed in previous chapters are useful for summarizing the information extracted from the P&IDs. They provide a simplified diagram of the system being troubleshot, limiting the illustration to a high-level process flow schematic with all the instrumentation tag available. It could also include flows, temperatures, and pressures for specific cases. The main benefit of BFDs is their simplicity and that they avoid navigating through multiple P&IDs. When possible, it is preferable to have BFD sketched on a single page for simplicity. Figure 4.1 provides an example of an ethane feed vaporization system integrated with an ethylene refrigeration unit. This single-page BFD summarizes the information extracted from five P&IDs. It also provides the basis for a simulation flowsheet in this particular evaluation [4].

Mechanical drawings of process vessels and heat exchangers are frequently required in troubleshooting. Besides supplying detail dimensions and geometry, they provide material of construction and operating limits, such as maximum allowable working pressure (MAWP). While basic dimensions of process vessels can be obtained from P&IDs, locations of level taps, thermowells, and pressure taps can

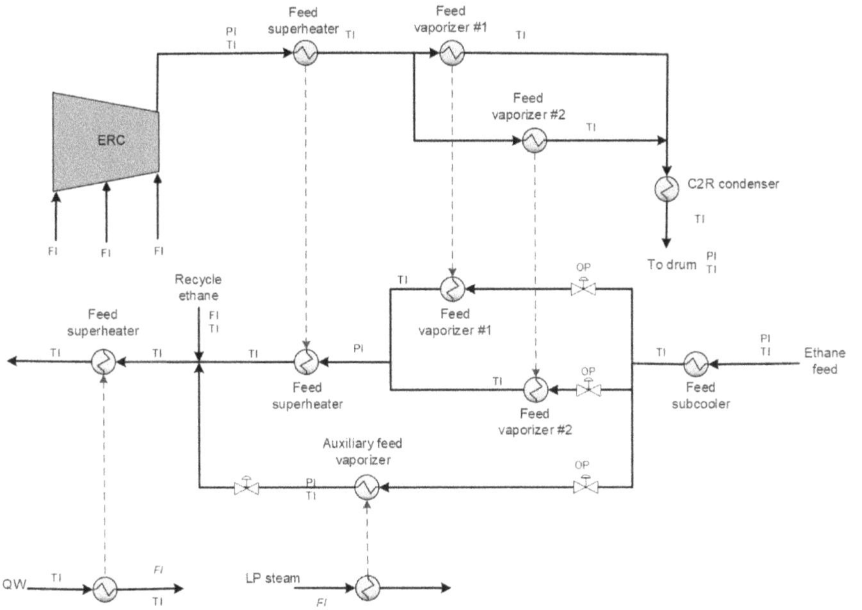

FIGURE 4.1 BFD Example [4].

Reproduced with Permission from figure 11, Ethane Feed Vaporizers Startup Difficulties, André Bernard, 2024 AIChE Spring Meeting – Ethylene Producer Conference

only be obtained from the mechanical drawings. Although P&IDs would report thermowell location relative to tray numbers in distillation towers, experience would show that understanding the exact location within the tower may explain some discrepancies between process simulation and measured value. It is typically recommended to measure tray temperature by having the thermowell in the liquid phase with the bottom of downcomers being an ideal location [5]. However, for control purposes, this location may result in lags and compromise on accuracy versus speed. The thermowells might have been positioned at optimum locations during the original tower design but gaps in temperature measurements from simulation may occur, following a retrofit. As illustrated in Figure 4.2, retrofitting conventional trays with high-capacity trays having downcomer dynamic seals would cause the thermowell to read the temperature in the froth of the tray below while before the retrofit it used to read the liquid from the tray above. Understanding those differences is critical as otherwise the troubleshooter may mistakenly reconcile the temperature gap by adjusting fractionation efficiency in the simulation.

Level tap locations and correlating the actual level against the level instrument reading is critical for evaluating residence time from low to high level. The criticality is even more important when high-level trips are associated with operation of knockout drums. Compressor inter-stage drums are good example where high-level trips typically protect the machine. Example 3.3 discussed how transient conditions during a process upset caused a PRC to trip on high interstage drum level. In this

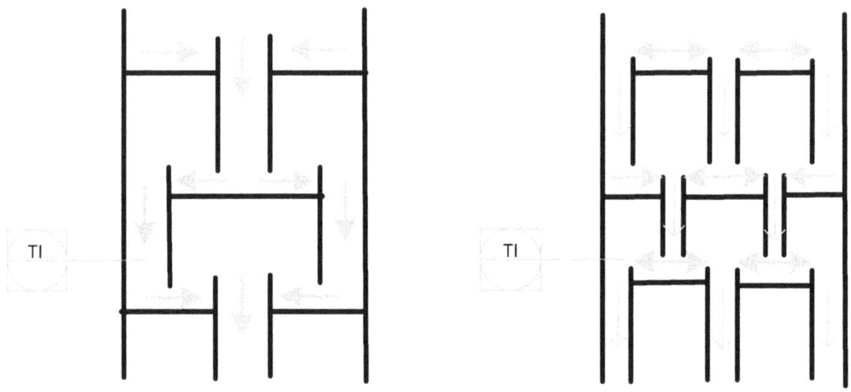

FIGURE 4.2 Thermowell Location versus Tray Retrofit.

example, good understanding of the level reading versus the actual liquid level in the drum was required not only for evaluating the sequence of events but also for providing robust recommendations.

Calculating the liquid inventory contained in vessels is straightforward on vertical vessels for which the level taps are located above the tan line. In this case, the volume of liquid contained between the level tap is directly related to the height of liquid above the bottom tap multiplied by the cross section of the vessel. If the level tap is underneath the tan line, the volume contained in the head of the vessel must be taken into account.

Exact location of inlet and outlet nozzle is important for conducting accurate hydraulic evaluations. Internals details from distributors, baffles, and mist eliminators in knockout drum and distillation towers will provide details that are part of the clues required for reconciling actual to expected performance. The momentum of fluids entering process vessels is closely related to the nozzle size and the presence or absence of distributor device.

While heat exchangers datasheets may sometimes suffice in providing the required information for configuring the exchanger geometry in rating software, mechanical drawings may sometimes complement missing or unclear information.

Mechanical drawings for rotating equipment provide more value to mechanical engineers and less to process engineers who would rely more on datasheets and characteristic performance curves.

As for any engineering documents, good MOC practices would ensure updating mechanical drawings resulting from any changes in the unit.

Equipment datasheets differ based on the type of equipment. They complement mechanical drawings for heat exchangers, process vessels, and rotating equipment.

Instrumentation datasheets are critical as they provide the type, calibration, and range of instruments. In many cases, flows are measured by pressure differential devices, such as orifice, venturi, or others, which correlate the volumetric flow to the pressure drop across the measuring device. The flow is, however, frequently displayed as a mass flow, based on a specified pressure, temperature, and density. Unless compensated for pressure, temperature, and density, significant deviations between the actual and measured mass flow are expected as process conditions depart from the calibration conditions. In such a case, the troubleshooter may have to include compensation for closing material balances around units. Level indicators do not always include the full range between the lower and upper taps. This information is normally specified on the level instrument datasheet and can then be used for correlating the actual liquid level. As for flow measurements, levels based on pressure differential between taps are affected by process conditions. The static head is dependent on density. As such, deviations in pressure, temperature, and/or fluid composition will likely deviate from calibration of the level instrument. Pressure and temperature are typically straightforward and just limited by the calibration range.

Performance characteristic plots for rotating equipment are frequently included with datasheets. They are typically provided as head and efficiency curves, which are critical for assessing the performance of the equipment.

Less common are performance characteristic plots of static equipment. However, as described in Chapter 2, distillation tray and catalyst operating windows are useful tools for assessing operation within the design capability of the equipment. Such documents, although typically not official engineering documents, should be treated as such and properly filed.

Piping layouts. Evaluating hydraulics between operating units requires knowledge of the piping layouts. Depending on sizes and plant vintage, this information may be available in different formats. In recent years, the use of three-dimensional (3D) models has somewhat eliminated the need to perform scale piping drawings as they can be automatically generated from 3D models [6]. Prior to the advent of 3D models, plastic scale models were used in the 1970s and 1980s. They provided an easy and quick way to get a draft isometric sketch with decent accuracy. Some plants do maintain isometric drawings of each pipe displayed on the P&IDs, which are a good reference for configuring accurate hydraulic calculations. In the absence of those documents or for a quick estimate of a pipe segment layout, an approximate sketch can be put together by walking the pipes in the plants. Walking the pipes does not mean walking on the pipe as many are elevated on pipe racks. It means to find reference points providing a safe walking passage that can be approximated to the length of pipe segments. As a rough guess, a large step (depending on people size) can be approximated to 3 ft. Elevation can be established from known reference points. A rough isometric sketch can be put together in the field with approximate pipe segments. A high-level evaluation can also be conducted using plot plan with location of origin and destination of pipe segment being evaluated. It is important, however, that the overall change in elevation between the origin and destination is well known and that fittings influencing pressure drops are being accounted. In estimating a pipe segment for hydraulic evaluation, it must be remembered that pressure drops are not highly influenced by inaccuracies in pipe lengths but are strongly dependent on diameters, elevations, and fittings, causing pressure drops, such as valves, orifices, or anything causing a change in diameter.

Engineering standards. Most companies have their own engineering standards, which are frequently derived from industry codes, standards, and local regulations. Troubleshooting involves identification of deviation resulting in undesired consequences. As such, design and/or operation practices deviating from engineering standards must be identified. Some of those deviations have been grandfathered due to the vintage of the plant and/or approved through variances after being thoroughly reviewed with risk mitigation strategies in place.

Documentation controls are management processes which ensure that documents are up to date but most of them are an accurate representation of what is operating in the field. This covers a wide range of engineering documents such as mechanical drawings, equipment datasheets, and P&IDs, to name a few. Chemical plants undergo significant changes over the years from flowsheet alteration to equipment retrofit. Quality certification such as ISO 9000, although not ensuring accuracy, ensures that a structured process is in place for handling documents. If executed with diligence, it raises the degree of confidence in the accuracy of engineering documents. It however comes with some challenges such as storage of the official version; management of red line copies being modified by ongoing projects and replacement of the existing documents with the revised as-built version from finished projects to official document storage. Although this sounds simple, the transfer of documents frequently involves different internal or external organizations (i.e., project teams, engineering contractors, and/or licensors). Given the criticality of some engineering documents such as P&IDs, process must be in place for people to report discrepancies observed in the field relative to the official control documents. This behavior must be encouraged as engineers and process operators are the best sets of eyes to pick on discrepancies as they frequently refer to those documents but also are well acquainted with the actual physical layouts of the units in which they are working.

Although documentation control is not a troubleshooting activity, it does have a significant impact on the effectiveness of troubleshooting. Inaccurate documents may result in unnecessary iterations and potentially unnecessary outages. For instance, obsolete pump or compressor head curves may suggest significant gap in performance that may justify an outage for inspecting and repairing the equipment. Now, a change in performance should be noticeable when modifying the equipment. However, poor MOCs and lack of performance monitoring would not raise attention until the equipment becomes a bottleneck. Miscommunication between disciplines during project executions is not uncommon and can result in alterations believed inconsequential on process equipment, while in reality could prematurely constrain capacity. As an example, mismatch in head curves between process and mechanical engineering was found while troubleshooting performance of a pump. Disagreement between both disciplines was reconciled after discovering that the "as built" head curve was never communicated to process or properly transfer to the equipment file. Alterations for improving reliability had been made without consultations with process. A slight change in process loads during project execution, believed to fall within the capability of the pump prior to the mechanical alterations, had not been communicated, as it was considered inconsequential. These changes resulted in a capacity bottleneck.

4.1.1.2 Operation, Maintenance, and Process Safety Documentations

Engineering documentation provides all the information relative to the process flow scheme, operational characteristics of individual equipment, and explains how a unit was intended to be operated. Operation procedure and maintenance records describe how it is actually operated. Process safety documentation provides history on incidents and gaps potentially threatening reliable and safe operation of the plant through PHAs reports.

Operating procedures are controlled documents which specify steps for the different tasks performed in the plant. Operating procedures can describe activities as simple as collection of samples, pump startup, and periodic flushes, which involve a relatively small number of steps. They can also be complex as for plant startup and shutdown, which involve numerous steps that may occur in parallel. There are also operating procedures for addressing abnormal conditions, such as power failures, loss of critical equipment, and other abnormal situations specific to the process. Operating procedures can be viewed as *this is what we say we do*, while deviations from operating procedures would be *this is how we do it*. The gap between the above two statements should raise awareness as it is an indication of inadequate procedure or acceptance of deviations. In either case, failing to address this gap could result in a potential incident. Literature [7] reports a case of pipe rupture in a hydrocracker caused by a significant temperature exceedance from the maximum allowable operating temperature. The operating procedure was calling for shutdown of the unit when exceeding a temperature of 425 °C. The rupture was estimated to have occurred at a reactor and pipe temperature of 760 °C. Previous exceedances were experienced and led the operator to believe that those excursions were acceptable. Exceeding operating conditions prescribed by operating procedures should be address as a near missed incident. A proper incident investigation would conclude if the operating window can be widened after following a rigorous management of change process, or if not acceptable, reemphasize the importance of following the procedure.

Operating procedures are part of the method category in the cause-and-effect diagram discussed in Chapter 1 and illustrated in Figure 1.4. However, they have a close tie to people as it is up to people to follow them or not. Operating procedures are an important piece of data in the search for deviations. Whether justified or not, deviations from procedures represent a deviation from normality in troubleshooting. Reasons why process operators deviate from procedures are discussed in Chapter 5. It is important to capture and understand those reasons as they may lead to changes in the procedure. However, deviations in procedures should not be discovered because of troubleshooting activities. They should be handled through the management of change process.

Process descriptions are documents that could be integrated within the operating procedures. However, it is best to have them as separate documents as they can create confusion in a step-by-step operating procedure. Process description would typically include a description of all components involved in an operating unit, including controls and SIS. Normal operating conditions and alarm settings are also provided as reference.

Operating procedures and process description provide a good reference in troubleshooting, which allows understanding how the unit is intended to be operated. Reviewing historical data will highlight deviations that might have become habits, which may or may not be appropriate. Nonetheless, those deviations must be identified and assessed relative to safe and reliable operation.

Operator logs document the different tasks accomplished by process operators throughout their working shifts. These logs are typically done digitally nowadays and are available to process engineers and unit leadership. Although some steps may be missing, depending on how busy process operators may have been during their shift, operator logs provide valuable information for troubleshooting. A review of operator logs should be combined with discussions with process operators to bridge the gap between what has been reported and what has the process data suggested.

Work permits are documents issued by process operators who are responsible for activities other than normal operations occurring in their units. Those documents are useful for reactive troubleshooting associated with incidents involving work in an operating unit. Those incidents are caused by omissions or violation of the requirements stated on the work permit. As an example, literature [8] reports a case involving fatalities due to omission of stating on the permit nitrogen purge operation in the vicinity of the work. A worker noticed a roll of duct tape inside a close by vessel, where he attempted to take it out and lost consciousness. His co-worker tried to rescue him and lost consciousness as well. Both workers were pronounced dead as they arrived at the hospital. Review of the work permit during the incident investigation revealed that a check box *Nitrogen Purge or Inerted* was checked as N/A instead of yes. Furthermore, though in good faith, by attempting to remove the roll of duct tape, the two workers were executing a vessel entry for which they did not have the proper permit.

Maintenance records provide history on how frequently process equipment are unavailable due to maintenance. These records should also provide the nature of the work done on the equipment. Maintenance strategies differ based on criticality of equipment and whether it is based on preventive or predictive maintenance. Nonetheless, it is not unusual for maintenance requirements to become reactive when equipment fails prematurely. An increase in maintenance frequency is an indicator that something has changed. Although it may not be directly related to the equipment being maintained, the nature of the work required is associated with symptoms (or deviation). Some of those symptoms can be associated with fouling, mechanical components failures, and/or excessive vibration. An increase in maintenance frequency can be caused by a change in the equipment exhibiting the symptoms or caused by external factors. In troubleshooting, the maintenance records can be used to establish the history and frequency of maintenance over time. The timeline of changes – either on the components being part of the equipment or external, such as operating conditions, new chemicals, maintenance, and performance history of upstream units – should be captured and compared. This comparison will help in establishing potential cause and effect.

As an example, two centrifuges were used for separating yeast from a fermenter effluent, with the yeast-rich stream recycled back to the fermenters. Due to an increase in fouling, the cleaning frequency of the centrifuges started to increase quite significantly. Fouling of the centrifuges was associated with imbalance of the rotating element which resulted in premature deterioration of some mechanical components. Yeast was separated from the fermented liquor by centrifugal force which pushed the yeast solution to the external circumference of a rotating cone, where it exited through nozzles symmetrically located around the external circumference of the cone. The yeast solution concentration was a function of the nozzle size, with small nozzle creating a denser solution (i.e., higher yeast concentration). The yeast solution from the centrifuges was collected in a tank from where it was recycled back to the fermenter by a centrifugal pump. Due to limited capacity of the centrifugal pump, operation made some alteration to the centrifuge nozzle configuration. The intent of those modifications was to generate a denser yeast solution, which would reduce the volumetric flow rate required by the centrifugal pump. The troubleshooting initiative was initially focused on reducing plant downtime and improving production. Reviews of maintenance records showed that downtime of the two centrifuges was a major cause of loss of production and maintenance costs. Troubleshooting of the machine performance uncovered a few deficiencies. First, a single centrifuge operation was insufficient for sustaining plant rates. As such, significant rate cuts were required whenever one of the centrifuges needed to be brought down for cleaning and maintenance. Second, in an attempt to make the yeast solution denser, operations instructed maintenance to put blanked nozzles in a few locations. These blank nozzles, although installed symmetrically, affected the balance of the centrifuge due to preferential formation of a yeast sludge on locations where blanked nozzles had been installed. A verification was done with the vendor regarding the use of blanked nozzles, which is a practice they did not recommend on this application due to the presence of fibrous material, which promoted yeast agglomeration and explained the accumulation of yeast sludge around the blanked nozzle's location. Furthermore, the presence of fibrous material was contributing to fouling of the machine. The change in nozzle configuration and the presence of fibrous material were the two root causes believed responsible for the high frequency of maintenance required by this machine. While fibrous material was already present from day 1, the change in nozzle configuration was more recent and the primary cause for the change in behavior. Recommendations included reverting to the original design nozzle configuration. Conversion of an upstream tank into a settler for gross removal of the fibrous material was also recommended. Although the centrifuge already had an upstream strainer, a settler would offload the strainer and reduce the overall carryover of fibrous material to the centrifuges. Operation at lower rates was recommended until a larger centrifugal pump could be purchased and installed. Management pushed back on implementing these changes until the new centrifugal pump became available, as they did not want to reduce rates. However, based on OEE described by Equation 3.1, they were told that the poor availability of the centrifuges would offset the potential gain in capacity at plant rates. Besides preventing premature failure of the centrifuge, the recommended low-rate operation would be equivalent to typical plant rates after adjustment for equipment downtime. This example demonstrates that a review of the maintenance record put the troubleshooting team quickly on the right path.

Inspection reports are another source of valuable information. Although the content is generally focused on the mechanical aspect, the information may provide clues on how process conditions may contribute to the deterioration rate. Inspection reports frequently include as found pictures (i.e., before cleaning) of equipment internals from when it is taken down for maintenance. These pictures are valuable pieces of information and may help in establishing potential root causes affecting the run length of the equipment. Clues such as the presence and location of fouling materials, signs of wear due to erosion, and/or unexplained discoloration of mechanical components helps in understanding the mechanism responsible for performance degradation of equipment.

Incident investigation reports provide history of incidents previously experienced. Those reports document the nature of the incident, the most likely root causes, and provide recommendations. They are a valuable reference, which should be reviewed early in the troubleshooting initiative, as they would save valuable time and effort if similarities exist with previous incidents. Example 3.3 discussed in Chapter 3 which look at the trip of refrigeration compressors following the trip of the crack gas compressor in an ethylene plant is an example where previous occurrence of such incidents had occurred. However, in this case, due to operation suspicions that changes implemented by a revamp project were the most likely root cause, the process engineer conducted a detailed evaluation of the sequence of events first. He reviewed historical incident reports after completing his evaluation as a validation for supporting his conclusions as the pre-revamp configuration would have been subject to the same transient conditions. Nonetheless, in this case, review of previous incident investigation reports confirmed his conclusion, but none of the previous evaluations had clearly identified the most likely root cause correctly as such, while early review should have been conducted, it would not have prevented the detail evaluation done.

PHAs of a plant are documents that are typically updated every 5 years. They include numerous failure scenarios with their risk ranking and recommendations on elements for which the residual risk level dictates mitigations and/or improvements. Table 3.2 illustrates an example of risk matrix where the residual risk level is based on the combination of likelihood and consequences. Although it is not expected to find incomplete action items on scenarios with residual risk level high and above, it is not unusual to find in incidents investigation conducted by the CSB report of unaddressed PHA or insurance underwriters' recommendations.

Review of PHAs in troubleshooting may help if the problem being investigated has similarities with the scenarios evaluated in the PHA. This may help evaluating the effectiveness of the mitigation documented in the PHA or re-emphasize the need for implementing PHA recommendations that are still outstanding. Findings from the troubleshooting initiative may also suggest modified or different mitigation strategies for bringing the risk level to acceptable.

HAZOP have similarities with PHA but are typically conducted as part of projects related to the construction of new plants or revamp of existing facilities. As for PHA, they provide a good reference relative to failure scenarios that have been evaluated with a list of recommendations for reducing the risk level to acceptable.

Material Safety Data Sheets (MSDSs) should be available for any chemicals produced or procured by a chemical facility. MSDS are useful relative to the physical and chemical properties of the substance. While their main purpose is for safety, they are providing information that can be useful for troubleshooting.

Certificate of analysis and product specifications. Chemical plants convert raw materials of less value into saleable products of higher value. Both saleable products and raw materials must meet quality specifications. In the value chain, products of one plant are raw material for downstream plants. The level of contaminants tends to be higher upstream the value chain than downstream. Nonetheless, most plants have specifications on their raw material. Exceedance in contaminants may cause operational problems such as catalyst poisoning, undesired secondary reactions, and/or impurities for which the separation unit has not been designed for. Besides raw materials, most chemical plants use several chemicals for preventing corrosion, fouling, and/or inhibiting undesired reactions. There are also chemical additives needed in the formulation of the final products. All of those chemicals are a potential source of entry for contaminants, which is prevented by agreed quality specifications between vendors and users. Understanding of quality specifications is important in troubleshooting when dealing with contaminants and/or the sudden presence of undesired chemical reactions that can have been triggered by the presence of new contaminants. In some cases, especially for feedstocks, it is not unusual for suppliers to provide a certificate of analysis. For specialty chemicals, it is more difficult due to the proprietary nature of the formulation used. In those instances, and when suspecting those chemicals as a point of entry for contaminants, the troubleshooting initiative should include the supplier as they can confirm or rule out their chemicals based on identification of the contaminants being suspected. Note that as for any troubleshooting exercise, the sudden presence of contaminants is the result of change. This could be caused by off-specification chemicals, in which case honesty and collaboration from supplier are critical. Or, in many cases, it is caused by introduction of a new chemical due to change in product or supplier.

It is important to note that while the supplier may provide a certificate of analysis, the analysis reflects the quality of the product at the point of origin. Contamination through transportation is not unheard of. As such, analysis of feedstock as received at the plant is a good practice when applicable. This can be easily done on materials transported by tank truck and railcars but more difficult to control in pipeline transfer.

4.1.1.3 Process Data

Engineering documentation provides the information required for understanding the process flowsheet and equipment characteristics. Evaluating performance requires understanding the conditions imposed on process equipment with the actual outcome from those conditions. These can be categorized as inputs, which are conditions that are within the control of process operator and involve SPs and valve settings. The outcomes or outputs are consequences from the inputs, as measured by process variables.

Raw process variables are parameters that are measured in the plant from transmitters that correlate a physical property to a 4–20 mA electrical signal. The current is then digitalized and inputs a distributed control system (DCS) where the signal is correlated to the value of a process variable. Similarly, the DCS digital output signal can be converted into a 4–20 mA signal which supplies final control elements such as control valves. The transition from analog control panels to DCSs in the late 1980s brought process data in the office of process engineers with the use of data historian; it can archive all the data from the DCS. The process data from the DCS can be qualified as raw process data and includes primarily flows, temperatures, pressures, and liquid levels. Online analyzers could also feed the DCS but are discrete variables updating values based on the analyzer sampling and analyses cycle. There are many other process properties that can supply information to the DCS providing that they can be measured and correlated to a 4–20 mA signal.

Data historians typically provide graphical trending of process variables in the time domain, which is extensively used by unit process engineers. A quick review of key variables would show instability problems. Comparing process variables (PVs) to associated SPs could reveal if a controller is operated on automatic or manual mode as the PV typically equals the SP when operated on manual with the controller configured for PV tracking. A constant controller output would also suggest manual operation. The trends would also provide times at which a controller SP was changed and how the downstream unit would have responded to the step change. Trends are also useful to diagnose control oscillations, which are often symptomatic of inadequate controller tuning. In some cases, re-establishing stability in manual followed by a switch to automatic mode may suffice for stabilizing the controller. Such a case would be symptomatic of abnormal disturbance that exceeded the capability of the controller in re-establishing stability.

Trending raw data in the time domain is however less effective when troubleshooting performance issues with operating units. As discussed in Chapter 2, factors influencing performance, such as fouling, flooding, and efficiencies, are not clearly noticeable when trending raw process variables in the time domain. Many process engineers spent countless hours trying to establish cause and effects between raw variables in process trends, while those cause and effects are driven by the fundamental principles discussed in Chapter 2. Process trends are, however, extremely useful in identifying time period where conditions deviate from expectations. Identification of those time stamps will narrow rigorous evaluation of a suspected process unit. As an example, the performance on the two vaporizers shown in Figure 4.1 exhibited a shift in behavior over time. As illustrated in Figure 4.3, a disturbance occurring around 28 days caused a shift in liquid level and outlet temperatures. Two relatively stable operating modes were observed on each side of the disturbance and identified as "Period 1" and "Period 2". Two instantaneous datasets of process variables were extracted from the data historian from those two periods for further analyses. As discussed later, under simulation, H&MBs were conducted for evaluating and comparing the performance of the two vaporizers.

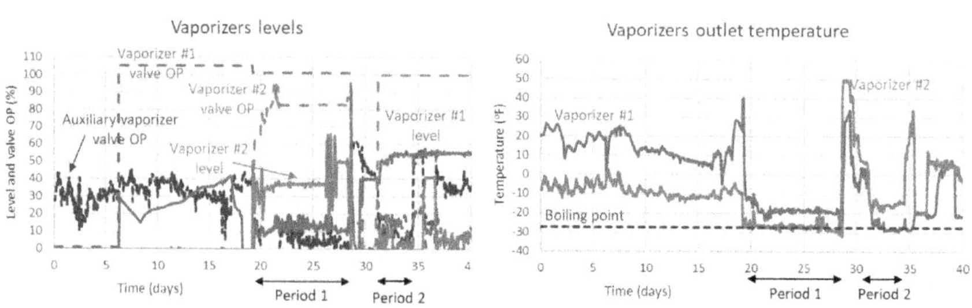

FIGURE 4.3 Vaporizer Raw Process Data Trend.

It is not uncommon to have a large number of local instruments that are not wired to plant DCS. These include local pressure gauges, temperature indicators, level gauges, and flow metering devices, such as rotameters. The value of some of those instruments is periodically recorded by process operators in log sheets, which are normally archives for limited period. Those log sheets provide another source of data for troubleshooting. Although available, the accuracy of local instruments is frequently questionable. For temperatures and pressures, their value or offset can however be verified using instruments process calibrators. The use of process calibrators is highly recommended when conducting a performance test on equipment, such as pumps, compressors, and/or heat exchangers.

Laboratory analyses are frequently performed for validating online analyzers or providing analysis of parameters that cannot be measured by online analyzers or in the absence of them. Some of the analyses are done for validating product quality specifications while others are intended as operating variables and may drive operational adjustments when trending outside control limits. Water chemistry is an example where periodic analyses will drive changes in chemical injection rates for maintaining water quality within operating limits. Laboratory analysis results are historized in paper copy or digital database. Unlike process variables continuously measured from the plant and archive in data historian, laboratory analysis results are discrete variables. Analyses frequency can vary from hours to once per shift or daily. The use of control charts is highly recommended if the analyses results are intended to drive operational changes. As illustrated in Figure 4.4, a control chart is a trend of a process variable in the time domain. The trend of the variable is compared against a target value and control limits defined as $\pm 1\sigma$, $\pm 2\sigma$, and $\pm 3\sigma$, with σ being the acceptable standard deviations. SPC rules suggest that no change is required if the variable distribution is balanced on each side of the target and within the $\pm 1\sigma$ control limits. However, the following trending conditions would suggest transitional behavior requiring operational adjustments: More than eight successive data points above or below target, more than four out of five successive data points above or below target $\pm 1\sigma$, more than two out of three successive data points above or below target $\pm 2\sigma$, and one data point above or below target $\pm 3\sigma$.

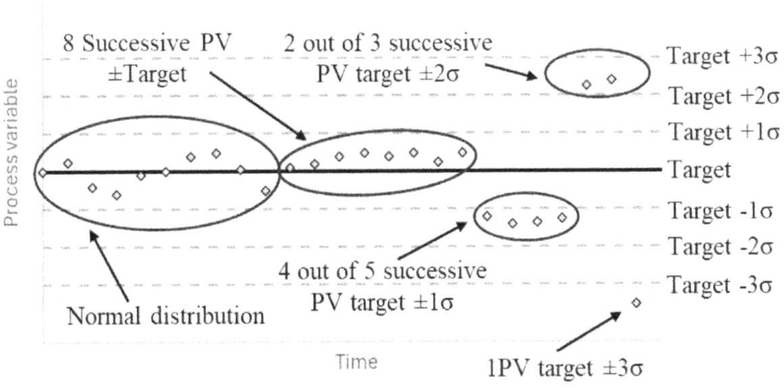

FIGURE 4.4 Typical Control Chart.

The above discussion refers to laboratory analyses that are done as part of the routine operation of the plant. In troubleshooting, some additional analyses could be requested for validating potential root causes. This may include streams or tests that are not normally performed as part of the normal routine.

During planned or unplanned outages, collection and analysis of solid deposits from fouled equipment provide information on the nature of the foulant material and the potential reaction mechanisms. This helps in identifying the fouling precursors responsible for those undesired reactions. However, unlike raw process data from data historians and periodic laboratory analyses, this data represents the cumulative response over an extended period to operating conditions and the presence of reactive components resulting in the accumulation of undesired material within the process unit. Nonetheless, this knowledge is critical in the development of mitigation strategies intended to reduce the rate and accumulation of fouling material.

The quality of stream analyses strongly depends on sampling methodologies. This is especially important for multiphase systems. As discussed for settlers in Chapter 2, for two-phase liquids or liquid/solid mixture, sampling must be conducted isokinetically to prevent withdrawing one phase over the other. Cleanliness of sampling containers is critical as well as proper purging of pressurized sampling bombs to prevent contamination. Depending on mixtures being sampled and analyzed, preservatives are sometimes required for preventing chemical reactions, which will affect the results. On liquid samples, this may also involve inerting with nitrogen to prevent reactions with oxygen. Most of those requirements are typically specified in standardized methods.

In troubleshooting, laboratories are assets that can provide more than just stream analyses. Evaluation of stream properties is often overlooked and is frequently responsible for design shortfall. As an example, process simulators can provide poor viscosity estimate on oil mixtures defined by pseudo components. Not only would that affect hydraulic calculations but could also result in poor heat exchanger design. Less frequent is equilibrium data for which significant amount of data is available in literature. However, for some novel applications, determination of vapor/liquid equilibrium may need to be evaluated. Laboratories also provide a control environment for evaluating and/or correlating the dependency of different process variables against each other. For instance, the designs of liquid/liquid and liquid/solid separators discussed in Chapter 2 are typically based on Stokes' law, which assumes a droplet or particle diameter with densities typically obtained from process simulations. In the design of a new facility, those parameters are typically unavailable and can only be assumed based on design experience from similar processes and/or in the case of densities obtained from process simulations. When troubleshooting a facility, samples can be taken and settling experiments (often referred to as Jar Test) conducted in laboratory. Those experiments will provide an accurate settling time. Particle size distribution can also be accurately established as well as densities. The impact of other parameters, such as pH, temperature, and chemical additives, can be evaluated in a control environment, which besides revealing clues relative to poor settling performance will provide optimum operating conditions.

Fouling of process equipment is another area where laboratory experiments are crucial for understanding the fouling propensity of mixtures and the reaction mechanism involved. The use of HLPS has been discussed in Chapter 3 for evaluating the fouling propensity of mixtures. Batch reactors can also be used for evaluating the fouling propensity of mixtures. HLPS and batch reactors can also help for evaluating the effectiveness of chemical inhibitors and dispersants.

In reactor development, chemical reaction kinetics are first evaluated at lab scale. Lab-scale experiments can also be conducted for understanding or validating the effect of contaminants relative to reaction inhibition or catalyst poisoning.

More sophisticated laboratory-scale setups have been used for studying kinetics and coke formation in cracking furnaces.

Not all plant laboratories have all the capability for executing some of the testing described earlier. This work, when deemed essential, can always be outsourced.

Steady-state process simulations are primarily used in process design for generating process loads of operating units and H&MBs. As discussed in Chapter 3 and illustrated in Figure 3.13, they are the central point of origin for process data used in design projects. A design H&MB provides a good reference in terms of defining the original normality of the plant. However, deviations from design conditions are frequent and the ability to conduct process simulation reflecting operating conditions believed responsible for poor performance are extremely useful in troubleshooting.

To be of value, process simulations must be properly configured. The reader is encouraged to review the simulation section in Chapter 2 under distillation, which discusses in more detail proper selection of thermodynamic and property methods when configuring simulation models. Process simulation has been referred to as essential for troubleshooting distillation towers. They provide stage-by-stage properties and important process variables that cannot be physically measured. Some of those process variables are critical for assessing the operation of a unit relative to its design characteristics. The same can be said for other operating units, such as heat exchangers, compressors, and chemical reactors.

As for process design, in troubleshooting, process simulations provide process loads, which can then be compared against characteristics defining the reliable operating window of process equipment. Many of such characteristics have been discussed in Chapter 2. Process simulations can also be used to compare expected to actual performance of operating units. Although run within their reliable operating window, deviation in performance can be the result of several factors. Conducting sensitivity analysis around parameters known to affect performance will provide clues on the potential mechanisms responsible for those deviations. Differences between simulated and actual performance are then reconciled based on criteria known or suspected to be responsible for the differences. In the case of heat exchangers, deviation in performance would be associated with fouling. The extent of fouling would be quantified by adjusting the fouling factors in a heat exchanger simulation for reconciling the results with the actual observed performance. Drop in performance of distillation tower would be quantified by manipulating the fractionation efficiency in a simulation model for matching plant data. In the case of rotating equipment, head offsets and deviation in expected efficiencies would be manipulated for reconciling simulation results with observed performance.

The ethane feed system illustrated in Figure 4.1 is an example where process simulation can be used for developing H&MB in the absence of flowmeters, although the total ethane and ethylene flows could be calculated from existing flowmeters. The split between the two exchangers in this case was unknown and is critical for evaluating the performance of the exchangers. The outlet temperature of the ethylene (C2R) exchangers suggested subcooling on one and saturation on the other one. Similarly on the ethane side, one showed superheated and the other one saturation. The ethane being on the kettle side of the exchanger suggested that saturation temperature would be equivalent to a vapor fraction of 1, while on the ethylene side a residual vapor fraction was likely. The combined ethane stream being superheated by a downstream exchanger used for de-superheating the ethylene provides an additional piece of information for closing the heat balance. As such, the remaining degrees of freedom can be satisfied by splitting the ethane flow for meeting the outlet temperature of one vaporizer as described by Equation 4.1. The ethane flow to the auxiliary feed vaporizer can be evaluated by a heat balance and the ethane outlet temperature as summarized by Equation 4.2, where the heat duty Q is calculated from the LP steam flow and heat of condensation. The ethylene flow split can be set for meeting the ethane outlet temperature of the other vaporizer (Equation 4.3). Finally, the ethylene vapor fraction (measured on vaporizer #2) is manipulated for meeting the temperature of the ethane superheater outlet temperature as illustrated by Equation 4.4. Establishing the H&MB of this system requires reconciling Equations 4.1–4.4. This can be quite tedious on a sheet of paper or on a spreadsheet given phase changes and the iterative nature of the calculation. However, a process simulation of this system can be done where Equations 4.1–4.4 are reconciled through design specifications in a sequential modular simulation or by appropriate classification of free and fix variables in an equation-based model. This example shows that process simulation is an effective tool for estimating process variables that are not being measured:

$$Ethane\ Flow\ to\ Vaporizer\ \#1 = f\left(Vap\ \#1\ ethane\ outlet\ Temp\right) \tag{4.1}$$

$$Ethane\ Flow\ to\ Aux\ Vap = f\left(Q, ethane\ outlet\ Temp\right) \tag{4.2}$$

$$C2R\ Flow\ to\ Vaporizer\ \#2 = f\left(Vap\ 2\ outlet\ Temp\right) \tag{4.3}$$

$$C2R\ Vaporizer\ \#2\ Vfrac = f\left(Ethane\ Feed\ Superheater\ outlet\ Temp\right) \tag{4.4}$$

The above example illustrates a relatively complex system where lack of instruments makes it difficult for establishing an H&MB. Converging the simulation itself in this case can also be time consuming. This requires being strategic in the number of simulation cases to run. The raw data trend for the feed vaporizer was previously discussed as illustrated in Figure 4.3, where two periods showing different system behavior would deserve a deeper evaluation. Comparison of the H&MB generated by the process simulations revealed that 60% of the *C2R* was flowing through vaporizer #1 in period 1 compared to 14% in period 2. A disturbance occurring around day 28 seemed to have caused a change in *C2R* flow distribution with the exchanger outlet temperature shifting as well (i.e., subcooled vs. saturated). Although the simulation suggested a hydraulic imbalance, it did not provide the reason for it as rigorous hydraulic is rarely performed at this level. Nonetheless, it provided an additional clue directing the search toward the *C2R* isometric piping system. Figure 4.5

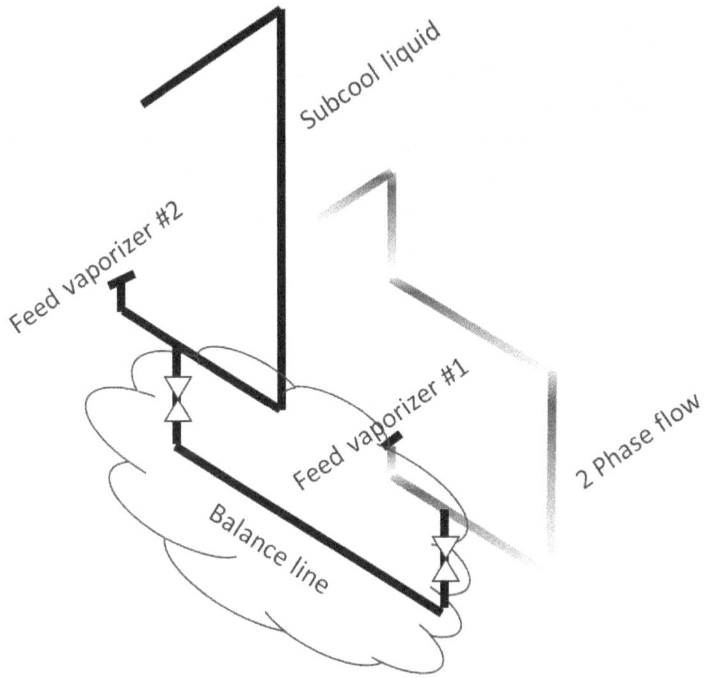

FIGURE 4.5 Feed Vaporizers Example *C2R* Outlet Piping.

illustrates the feed vaporizers' *C2R* outlet isometric piping. As described earlier, the *C2R* exits one of the vaporizers with a vapor fraction while the other exits at subcooled temperature. The isometric piping illustrated in Figure 4.5 shows vertical risers downstream the exchangers. As such, the static head of the riser characterized by two-phase flow is less than the one for which *C2R* is subcooled. The balance line shown in Figure 4.5 was not part of the original design and added later for resolving this problem. Therefore, *C2R* supplying the two vaporizers would follow the path of least resistance and explain the higher flow rate on the exchanger where *C2R* does not fully condense. The design H&MB and rigorous hydraulic calculations suggested a 50% flow split between the two exchangers. It is important to highlight that the design calculations were accurate but performed at steady state. In the startup transient, *C2R* was initially flowing through the vaporizers as a vapor until condensing would occur by commissioning ethane to the vaporizers. Unless perfectly balance, one of the risers would preferentially fill up with liquid, creating a higher back pressure on the associated exchanger. Operation initially assumed that one exchanger was performing poorly due to fouling. However, the simulation reflecting Periods 1 and 2 suggested that the *C2R* flow shifted from one vaporizer to the other after experiencing a disturbance around day 28. To add to the complexity, the design steady-state H&MB indicated partial condensing of *C2R*, which would make it extremely difficult to balance the two risers. As stated earlier, a balance line was installed for equalizing the static head of both risers, which helped in resolving this problem. This case is a good example showing the connectivity between engineering documentation (P&IDs, BFD, and isometric), raw process data, and process simulations for troubleshooting a problem.

As previously stated, due to complexity of the simulation, the time stamps evaluated must be strategic for minimizing the number of case study. However, in some cases, trends of performance indicators could be of interest. For example, monitoring the *UA* value of a heat exchanger over time as illustrated in Figure 3.8 requires calculating the heat duty from measured temperatures and flows. This calculation can be complicated if it combines latent and sensible heat along with pressure fluctuations. To assess a drop in performance, the *UA* value would be trended over an extended time period, which would require numerous simulations. This would be impractical and time consuming. An alternative would be to establish regressions of the dew point, bubble point, latent heat, heat capacities between superheat, subcooled temperatures, and saturation for the expected operating pressure range. This can be done with heater blocks in a process simulation configured to run the different conditions needed for establishing the regressions. The range of operating pressure is typically limited such that the above temperatures and heating variables are relatively linear. The regression can then be used in a spreadsheet for calculating the heat duty based on measured flow and temperatures. The *UA* value is then calculated from the calculated duty and LMTD.

Dynamic simulations. Abnormal conditions such as process upsets typically involve transient conditions that are not always intuitive. Steady-state process simulator provides conditions reflecting an equilibrium state. As such, a steady-state simulation does not have the capability to predict the drastic drop in temperature experienced when introducing liquid hydrocarbon into a nitrogen-filled vessel. This drop in temperature, caused by the partial pressure imbalance, is a transient state, which once reaching steady state will match the prediction of a steady-state simulator. Proactive and reactive troubleshooting frequently involves evaluation of system failures, which are associated with transient behavior. A sudden loss of reboiler duty on a demethanizer tower is described in Example 3.1, where a dynamic model was used for establishing the CET throughout the transient condition. The transient CET was then compared to the MAT of the vessel as illustrated in Figure 3.1, where it is demonstrated that it would drop below the MAT and potentially cause a vessel failure by cold brittle fracture. The same dynamic model was used to demonstrate that SIS would not be adequate for protecting the vessel. The above example is based on proactive troubleshooting, typically conducted as part of process safety risk evaluation where there is no actual plant data available from similar failure. In reactive troubleshooting, the evaluation is not hypothetical as it relates to an actual event. Consequently, plant data of the transient conditions experienced are available from the plant data historian. A reactive troubleshooting evaluation of a refrigeration compressor trip has been discussed in Example 3.3. In such cases, the dynamic model is tuned using plant data. Tuning of the model is extremely useful in establishing the sequence of events. It eliminates biases and inaccurate speculations which do not align with thermodynamic first principles. As for any models, it should be configured in alignment with the plant input/output variables; that is, input being those within the control of process operators and output being the outcome from the inputs. The inputs are then manipulated in the model such that the output values align with the plant values. This may require manipulating some of the unit operation characteristics in the model if the transient conditions were associated with change in unit performance. The dynamic behavior of a model is affected by material holdup (i.e., volume) and PID controller tuning. Matching the holdup can be tricky as piping is typically not included in models. The piping holdup is typically embedded in process vessel or mixer blocks. However, depending on proximity of process units being model, it may affect lags. Dynamic models are more sensitive relative to convergence than steady-state models, especially when modeling aggressive transient conditions. Good practice starts with simple models and adds complexity level as required.

Dynamic models typically originate from a steady-state model and are bound by the same requirements relative to thermodynamic and property methods for accurate results. They require additional inputs such as vessel holdup and controls. They require a higher skill set, but as illustrated by Examples 3.1 and 3.3 they provide good value in troubleshooting transient behavior and were found successful in disproving speculations.

Hydraulic calculations. As previously mentioned, rigorous hydraulic calculations are rarely done in a process model. If required, they can be done separately, as including them in process models adds unnecessary complexity that may affect convergence. Most process simulators can accommodate simple hydraulic through pipe blocks. However, complex networks of multiple users such as cooling water systems are not easily modeled in sequential modular (SM) simulator. The flow split fractions are specified in an SM simulation, while in reality it is driven by pressure differential and hydraulic resistance. Therefore, there is software that is exclusively designed for resolving flow pressure networks. Most of those software are typically good with single-phase liquid such as water. However, caution must be exercised when using them as some are not strong on properties and tend to result in significant gaps, especially with mixtures. Two-phase flows can also be problematic. It is therefore advisable to validate properties and flashing conditions with a proven SM process simulator.

Nonintrusive measurements are conducted either as part of periodic checks on equipment integrity or for validating suspected root causes. There are many process units and systems in a chemical plant that do not have local or online instrumentations for evaluating or monitoring performance. Although it is not always true, older plants tend to have less instruments than new ones. There are also process parameters that cannot be measured due to locations, extreme conditions, or being economically impractical. For example, there are no online instruments measuring froth height on distillation trays. TMT in fired equipment cannot be measured by thermocouples due to excessive temperature. Cooling water flows are rarely measured to process users as the numerous flow metering devices required would be too costly. As previously discussed, process simulations can compensate for some of the missing parameters. However, it also has limitations and, in many cases, requires field validations.

Portable ultrasonic flow meters are devices that are attached on the outside surface of pipes. Sound waves emitted from transducers positioned at an angle against each other are correlated to the pipe fluid velocity. They are frequently used for measuring cooling water flows to heat exchangers in cooling water systems. They can also be used for measuring gas velocities as well. Jlidi [9] discussed the use of ultrasonic gas flowmeters for monitoring flare gases. Although external clamp-on models typically used on portable devices are not as precise as fixed units with transducers exposed to the process fluid, they provide sufficient accuracy in the context of troubleshooting and high-level monitoring. As for most flow meters, their application is limited to single-phase systems.

The use of qualitative and quantitative gamma scans has been discussed in the distillation section of Chapter 2. Gamma scans can be done to monitor the progress of fouling in pack tower operating in fouling services. They are primarily used for confirming suspected problems related to MI, flooding, and/or distribution. The use of gamma scans can extend beyond distillation tower. For example, full vaporization at the outlet of a thermosiphon reboiler was suspected based on measured temperatures and process calculations. Gamma scans of the reboiler inlet and outlet piping confirmed those suspicions, which were caused by severe pluggage of the reboiler inlet resulting in low process fluid circulation. Gamma scans can also be used in process vessels for evaluating the location of the vapor/liquid/solid interface.

Thermal imaging is another technique that can be useful in performance monitoring and troubleshooting. Thermal imaging uses cameras that detect radiation in the far infrared (IR) range of the electromagnetic spectrum from emitting objects. Temperature of the object surface is correlated to the color of the image generated by the camera. It is important to understand that the temperature is representative of the metal surface and not the fluid contained inside a vessel. Barnard [10] discussed several applications of thermal imaging in the CPI, such as maldistribution in packed towers, heat exchangers, and solid accumulation in process vessels. Heat losses due to process leaks, damaged refractory in fired equipment, and passing steam traps are other anomalies that are discussed by Barnard. As for gamma scans, base line thermal imagery of normal performance helps in assessing deviations from normality.

Non-contacting IR thermometers (pyrometers) [11] can be used for measuring the external skin temperature of pipes, which can be useful in the absence of local or online temperature indications. Although it does not measure the actual process fluid temperature, it provides an estimate that can be useful for troubleshooting. If required, the process fluid temperature can be estimated from the pipe skin temperature using heat transfer correlations. As discussed in Chapter 2, tube metal temperatures (TMTs) in fired equipment are periodically measured by process operators or engineers using pyrometers. The high temperatures in this service exceed reliable operation of contacting instruments such as thermocouples and resistance temperature detectors. Although there are promising technologies using non-contacting infrared (NIR) cameras for allowing online 24/7 monitoring of TMTs [12], manual TMT measurement is still common practice in the industry.

X-rays are another type of non-destructive testing typically conducted in chemical plants. However, their use is more geared toward process vessel MI. From a process perspective, some mechanical anomalies may cause undesired hydraulic restrictions impacting the plant capacity.

4.1.2 Performance Monitoring

It is said that students who know their marks performed better than if they did not. This concept applies well to athletes where the difference between a good and bad performance is based on a set of measurable metrics. One aspect of troubleshooting involves identification of symptoms causing deviations from normality. The concept of normality must therefore be clear and well defined. As for athletes, normality in a chemical plant is based on a set of measurable metrics, frequently called KPIs. The extent and level of details of KPIs vary from plant to plant. They could be limited to production targets, or be more detailed with inclusion of losses, energy balances, and/or environmental emissions. Performance negatively deviating from KPI targets is symptomatic of an operational problem. It can be stated that initiating troubleshooting activities as a result of KPI deviation is reactive. Proactive troubleshooting goes beyond KPIs and involves key units' performance monitoring. A good monitoring program should provide trends of deterioration and/or early warning that this will eventually impact plant KPIs.

Unless it is caused by human error or sudden failure of a piece of equipment, performance degradation does not occur overnight. As previously stated, normal performance degradation is expected over time. Proper unit monitoring and historical data should be used to compare degradation rates from one plant cycle to others. Higher rates of degradation from the previous plant cycle should raise awareness and initiate proactive troubleshooting. Unfortunately, many plants do not monitor the performance rate of decay and react to consequences.

An overall plant material balance should provide a high-level validation of feedstock conversion yields and recovery effectiveness. From a process monitoring perspective, this should represent a minimum requirement, which should be reviewed daily. Many plants focus strictly on the final product rate and frequently fail to close the balance with coproducts, fresh feedstocks, and losses. They report production records without being fully aware of the recovery effectiveness or reactor yields. Some plants even rely on advanced process control strategies for maximizing feed rates without consideration to losses. Why is it missed? One of the reasons is that chemical plants are typically subdivided in areas, with each area operating in silos with their own sets of performance metrics. Why is it important? Reactor yields impact the process loads of the recovery area. Low reactor conversion typically results in composition shift, which may require distillation towers to operate at higher reflux ratio, which increases the internal hydraulic loadings and could cause premature capacity constraint. The low conversion operation may satisfy the reactor areas metrics from the perspective of unit run length but affect the overall performance of the plant. A simple overall material balance would provide insights into those deviations and should raise awareness for deeper investigation when deviation from expected performance is observed. A high-level energy intensity balance (i.e., energy usage/mass of valued products) would provide further information on the performance of the unit. In the above reactor low conversion operation, higher reflux ratio operation would translate into higher energy consumption for achieving separation, which should be reflected in energy intensity calculations. Keeping historical records of those high-level material and energy balance would allow a useful comparison for assessing unit effectiveness. Plotting key metrics over time tends to be more effective than database of tabular data. The use of control charts similar to what is illustrated in Figure 4.4 for KPIs is an effective way in raising awareness that something is changing. This can be rationalized due to intentional changes in operating conditions or the results of undesired deviation in performance.

Unexplained deviations in the overall material and energy balance should trigger a deeper evaluation of individual process units. The evaluation should start with unit-specific H&MB along with comparison of process loads relative to operational boundaries. Comparison with historical data would help in diagnosing a potential loss of performance.

Many facilities run offline or online optimizer programs, which provide rigorous calculations of an operating unit. Operation can always leverage on the output of those programs. However, while those programs may provide value, a simple overall material and energy balance is a simple check on the output provided by the optimizer. Optimizers are complex programs, which are often configured by external specialists. In-house expertise is often limited if even available.

Plants who have good monitoring programs would typically excel at troubleshooting as they already have most of the data required for initiating a troubleshooting initiative. This speeds up the data gathering step. The availability of process models whether for unit monitoring or process design avoids building them from zero. Building process models is relatively time consuming and unfortunately often skipped in the interest of time. However, this results in unvalidated speculation on the most likely root causes. Plants having poor monitoring programs would tend to feel the pressure of time, skip data gathering, H&MBs, and understanding the expected reliable operating range of process units. They become more focused by the troubleshooting work process than the relevance and quality of the data used for establishing root causes. More organized plants use the troubleshooting work process for structuring the data, which helps them in finding a logical path to the most likely root cause.

Nomenclature

BFD	block flow diagram
C2R	ethylene refrigerant
CET	critical exposed temperature

CPI	Chemical Process Industry
CSB	U.S. Chemical Safety and Hazard Investigation Board
DCS	distributed control system
H&MB	heat and material balance
HAZOP	hazard operating studies
HLPS	hot liquid process simulator
IR	infrared
KPI	key performance indicator
LP	low pressure
MAT	minimum allowable temperature
MAWP	maximum allowable working pressure
MOC	management of change
MI	mechanical integrity
MSDS	Material Safety Data Sheets
NIR	non-contacting infrared
OEE	overall equipment effectiveness
OP	output
P&ID	pipe and instrumentation diagram
PHA	process hazard analysis
PFD	process flow diagram
PID	proportional-integral-derivative
PV	process variable
Q	heat duty
SIS	Safety Instrumentation System
SM	sequential modular
SIS	safety instrumented systems
SP	set point
TMT	tube metal temperature

REFERENCES

[1] G. Nasby, "Using Process Flowsheets as Communication Tools," *Chemical Engineering Progress*, vol. 108 no. 10, pp. 36–44, October 2012.

[2] CSB, "Fatal Equipment Rupture, Explosion and Fire at the KMCO Chemical Facility, Crosby, Texas, Incident Date: April 2, 2019, No. 2019–02-I-TX," *U.S. Chemical Safety and Hazard Investigation Board (CSB)*, December 2023.

[3] "Process Safety Beacon, Are Your P&IDs Up to Date?," *Chemical Engineering Progress*, vol. 120 no. 7, p. 15, July 2024.

[4] A. Bernard, "Ethane Feed Vaporizers Startup Difficulties," in AIChE Spring Meeting, Ethylene Producer Conference, New Orleans, LA, 2024.

[5] H. Z. Kister, Distillation Operations, McGraw-Hill, 1990.

[6] J. R. Beale, P. Bowers and P. Smith, *The Piping Guide to Piping design*, Gulf Publishing Company, 2010.

[7] "Process Safety Beacon, Conduct of Operations," *Chemical Engineering Progress*, vol. 111 no. 6, p. 26, June 2015.
[8] "Safe Work Practices," *Chemical Engineering Progress*, vol. 111 no. 4, p. 20, April 2015.
[9] K. Jlidi, "Introduction to Ultrasonic Flare Gas Flowmeters," *Chemical Engineering Progress*, vol. 118 no. 7, pp. 21–25, July 2022.
[10] H. Baranard, "Use Thermography to Expose What's Hidden," *Chemical Engineering Progress*, vol. 113 no. 1, pp. 21–27, January 2017.
[11] G. Freemer, "Measuring Temperature by Indirect Means," *Chemical Engineering Progress*, November 2011, vol. 107 no. 11, pp. 24–27.
[12] S. H. Karsten Schulze, "Improving TMT Measurement Accuracy Using IR Cameras on Ethylene Cracking Furnaces," in AIChE Spring Meeting, Ethylene Producer Conference, New Orleans, LA, 2024.

5 Troubleshooting – The People Aspect

5.1 THE PEOPLE ASPECT

As previously discussed, the human factor is responsible for a significant portion of problems encountered in chemical plants. Technology and procedures can only do so much. Ultimately, execution falls in the hands of people. The 2010 and 2020 decades are of particular interest when the work force experienced a generational change from baby boomers retiring and millennials taking over. Many baby boomers started working in chemical plants when analog control panels were the human/plant interface. PID controllers were hardware boxes laid out across a wall, with indicators and paper-rolled charts. Hardwire configurations between transmitters, controllers, and final control elements limited opportunities for automation. The emergence of DCSs in the mid-1970s, which gained popularity during the 1980s and 1990s, unleashed ample opportunities for more complex automation. DCS capabilities provided the basis for implementation of advance process control (APC) and optimization [1]. Being intelligent active software layers above regulatory control [2], they are more effective than skilled process operators manually optimizing operation through individual controller SP changes. The advent of DCS, APC, and online optimization significantly contributed to reducing the workload of process operators. The digitalization of control systems also brought real-time and historical process data to process engineer's computers. Those benefits, however, came with some drawbacks as they are typically configured by external experts. As such, complex APC applications may be perceived as black boxes [3] by process engineers and operators. The reduction of process operators' interventions comes at the expense of losing the insight gained over years of manipulating process variables and witnessing the outcomes.

The generational change of the workforce in the CPI is associated with different skill set and culture. Baby boomers have been characterized by the need to know how everything works in a plant down to each nut and bolt, while millennials know how to use the tools without necessarily needing to know how it actually works [4]. Baby boomers' deep knowledge of the plant gave them the ability to take over during an upset and bring conditions back to normal or safe park. In many cases, that knowledge has been imbedded in automated or partly automated response sequences. This evolution has resulted in changes in process operator training and headcounts. Technical support roles from process and process control engineers have also evolved.

The above discussion demonstrates that automation has largely changed how people interface with the process. However, people's behaviors and personalities are aspects that have not necessarily changed and are often overlooked in troubleshooting. It is fair to assume that two different operating crews, which had the same

DOI: 10.1201/9781003449676-5

training, would react differently to an abnormal event that had never been previously encountered. The human factor therefore introduces an element of variability that should be considered in troubleshooting.

5.2 PROCESS ENGINEERING PERSPECTIVE

Unlike process operators who may work in an operating unit for their entire career, a process engineer typically rotates roles and/or transitions to management and business positions. Some process engineers' career progression remains in technical roles and advances to positions requiring a higher skill set, such as process design or becoming subject matter experts (SMEs). Role rotation has been reported as a retention strategy with millennials [5]. Because of role rotation, unit process engineers would have little exposure to infrequent major activities such as plant turnaround, startup, and shutdown. Their knowledge base relative to process upsets remains limited to what they might have been experiencing while in the role.

They typically work closely with operation supervisors, who are typically former experienced process operators. Operation supervisors' seniority and knowledge of the unit can be intimidating to young process engineers. It is not unusual, unfortunately, for process engineers to try to emulate the behavior of operation supervisors or experienced process operators. In other words, they rely on intuition rather than leveraging on fundamental principles they learned in school. Intuition is defined as the ability to understand something based on gut feel rather than analyzing facts. Joice C. Hall (founder of the Hallmark Company) called intuition "the vapor of past experience" [6], meaning that good intuition comes from past experiences. Operation supervisors and process operators have significant experience, and, although, as discussed later, may follow a structured reasoning process, they must frequently make decision in real time relative to unexpected events. Therefore, good, experienced operators would frequently rely on intuition. Conversely, process engineers would not provide value to operation if they follow the same thinking process. Besides not having the experience that would give them good intuition, their role is opposite to that, where they should be analyzing facts and data for establishing the most likely root cause and appropriate response to an undesired event. Process engineers rarely have to make decision in real time and as such work in an environment giving them time for conducting detailed troubleshooting evaluations. In schools, problems were formulated with a clear question asking for clear answers with all the required data provided as part of the question. It is normal for young process engineers to struggle in troubleshooting plant issues as the problems do not come as a clear question with all the data. As discussed in Chapter 1, one aspect of troubleshooting is about formulating the question or problem definition. The data required for solving the problem does not come with the question either. As discussed through this book, conducting H&MBs and calculating process loads on suspected equipment provides the foundation required for understanding deviations from normality. This is far from following an intuition thinking process, but this is where and how process engineers can effectively contribute to troubleshooting. Using sports as an analogy, process engineers can be seen as sideline observers who watch and analyze game videos. They identify errors and/or weaknesses and communicate strategy adjustments to the coaching staff. In this analogy, process operators are the players and operation supervisors the coaching staff. One other challenge for unit process engineer is in prioritizing their work due to the dynamic nature of an operating unit. It is common to describe work in an operating unit as what is important today as nothing to do with yesterday's or tomorrow's crisis. This can be disruptive and negatively impact the ability of process engineer in focusing on what matters.

The previous chapters of this book focus strongly on fundamental principles and data, which, as described earlier, is where process engineers can largely contribute to troubleshooting. However, beyond having abilities for resolving technical problems, process engineers must be able to interact effectively with operation supervisors and process operators. Norman Lieberman [7] states that troubleshooting should start by a discussion with process operators. Besides the extensive knowledge they have gained by operating a unit for several years, those present during an abnormal event have firsthand experience. They can share what went through their mind while trying to bring the unit back to stable conditions or safe park. They provide important piece of information that can be later reconciled with a technical evaluation. Those discussions are not always easy depending on site culture and history. Engineers generally have higher education than process operators and in many cases operation supervisors. Although they typically have less experience, their initial conduct with operation could make them either successful or irrelevant to the organization. Pride of having completed an engineering degree is not a reason to disrespect people that may have lower education but have a wealth of knowledge through many years of experience. Arrogance shall be avoided if a new engineer wants to gain trust from operating crews. The last two sentences may sound odd, but too many young engineers destroy their career by being incapable of effectively communicating with operations. It is important to understand that success is achieved through the help of others [8]. Trust and respect are gained by respecting and trusting others. Although difficult, that means being a good listener and remaining calm when a process operator is expressing frustration. In stressful time, the first five minutes of a conversation may seem confrontational with expression of criticisms. A good strategy would be to listen, acknowledge how the person feels and the problems they are reporting. The objective here is not about being right but is to gather information and avoid getting into unnecessary debate. There will be a point in the conversation where the individual will become more rational and starts focusing on the problem instead of his frustration. As much as the criticism may hurt, those situations are opportunity for engineers to show leadership. Leadership is often confused with authority and formal position of power within an organization. However, you do not need to hold a position of power to show leadership. William A. Cohen [8] states that leadership has to do with the ability to win other people's mind and influence them in doing something that they may not otherwise do for reaching goals and objective. Max De Pree [9] defines good leaders as good listeners and servants of their followers because they remove obstacles preventing them from being effective at their jobs. The above statement summarizes two key elements for being successful in troubleshooting, which are gaining key information relative to an abnormal situation by being a good listener, and winning trust and support from operation in resolving problems and implementing corrective actions.

People are not like machines as they have different personalities with transitional health and emotional states. The emotional state of a process operator would be quite different if engaged in a friendly discussion outside work as compared to being overwhelmed with questions shortly after a plant upset. Albert J. Bernstein and Sydney Craft Rozen [10] stated that people tend to revert to primitive behaviors when in a stressful situation. They discuss numerous factors such as impulsiveness, defensiveness, and irritability, to name a few. Those emotional states are expected when questioning process operators following a plant upset. By expecting those behaviors, the troubleshooter has to remain in control by staying calm, remaining in listening mode, and avoiding fighting back. The objective of a discussion following an abnormal condition is to gather information, not provide advice or criticism. Most people do not like making mistakes and hate being told that they made a mistake. Procedures could be technically accurate but difficult to execute, which may explain deviations from them. Curiosity of the troubleshooter rather than criticism must be driving questions – trying to understand the sequence of events from an operator perspective and state of mind. People do what they do, at the time they do it for reason that make sense to them [11]. Besuijen [12], a former operator himself, who became a trainer, discusses physiological factors affecting operators' decision-making under stressful situations, such as abnormal conditions. Those factors cause the brain to revert to primitive response and override rational thinking leading to freezing, tunnel vision/fixation, and can shift time perception. Two often-overlooked physiological factors, as stated by Besuijen, are the impact on coordination and speech. Shaking hands is a common consequence of executing tasks under stressful conditions. This may result in erratic inputs to DCS caused by excessive mouse clicking, erratic mouse movement, and/or typing too many numbers. Some, but not all, DCS systems provide robustness against large changes in SPs or outputs. Some are also configured with ramps, which mitigates erratic inputs. Nonetheless, those erratic moves can be detrimental to an effective recovery. Alteration or speech impairment caused by stress is another factor, which will impact communications between the control room and the field. Denial in the initial seconds of a plant upset is quite a common reaction reported by process operators. The above factors are too often not given sufficient considerations when conducting investigations following a plant upset. It is also challenging to get process operators to share how those factors might have influenced their response as it can be perceived as a sign of weakness. This is why process engineers conducting troubleshooting must establish a strong trust relationship with operators so that they can become aware of those factors. Doing so will help in establishing effective response and providing proper training will better prepare process operators in responding to similar events. Advice and recommendations are to be communicated once a troubleshooting evaluation is complete. This should be done in a relaxed and collaborative environment.

Although it was mentioned earlier that process engineers are rarely involved in real-time troubleshooting, there are circumstances, such as startup and commissioning of new and/or modified units, in which they are physically present in the control room and expected to actively participate. However, it has been the experience of the author that expectations or role definition of process engineers during those circumstances are unclear. If something goes wrong, they are expected to have answers. As such, and depending on their experience and personality, they either become quiet spectator or engage in discussion purely based on speculation. As stated earlier, engineers provide value by analyzing facts, which can be challenging in real-time troubleshooting due to time constraint in decision-making. Those circumstances were discussed in Chapter 3 and involve upfront preparation and proactive troubleshooting. Startup and commissioning procedures involve numerous steps which can be segregated by systems and documented by separate procedures. For example, commissioning the refrigeration system of an ethylene plant is one activity that precedes hydrocarbon feeds to cracking furnaces followed by startup of the crack gas compressor (CGC). The refrigeration system of conventional ethylene plant includes two different systems: ethylene (C2R) and propylene refrigeration (C3R). The C3R system is typically started first followed by the C2R system. Once the refrigeration system is on stream and stable, operation can proceed with hydrocarbon feed to the cracking furnaces followed by startup of the CGC. Each of the above system has its own startup procedure. In most cases, completion of one procedure can be followed by a pause prior to execute the subsequent one. Doing so allows confirming that the startup sequence is on target or that any deviations from predicted value can be reconciled and do not represent a threat for progressing the startup sequence to the next procedures. Process variables such as compressor power, operating point on the head curve, and steam balances would typically be compared to targets. Some of those process variables can be seen by process operators, but their focus is on executing steps within a procedure, maintaining unit stability and responding to alarms. They may not realize that deviations from target may hurt execution until it becomes a problem. When facing a problem, they may have suspicions based on intuition and mental models, and too often have to rely on that for resolving issues as they occur. This is where process engineering can help, support, and minimize startup delays caused by ineffective troubleshooting. Bernard [13, 14] discussed how tools such as process simulations and spreadsheets can be developed up front for monitoring and providing technical guidance during execution of the startup sequence of revamp facilities. The same strategies can apply to existing facilities. In today's digitalization environment, spreadsheets can be developed for automatically extracting data from plant data historians, exporting data inputs, and importing results from simulations. The spreadsheets can also be configured with process calculations, time plots, and equipment performance diagrams. Such tools allow a process engineer to detect quickly operation moves that might push equipment outside their reliable operating window. This can then be communicated quickly with direction relative to appropriate operating conditions. This is especially useful in revamped circumstances as a process operator can no longer operate the unit based on experience from pre-revamp configuration and so need guidance. In unmodified units, process operators may have years of experiences, but as mentioned earlier, startups are infrequent events for which they have limited exposures. It is not unusual for process operators to print their graphics before shutting down a unit and used them as reference on startup. Their thinking is typically linear, which, at low rates, can bring equipment operating conditions below turndown. In this way, the input/output tables discussed in Chapter 3 would provide value. Additionally, having a process engineer on the sideline conducting load calculations would avoid operating outside the reliable operating window of process units.

This is an effective way of involving process engineers and getting value from their skills in those circumstances. They do not necessarily need to build the tools but should be trained to use them effectively when needed.

Process engineers should be encouraged to participate in OTS rehearsal for startup or commissioning activities. As previously stated, they should be active members of the technical team developing the procedures, which would make them more effective during the actual event. Monitoring tools can be developed or tested during those rehearsals.

Communication is another aspect of process engineers' interaction with operation and management. Findings from troubleshooting initiatives must be communicated. Depending on the magnitude of the evaluation, communication may take the form of a written report by a single engineer, email, and/or presentation. A complex evaluation may include a team where each member would contribute a section of a global report. It is advisable to complement the release of a report with an oral presentation, which gives management and operation opportunities for questioning the findings and suggesting additional elements. The written reports become a reference document providing historical findings that will become useful in case of reoccurrence or problems presenting similarities. It can also serve as training reference. Science and engineering reports are formal documents, which are typically written in an impersonal style. Although some advice using the second person, as a mean to engage the reader more effectively than by using the third person [15], most references agree on accuracy, clarity, and conciseness. Another key element, whether written or oral communication, is to know your audience [16]. A troubleshooting evaluation will likely include significant amount of technical details, for which people may not have the requisite knowledge or interest. It is often a challenge to leave technical details out of a communication given the sense of pride and desire to share how a difficult problem has been resolved. Nonetheless, the communication should focus on facts, assumptions, high-level methodology, conclusions, and recommendations. Detailed calculations and simulations should still be provided but as appendices, as they can become useful historical references if similar problems were to occur in the future.

It is important to clearly state assumptions, as they represent gaps in an evaluation that is missing factual data. Those assumptions can be tested in different ways, such as sensitivity analysis with process simulator or based on similarities documented in literature. Bloch [17] states the difficulty for managers in making proper decision from conclusions based on shaky data and inaccurate facts. He discourages the use of opinions and partial facts, indicating that this practice is frequently motivated by the desire to hide a truth that is frequently related to human factors. Effective troubleshooting cannot be effective without transparency, and management cannot make good decisions without being provided with accurate information.

5.3 PROCESS OPERATORS

Although they may apply or request transfer, process operators typically work in the same units sometimes for their entire career. This gives them a lot of insight on the process unit's behavior. They rely a lot on mental models and intuition, which are enhanced through experience, training, and thinking patterns. Although many of the troubleshooting work processes are described by formal activities associated with meetings and documentations, the foundation of those processes relies on dimensional questions, which should become second nature. For instance, the KT [18] problem-solving methodology described in Chapter 1 originates from studying the thinking process and decision-making of successful Air Force officers. Similar to process operators, pilots flying airplanes must resolve problems and make decision in real time. The time constraint for decision-making also brings an element of stress, which has been discussed earlier and may also affect judgment. Dimensional questions *What, Where, When, and Why* along with association to cause and effect should become second nature in problem solving. Unlike process engineers, process operators are not expected to understand the fundamental principles that explain a sudden failure. Their role is to recognize the failure and assess if they could keep the unit on stream safely and reliably through adjustments of operating conditions or if the failure requires to bring the unit to safe park. This decision is not always straightforward and is frequently associated with conflicting priorities. It is the instinctive nature of process operators to keep the unit on stream and protect the integrity of equipment. This, in some cases, could be a detrimental path forward for the entire plant. Bernard [19] evaluated a site-wide power failure in an ethylene plant that resulted in a loss of the steam system. This caused coking and damages on the radiant coil of cracking furnaces. The loss of the steam system was attributed to a trip of the boiler feed water pumps triggered by a low level on the drum supplying the pumps. The evaluation concluded that excessive use of boiler feed water to the cracking furnaces was largely responsible for the trip. The cracking furnace operators during those events were focused on bringing the furnaces to a safe park. This involved pulling hydrocarbon feed and keeping steam flowing through the radiant coil to avoid coking and thermal stress that could damage the coil. Due to numerous furnaces, this involves many steps from the field and the control room. A normal transition involved removing hydrocarbon feed, adjusting firing and pinching back on boiler feed water supply to the furnaces as the presence of bypass orifices was known to cause oversupply of boiler feed water to the steam drum. The oversupply was managed by an overflow, which disposed excess boiler feed water to drain. In the chaos of those events and workload of the field operator, none of those valves could be pinched back. While trying to protect the furnaces, the strategy failed as the loss of the boiler feed water pumps caused all the furnace to experience a full trip associated with severe coking and thermal stress. Compromising the integrity of a few furnaces by initiating a full trip would have prevented damage on a larger number of furnaces and enabled a speedy re-start of the plant. While other recommendations were issued from this evaluation, it highlighted that without clear rule of engagement in such situations, and as well-intentioned process operator might be, they may not follow a path minimizing consequences of the overall plant. In this particular case, power failures were unusual events for which procedures had never been tested or implemented. Furthermore, given the short duration, such procedure had to be implemented by memory. The reactive troubleshooting evaluation of those events revealed weaknesses and deficiencies on backup and redundant systems that were later addressed.

Some known abnormal condition scenarios are typically documented through procedures. Written procedure must be easily accessible to process operators. However, most process operators' interventions must be initiated quickly in response to an abnormal event. Time spent in retrieving and reading the document is a luxury that is rarely available. As such, responses are likely initiated by memory and intuition. Rob Fisher [20] suggests that tasks having more than seven steps and performed from memory have an exponentially higher error rate. While he discussed other weaknesses in procedure writing, he defines three performance error modes having different error rate. The first one is skill-based task, which he defines as actions done out of habit. Driving back and forth to work, washing hands are day-to-day tasks that fall under that category and have a low error rate of about 0.1%. Rules-based is the second error mode, which requires to know the rules, when and how to apply them. Those rules refer to operating procedure, safety guidelines, and work instructions. However, those are not always available in a written form and are frequently accomplished by memory. The error rate of this mode is reported at about 1%. The third performance mode is knowledge-based, where overconfidence may result in deviations from procedures and/or recommended operating parameters. The knowledge-based mode is also characterized by people lacking knowledge and second-guessing the path forward. This performance mode is reported to have an error rate of 10–50%. Response to abnormal conditions often relies on the knowledge-based mode. However, by leveraging on proactive and reactive troubleshooting evaluations, systems and procedures can be improved. With proper and frequent training strategized toward response of the most disruptive events, the performance error mode can be upgraded to rule-based with an error rate of about 1%. Effective procedure should aim at simplifying and minimizing the number of steps. Partly automated trip response in DCS, besides driving the operator through the procedure, brings simplicity and consistency in execution. The ability to test and validate automated trip response is the biggest hurdle for implementing such strategy. The performance of such automated application can only be evaluated when experiencing the abnormal condition. This challenge can be overcome by testing those procedures with an Operator Training Simulator (OTS) system. OTS rehearsals would also provide training benefits for process operators in the execution of those procedures.

OTS. The description and value of OTS relative to startup and commissioning of new plants have been discussed in Chapter 3. Nonetheless, the primary function of OTS is for training process operators. The rise of automation, improved reliability, and longer run lengths of chemical plants has limited process operators' interactions with the units. OTS provides great value in maintaining process operator hands on skills. It also helps in speeding up new process operator progression and competency in response to the generational change previously described. Classroom training and reviewing procedure is knowledge that will be lost if not applied. Furthermore, while classroom training and procedures might be technically accurate, they could sometimes be confusing to understand until practiced through execution. An OTS provides hands-on experience for practicing execution of procedures in an environment free of consequences, where error can be reviewed and corrected. Doe [21] and Besuijen [22] described how OTS are used for the progression of process operators, their certification, and training against abnormal conditions.

5.4 SUBJECT MATTER EXPERTS

SMEs have deep technical knowledge in specific fields. They are specialists, not generalists. Their expertise can be wide, like fired equipment in general or narrow down to single components like burners in fired equipment. They are not necessarily engineers; some can be technologists or former process operators. Their knowledge and experience allow them to associate symptoms with potential root causes quickly. They also know what kind of testing or calculations are required for converting assumptions into facts. They can therefore be extremely useful for troubleshooting. SMEs may come from within the organization or contracted externally.

As for process engineers and process operators, they are human beings with their own personalities. Some have better people skills than others. Some have strong personalities, which can become intimidating to other team members. Their deep knowledge can also cause biases with opinions based on past experiences but not supported by facts related to the current problems. As previously discussed for process engineers, good SMEs have to be good listeners for gathering relevant facts and data. Ego would not make them successful nor respected. Adopting a mentoring/teaching approach with process engineers and operators would gain them respect and trust. They should also be supportive and collaborative with the work process selected by the troubleshooting team. SMEs should not facilitate the troubleshooting work process, as they would be biased by their own knowledge and experience.

5.5 MANAGEMENT PERSPECTIVE

As for process engineers and process operators, managers have different personalities and backgrounds. They have formal authority and decisional power. Many are competitive by nature and excel in bonding with people of influence. This often explains their rise in the corporate world. They know what matters to be successful and do not get hung up on details. Their competitive nature can be divided into two distinct management styles. Bernstein and Rozen [23] described the ultra-competitor personality as someone for who winning is all that matters. They see any discussion or negotiation as I-win-you-lose situation. They are typically short-term minded and focus on what would benefit them personally. Because of their nature, people would be prone to tell them what they want to hear rather than presenting them with the actual facts. This creates a toxic work environment that is counterproductive in troubleshooting. Bernstein and Rozen suggest a few strategies that may help when dealing with this personality type. Those strategies would not significantly affect the workflow in formal troubleshooting activities but will help in making them effective. Managers should be invited to troubleshooting meetings. However, because they have dominant personalities, it is imperative that meeting invites include agendas with agenda items timelines. Parking lot can be used for documenting their interventions if irrelevant to the discussion. Action items can be delegated to qualified team members when relevant. Given that they typically have a dominant personality type, they could easily hijack a meeting. The parking lot is a way to acknowledge their thoughts and criticism without confrontation and causing excessive disturbance to the meeting. It is not unusual that they would either change their mind or forget about what they were saying shortly after a meeting. Irrelevant comments are often a way for them to exert a form of dominance. Meetings should be documented by updating forms associated with the work process being followed along with minutes of meetings. This should include all the facts and data supporting the troubleshooting initiative. Avoid relying on the managers' opinion and trust the proper source of information based on facts and engineering calculations. They could try to develop a script that makes them look good but hide the truth. Again, those opinions can go on parking lot. The parking lot should be kept anonymous such that they will not look bad. Recommendations should be documented in a final report. In the end, the manager makes the final decision for implementing recommendations. Given their competitive nature and desire to look good in front of their superior, there is a high likelihood that they would endorse the work once completed. There is no advantage for them to go against the team they are supposed to lead. However, team members should not expect any praise. This discussion is about providing tools for dealing with such individuals and is not intended as career guidance, which is beyond the scope of this book. However, this management style is not unusual and documented in literature. It is responsible for high turnover of employees and is quite detrimental to the effectiveness of an organization.

There are managers who have that personality type but became aware of their behavior. There are strategies that might help them becoming more effective. Recognizing the behavior is a big first step. They should frequently remind themselves which team they are on, spending more time with their subordinates and looking for opportunities to praise the team. They should surround themselves with experts that they respect, trust, and pass ideas by them first. Decision-making should also be focused on the long-term benefits.

Another managing type that can be difficult to deal with are highly skilled engineers promoted to managing positions. Their knowledge and experiences can put them in a position where they do not trust any of their subordinates. They often believe that they can do the work better, which can be true. However, by getting involved in the details they are no longer doing their managing duties and frustrates their subordinate engineers by the lack of trust. There was a time in some organizations when good technical engineers were quickly promoted through the management ladder. However, transitioning to management duties is not a promotion but a career change. Unless such individuals start trusting their subordinate and focusing on management duties, they might as well stay in a technical role as they would otherwise be poor managers.

Newly promoted managers often have the false idea that they must know everything and make all the decisions. Glen Sather, a former hockey player who became a successful team manager with the Edmonton Oilers and New York Rangers in the National Hockey League (NHL), once said in an interview, that the only skill you needed to be a successful manager in professional hockey was to know who can and who cannot play the game. Different personalities, skills set, and experiences are typically found within an operating crew. Some have less confidence than others and just want to do their work shift with minimum disturbances, while others excel and take pride in resolving complex problems. Factor influencing the productivity and effectiveness of process operators goes beyond training; it also relates to organizational culture and work environment. Management must provide the leadership required such that process operators feel respected and trusted. Similar as for successful sports team, management has the responsibility to create a work environment that enables ownership, pride, and commitment to success. Sather added that you hire people for whatever you do not know. This also means, as stated earlier, that you get success through the help of others. As per the previous discussion, being a formal manager does not entail that you are a good leader. Managers with good leadership skills, however, will likely be more successful. Success of the team takes precedence over their own. As such, they are not worried about surrounding themselves with competent people that they trust. They do not hesitate to delegate leadership duties to their subordinate when appropriate. DePree [9] refers to this form of delegation as roving leadership. For instance, in real-time troubleshooting, a good manager would not question or overrule direction provided by an operation coordinator in front of process operators. They would not blame people but rather delegate resources and provide support in troubleshooting activities. The support and trust provided would give them accurate facts, data, and recommendations helping them making appropriate decisions. Cohen [8] ranked the top 13 elements that motivate people at work. The top five are: (1) *working for people who treat me with respect*, (2) *Interesting work*, (3) *Recognition for good work*, (4) *Chance to develop skills*, and (5) *Working for people who listen if you have ideas about how to do things better.* While they all are important, (1) and (5) are highly relevant to troubleshooting. People need to know they are trusted if they are expected to communicate what went through their mind during a process upset without fear of being reprimanded or blamed. They also want to be listened if they have opinions or ideas. Those themes were discussed relative to interaction between process engineers and operators. However, trust and good communications start with management.

Collins and Devanna [24] define the 4 Ms of manufacturing management as manpower, machinery, materials, and money. They state that the ability to deliver output depends on a mix of those four elements. Although troubleshooting may involve additional cost, the money aspect falls under business management and not so much under technical troubleshooting. Manpower and technology are more relevant to troubleshooting. It has been previously stated that the purpose of a chemical plant is to convert raw material into product streams of higher value. To be economically profitable and viable, operation of a chemical facility is associated with constraints and requirements that are interconnected such as capacity, budget, scheduling, inventory, quality, and regulations. As discussed in Chapter 3, deviations from those metrics are consequences triggering troubleshooting initiatives. In many circumstances, deviations from some of those metrics can be hidden or imbedded into one aspect that is not necessarily the root cause. For instance, a quality issue can be hidden by operating the plant at lower rates, or a capacity problem can be hidden or inconsequential due to high inventory. Another example is recovery from an upset that may not affect any of those metrics. Plant managers and unit leaders are accountable and evaluated based on the above metrics. Ideally, each abnormal condition and near miss should be evaluated. The degree of rigor and mobilization of resources is often constrained. Managers must therefore prioritize and strategize the deployment of resources based on the criticality of the event or problems being experienced.

5.6 FACILITATORS

Depending on the magnitude and complexity of the problem being evaluated, the troubleshooting team can vary from a single individual to a team involving participants of different backgrounds. Process engineers frequently conduct informal troubleshooting on their own. However, larger problems typically involve group meetings. Those meetings can easily go sideways when not properly structured. For instance, SMEs have difficulties in seeing the problem beyond their area of expertise. Managers might feel the urge of resolving the problem quickly without fully understanding the actual root cause. Their position of authority might also intimidate or influence team members. Differences in opinion between team members may result in nonproductive arguments. For the above reasons and others, there are significant benefits in getting a facilitator for leading those meetings. By definition, a facilitator is a neutral guide who leads a group of people navigating through a work process with the objective of finding a solution. KT [25] states that in performing this function, a facilitator must stay neutral, while guiding the team to a consensus. The facilitator expertise lies in structured problem-solving work process, not the technology being evaluated; support every team member and stimulate the team creativity; and encourage everyone to participate and contribute. It is not unusual in team discussion that people with introverted personalities remain quiet, either for fear of saying something foolish or believing that their opinion may not matter. It is the role of the facilitator to get those individuals to open up. Whether relevant or not, comments outside the box may break biases and open the door to unexplored opportunities.

KT [25] state five challenges which can be overcome with the involvement of a skilled facilitator.

1. *Thinking Inside the Box.* Most people have biases due to their role and experience. A facilitator chairs the meetings and leads a structured troubleshooting work process.
2. *Technical Complexity.* Discussions among SMEs may generate many theories, which could cause the team to diverge from the solution by being hung-up on details that may not matter. A facilitator has the ability to extract the fact, parking lot opinions or create action items by delegating separate evaluations for validating assumptions/opinions into facts.
3. *The Solution That Worked Before.* The use of historical events has been discussed in the previous chapter. It is extremely useful information which provides important clues. The troubleshooting team could easily assume that what worked before will work again. However, by going through the KT dimensional questions, the team may find slight differences for which past solutions might not be successful.
4. *Poor-Quality Information.* The success of a troubleshooting work process lies in the quality of the data collected. The facilitator cannot guarantee the quality of the data. This role belongs to process engineers and operation. Individual unexperienced in troubleshooting may need guidance for collecting data and conducting evaluations. A facilitator provides this type of guideline by involving the right people and assigning action items. The data must fit the structured work process followed by the facilitator.

5. *Lack of Common Understanding of the Issue.* This refers to problem definition. While this should be straightforward, it is not unusual for team members to argue around what the problem definition should be. Some often confuse the problem with the suspected root cause. Poor problem definition can mislead the troubleshooting effort. An experienced facilitator can guide the team in understanding and agreeing on the problem definition.

The presence of a facilitator does not exempt process engineers and SMEs from conducting the evaluation described in this book. H&MBs and equipment process loads are important pieces of the puzzle that facilitator will strategically document in the work process they are following. Troubleshooting work process is often described as detective work, where each piece of evidence must align for supporting a probable root cause.

Nomenclature

APC	advance process control
C2R	ethylene refrigeration
C3R	propylene refrigeration
CGC	crack gas compressor
CPI	Chemical Process Industry
DCS	distributed control system
H&MB	heat and material balance
KT	Kepner–Tregoe
NHL	National Hockey League
OTS	operator training simulator
PID	proportional-integral-derivative
SME	subject matter expert

REFERENCES

[1] B. G. Houk, D. L. DeBari, S. Bitar, D. Bartusiak and K. Ribe, "Open Process Automation for Distributed Control Systems," *Chemical Engineering Progress*, vol. 116 no. 9, pp. 45–51, September 2020.
[2] P. Nordh, "Understand Advanced Process Control," *Chemical Engineering Progress*, vol. 112 no. 6, pp. 69–72, 27 August June 2016.
[3] R. Sadeghbeigi, "3.4 Advance Process Control (APC)," in *Fluid Catalytic Cracking Handbook, An Expert Guide to the Practical Operation, Design, and OPtimization of FCC Units*, Elsevier, 2020, pp. 42–45.
[4] P. Welander, "How Many People Will Be Running Your Plant?," *Control Engineering*, vol. 61 no. 10, p. 52, October 2014.
[5] S. Jenkins, "CPI Workforce Issues See Positive Movement," *Chemical Engineering*, vol. 124 no. 12, pp. 14–17, December 2017.
[6] R. Rowan, *The Intuitive Manager*, The Berkley Publishing Group, 1986.
[7] N. P. Lieberman, *Troubleshooting Process Operations*, Third Edition, PennWell Publishing Company, 1991.

[8] W. A. Cohen, The Art of the Leader, Prentice Hall, 1990.
[9] M. DePree, *Leadership is an Art*, Dell Publishing, 1989.
[10] A. J. Bernstein and S. C. Rozen, Dealing with All Those Impossible People at Work, Dinosaur Brains, Ballantine Books, 1989.
[11] S. Dekker, *The Field Guide to Understanding Human Error*, CRC Press, 2014.
[12] R. Besuijen, Troubleshooting Tactics. How Process Operators Make Critical Decisions, Ron J Besuijen, 2022.
[13] A. Bernard, "Post-revamp startup planning and support," *Petroleum Technology Quarterly* Q4, vol. 12 no. 5, pp. 143–151, 2007.
[14] A. Bernard, "Planning an Olefins Unit Startup with an Operating Training Simulator as Part of a Plant Revamp Commissioning Strategy," in AIChE Spring Meeting, Ethylene Producer Conference, Houston, TX, 2023.
[15] R. W. Bly, "Become a Better Writer," Chemical Engineering Progress, April 2017.
[16] R. W. Bly, "Give a Great Presentation," Chemical Engineering Progress, October 2018.
[17] H. P. Bloch, "Safe Fluid Machines Are Our Responsibility," Hydrocarbon Processing, April 2019.
[18] C. H. Kepner and B. B. Tregoe, The New Rational Manager, Kepner-Tregoe, Inc, 1981.
[19] A. Bernard, "Analyses of a Rapid loss of a Steam System During a Power Failure," in AIChE Spring Meeting, Ethylene Producer Conference, Atlanta, GA, 2005.
[20] R. Fisher, "Effective Procedures and Process Safety," Chemical Engineering Progress, March 2024.
[21] K. Doe and D. Copparrd, "Process Simulation for Operations Training and Other Benefits," in AIChE Spring Meeting, Ethylene Producer Conference, Houston, TX, 2001.
[22] R. Besuijen, R. Young and M. Guaraldo, "Operator Training Simulators – Nova Chemicals' Experience for Training, Certification and Other Applications," in AIChE Spring Meeting, Ethylene Producer Conference, Houston, TX, 2012.
[23] A. J. Bernstein and S. C. Rozen, Neanderthals at Work, Ballantine Books, 1992.
[24] E. G. Collins and M. A. Devanna, The Protable MBA, John Wiley & Sons, 1990.
[25] Kepner-Tregoe, "The Value of a Facilitator in Organizational Problem Solving," [Online]. Available: https://kepner-tregoe.com/resources/whitepapers/the-value-of-a-facilitator-in-organizational-problem-solving-and-decision-making/. [Accessed 1 January 2026].

Index

D

E

F

Q

R

S

T

For Product Safety Concerns and Information please contact our EU representative GPSR@taylorandfrancis.com
Taylor & Francis Verlag GmbH, Kaufingerstraße 24, 80331 München, Germany

www.ingramcontent.com/pod-product-compliance
Ingram Content Group UK Ltd.
Pitfield, Milton Keynes, MK11 3LW, UK
UKHW022218190926
475702UK00003B/427
* 9 7 8 1 0 3 2 5 8 3 4 2 6 *